W9-ACU-985

CONTROL AND DYNAMIC SYSTEMS

Advances in Theory and Applications

Volume 33

CONTRIBUTORS TO THIS VOLUME

JAMES L. FARRELL
TIMOTHY JOSEPH FREI
KARYN S. HAALAND
RONALD A. HESS
NORIYUKI HORI
KIMIO KANAI
GREGORY W. MEDLIN
L. MEIROVITCH
WALTER C. MERRILL
PETER N. NIKIFORUK
Y. SHARONY
D. D. SWORDER

CONTROL AND DYNAMIC SYSTEMS

ADVANCES IN THEORY AND APPLICATIONS

Edited by
C. T. LEONDES

School of Engineering and Applied Science
University of California
Los Angeles, California

College of Engineering
University of Washington
Seattle, Washington

VOLUME 33: ADVANCES IN AEROSPACE SYSTEMS DYNAMICS AND CONTROL SYSTEMS
Part 3 of 3

ACADEMIC PRESS, INC.
Harcourt Brace Jovanovich, Publishers
San Diego New York Boston
London Sydney Tokyo Toronto

This book is printed on acid-free paper. ♾

COPYRIGHT © 1990 BY ACADEMIC PRESS, INC.
All Rights Reserved.
No part of this publication may be reproduced or transmitted in any form or by any means, electronic or mechanical, including photocopy, recording, or any information storage and retrieval system, without permission in writing from the publisher.

ACADEMIC PRESS, INC.
San Diego, California 92101

United Kingdom Edition published by
ACADEMIC PRESS LIMITED
24-28 Oval Road, London NW1 7DX

LIBRARY OF CONGRESS CATALOG CARD NUMBER: 64-8027

ISBN 0-12-012733-4 (alk. paper)

PRINTED IN THE UNITED STATES OF AMERICA
90 91 92 93 9 8 7 6 5 4 3 2 1

CARLSON
QA
402.3
.A1A3
V.33, pt.3

CONTENTS

Decoupled Flight Control via a Model-Following Technique Using the Euler Operator

Peter N. Nikiforuk, Noriyuki Hori, and Kimio Kanai

Advances in Digital Signal Processing

Gregory W. Medlin

Methodology for the Analytical Assessment of Aircraft Handling Qualities

Ronald A. Hess

Identification of Pilot–Vehicle Dynamics from Simulation and Flight Test

Ronald A. Hess

Strapdown Inertial Navigation System Requirements Imposed by Synthetic Aperture Radar

James L. Farrell

Trajectory Estimation Algorithm Using Angles-Only, Multisensor Tracking Techniques

Timothy Joseph Frei

A Perturbation Approach to the Maneuvering and Control of Space Structures

L. Meirovitch and Y. Sharony

CONTRIBUTORS

Numbers in parentheses indicate the pages on which the authors' contributions begin.

James L. Farrell (177), *Development and Engineering Divisions, Westinghouse Defense and Electronic Systems Center, Baltimore, Maryland 21203*

Timothy Joseph Frei (199), *Space and Technology Group, TRW Inc., Redondo Beach, California 90278, and Department of Mechanical, Aerospace, and Nuclear Engineering, University of California, Los Angeles, California 90024*

Karyn S. Haaland (35), *Department of Applied Mechanics and Engineering Sciences, University of California—San Diego, La Jolla, California 92093*

Ronald A. Hess (129, 151), *Department of Mechanical Engineering, Division of Aeronautical Science and Engineering, University of California, Davis, California 95616*

Noriyuki Hori (59), *Department of Mechanical Engineering, McGill University, Montreal, Quebec H3A 2K6, Canada*

Kimio Kanai (59), *Department of Aerospace Engineering, National Defense Academy, Yokosuka City, Japan*

Gregory W. Medlin (89), *Department of Electrical and Computer Engineering, University of South Carolina, Columbia, South Carolina 29208*

L. Meirovitch (247), *Department of Engineering Science and Mechanics, Virginia Polytechnic Institute and State University, Blacksburg, Virginia 24061*

Walter C. Merrill (1), *National Aeronautics and Space Administration, Lewis Research Center, Cleveland, Ohio 44135*

Peter N. Nikiforuk (59), *College of Engineering, University of Saskatchewan, Saskatoon, Saskatchewan S7N 0W0, Canada*

Y. Sharony (247), *Department of Electrical Engineering, Virginia Polytechnic Institute and State University, Blacksburg, Virginia 24061*

D. D. Sworder (35), *Department of Applied Mechanics and Engineering Sciences, University of California—San Diego, La Jolla, California 92093*

PREFACE

Developments in the many technologies which support the growth of aerospace systems have undergone an incredible revolution over the past 25–30 years. For example, integrated electronic circuits have increased in density by about eight orders of magnitude since 1960. This makes possible many things in aerospace vehicle systems that were previously impossible. Advances in other areas such as sensor systems, materials, structures, propulsion, software, and systems integration techniques all very clearly and strongly suggest that it is now most appropriate to cover this subject in *Control and Dynamic Systems*. However, the proliferation of significant developments and advances has been so great that adequate coverage could not possibly be encompassed in one volume, thus, this volume is the last of a trilogy devoted to the theme, "Advances in Aerospace Systems Dynamics and Control Systems."

The first contribution, "Sensor Failure Detection for Jet Engines," by Walter C. Merrill, presents a comprehensive survey and analysis of analytical redundancy techniques to improve turbine engine control system reliability. In particular, Merrill notes that sensor redundancy is required to achieve adequate control system reliability. The three types of sensor redundancy are direct, analytical (the subject of this contribution), and temporal. Because of the requisite increasing complexity of modern aerospace systems, this article's comprehensive survey and analyses constitute a unique reference source. The next contribution, "Human Response Models for Interpretive Tasks," by D. D. Sworder and Karyn Haaland, deals with two significant areas in modern aerospace systems: (1) techniques for the proper allocation of tasks between human and machine (computers, sensors, etc.) in man–machine systems and (2) techniques for remote operations such as those designed for remotely piloted or teleoperated vehicles. As a result, this contribution is particularly welcome.

The next contribution, "Decoupled Flight Control via a Model-Following Technique Using the Euler Operator," by Peter N. Nikiforuk, Noriyuki Hori, and Kimio Kanai, addresses control configured vehicles (CCV) through model-following techniques. As is well known, CCV techniques are becoming increasingly prevalent in both commercial and military aircraft. In conventional flight control, flight path change is achieved by first changing aircraft attitude; then vertical and horizontal forces, which are generated as a result of attitude change, produce flight path changes. Among other purposes, CCV was introduced to control directly the vertical and horizontal forces. One means for achieving this is to have the aircraft performance follow the model of a decoupled system (the model-following technique). Since modern flight control systems tend to be designed with digital controllers, the discretization process introduced as a result of digital control must faithfully reproduce the continuous time system. The Euler operator, described in this contribution, achieves this very closely. Because of the growing importance and application of CCV techniques in modern aircraft, this contribution is essential to this trilogy.

Virtually all equipment on modern aerospace systems involves the utilization of digital computers, and, in many cases, digital signal processors. The next contribution, "Advances in Digital Signal Processing," by Gregory W. Medlin, treats this area of major significance to modern aerospace systems. Optimum digital filter design techniques are described in detail, and their effectiveness is exemplified through the presentation of comprehensive computer studies.

"Methodology for the Analytical Assessment of Aircraft Handling Qualities," by Ronald A. Hess, represents an essential element in this trilogy: Ultimately, the human factor, the pilot, is the key in modern aircraft systems. Hess, a leading figure on the international scene in this area, has provided a splendid contribution. The comprehensive review and analysis of aircraft handling qualities presented here will be a valuable reference for working professionals in this rather complex area. The next contribution, "Identification of Pilot–Vehicle Dynamics from Simulation and Flight Test," also by Hess, deals with the identification of pilot–vehicle dynamics as part of any simulation or flight test experiment. It is mandatory that researchers be able to extract as much information as possible regarding the pilot–vehicle system from each simulation and flight test experiment because of the very high costs of the tests. It is particularly useful to be able to identify pilot–vehicle dynamics as part of any simulation or flight test experiment. As noted in this contribution, the control theoretic model of the human pilot has become the fundamental mode of thinking on the part of researchers involved in analyzing pilot–vehicle systems. Thus, this model is the object of most identification efforts. These and other related issues are covered in this most welcome contribution.

One of the earliest applications of flight was to reconnaissance. Since those early days, the techniques for reconnaissance have progressed to an amazing level, for both aeronautic and space systems, as a result of technological advances. In one modern reconnaissance technique, synthetic aperture radar (SAR), an aircraft synthetically creates a much larger radar aperture by recording and processing radar ground returns over a given flight path. Because of the synthetically generated larger aperture, a higher angular resolution results, and photo-like radar images are generated. However, fundamental to this whole process is sufficiently precise knowledge of the aircraft's space–time history during the generation of the synthetic aperture. This is provided by inertial navigation systems (INS) in all SAR systems. The next contribution, "Strapdown Inertial Navigation System Requirements Imposed by Synthetic Aperture Radar," by James L. Farrell, specifies how such systems requirements are defined. Because of the significance of reconnaissance in modern aerospace systems, this article by one of the leading figures in the field is most welcome.

"Trajectory Estimation Algorithm Using Angles-Only, Multisensor Tracking Techniques," by Timothy J. Frei, presents fundamental algorithmic techniques for modern aerospace passive (infrared) sensor systems. All modern aerospace systems have sensor systems on board, and one of the principal components of such sensor systems generates target data. Radar (active) systems can generate three-dimensional target data, but passive sensors, such as infrared sensors, can only generate target angle data. However, the data from such passive systems which are based on two or several platforms can be correlated to generate such three-dimensional data. Because of the importance of these issues in a number of aerospace applications, Frei's contribution is an important addition to this volume.

The final contribution, "A Perturbation Approach to the Maneuvering and Control of Space Structures," by L. Meirovitch and Y. Sharony, deals with implementable optimal control techniques for complex aerospace structures. Such structures, particularly large space structures, tend to be highly elastic, and so the implementation of maneuvering and control systems for them can be enormously complex if not approached properly. After noting that such structures can be described by hybrid representations; that is, nonlinear differential equations for the rigid body mode and linear partial differential equations for the elastic modes, the contributors point out that the optimal control of such systems represents formidable implementation difficulties. As a result, implementable perturbation alternatives are presented wherein the elastic body modes (first order) represent a perturbation on the (nonlinear) rigid body (zero order) mode. Meirovitch is one of the leading authorities on systems techniques for aerospace structures, and so this article by him and his colleague, Y. Sharony, will be of great interest to those in the field.

This book is a particularly appropriate volume with which to conclude this unique trilogy. The authors are all to be commended for their superb contributions, which will most certainly be significant reference sources for research workers and practitioners on the international scene for many years to come.

CONTROL AND DYNAMIC SYSTEMS, VOL. 33

SENSOR FAILURE DETECTION FOR JET ENGINES

WALTER C. MERRILL

National Aeronautics and Space Administration
Lewis Research Center
Cleveland, Ohio 44135

I. INTRODUCTION

This article surveys the use of analytical redundancy (AR) to improve turbine engine control system reliability. Since 1950, hydromechanical implementations of turbine engine control systems have matured into highly reliable units. However, as shown in Fig. 1, an increase in control complexity has occurred and is expected to continue. This increased complexity has made it difficult to build reliable, low-cost, lightweight hydromechanical controls. On the other hand, microprocessor-based digital electronic technology allows complex control systems to be built with low cost and weight. However, these digital electronic controls do not have the maturity and, therefore, the demonstrated reliability of hydromechanical engine control systems.

Thus, in an effort to improve the overall demonstrated reliability of the digital electronic control system, various redundancy management techniques have been applied to both the total control system and to individual components. One of the least reliable of the control system components is the engine sensor. In particular, a study of fault-tolerant electronic engine controls shows that sensor redundancy will be required to achieve adequate control system reliability [1]. There are three types of sensor redundancy: direct, analytical, and temporal. Direct, or hardware, redundancy uses multiple sensors to measure the same engine variable. Typically, a voting scheme is used to detect failures. Analytical redundancy uses a reference model of the engine and redundant information in dissimilar sensors to provide an estimate of a

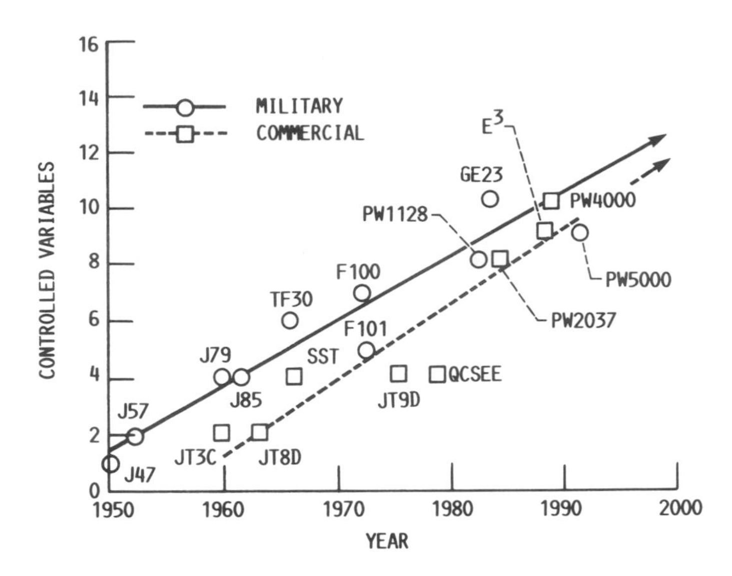

Fig. 1. Trends in control complexity of aircraft turbine engines.

measured variable. Estimates and measurements can be used in a variety of ways to detect failures. Temporal redundancy uses redundant information in successive samples of the output of a particular sensor to determine failures. Range and rate checks are simple and often-used examples of temporal redundancy.

Hardware redundancy is insensitive to failure magnitude since any detectable discrepancy between two like sensors indicates a failure. Thus, hardware redundancy handles hard (out-of-range or large in-range) failures as well as soft (small in-range or drift) failures. Analytically redundant schemes can distinguish failure type and, in fact, can be made sensitive to a particular type, such as soft failures. Range and rate checks are simple and reliable detection methods, but are limited to hard failures. Often range and rate checks are combined with analytical redundant schemes to cover both hard and soft failure types. As shown in [1], hardware redundancy results in more costly, heavier, less practical, and less reliable systems than do various analytical redundancy strategies. Since cost, weight, and reliability are important drivers in turbine engine control systems design, many researchers have investigated analytical redundancy strategies.

State-of-the-art digital electronic control schemes, such as that for the PW2037 engine [2], make use of a combination of hardware and analytical redundancy to provide adequate system reliability. Here, dual-redundant sensor measurements and a synthesized or estimated measurement are compared to detect sensor failures. This approach is comparable to that used in the aircraft control for the F8 digital fly-by-wire aircraft [3]. In each case a two-step approach is used. First the dual sensors are compared to determine if a discrepancy exists. Then a comparison is made to the estimate to isolate the faulty sensor. Operation continues with the good sensor. Here analytical redundancy allows system operation after both sensors have failed to further improve system reliability. Eventually, as AR-based techniques improve, additional reliance on AR strategies would allow single-sensor operation with the resultant savings in cost and weight.

The objective of this article is to survey the application of analytical redundancy to the detection, isolation, and accommodation (DIA) of sensor failures for gas turbine engines. This includes those approaches that use software implementations of temporal redundancy combined with analytical redundancy. Hardware redundant strategies are not covered. This survey first reviews the theoretical and application papers that form the technology base of turbine engine analytical redundancy research. Second, the status of important ongoing application efforts is discussed. Also included is a review of the PW2037 engine control system sensor AR strategy. This is the first operational engine to include AR-based strategies. Finally, an analysis of this survey indicates some current technology needs.

II. ANALYTICAL REDUNDANCY TECHNOLOGY BASE

In this section, papers that document the AR technology base will be reviewed. Seventeen papers are considered. The papers will be reviewed in essentially chronological order. The attributes documented in each paper, as discussed in this section, are summarized in Table I.

Wallhagen and Arpasi [4] presented the first (April 1974) use of sensor AR to improve engine control system reliability. A J85, single-spool, turbojet with two sensed variables and three controlled variables was tested at a sea-level static condition. The inputs were compressor variable geometry, fuel flow, and exhaust nozzle area. The sensors were a magnetic pickup for rotor speed and a high-response gauge transducer for compressor static discharge pressure. Failure detection was accomplished by comparing the rate of change of the sensed variables with predetermined limits. Four consecutive out-of-range rates declared a failure. Since each sensor was tested for catastrophic (i.e., hard) failure only, isolation is immediate. Failures are accommodated by replacement of the failed sensed value with a synthesized estimate. This synthesized variable is obtained from a tabulation of the synthesized variable as a function of the remaining engine variables. Different tables were stored for steady-state and acceleration conditions. No explicit dynamical relationships were included. The DIA logic was implemented in fixed-point assembly language on a

TABLE I. SUMMARY OF ATTRIBUTES FOUND IN THE LITERATURE[a]

Paper	Test-bed system	Reference model	Type of detection and isolation	Type of accommo-dation	Computer environment	Evaluation	Number of sensors	Comment
Wallhagen and Arpasi [4]	J85 turbojet	PS	Hard failures via rate-limit threshold	Replacement	Real-time minicomputer	SLS	2	
Hrach *et al.* [5]	TF30 real-time hybrid simulation	PS	Hard failures via rate-limit threshold	Replacement	Real-time minicomputer	Five operating conditions through flight envelope	4	Extension of [4]
Ellis [6]	Turbofan nonlinear digital simulation	PS, PL	Threshold check of weighted-average estimate	Estimates always used by control weighted-average modified	Digital simulation	SLS	6	
Wells and de Silva [7–10]	Nonlinear model of turbojet	PL with bank of filters	Bayesian hypothesis	None	Digital simulation	SLS	2	
Spang and Corley [11]	QCSEE real-time hybrid simulation	SC with Kalman filter	Threshold comparison of filter residuals	Substitution	Real-time digital control	SLS	7	
DeHoff and Hall [12]	Single-spool turbojet	SC	Maximum likelihood	None	Digital simulation	SLS	1	Theoretical study

Sahgal and Miller [13]	F100 nonlinear simulation	PL	None	Substitution	Digital simulation	Two conditions	4	
Leininger and Behbehani [14–16]	QCSEE nonlinear simulation	PL	GLR	None	Digital simulation	SLS	6	Theoretical and application study
Meserole [17]	F100 nonlinear simulation	PL	Detection filter	None	Digital simulation	SLS	15	Theoretical and application study
Leininger [18]	QCSEE simulation	Linear	*t* test	None	Digital simulation		SLS	Theoretical study of model uncertainty
Weiss *et al.* [19]; Pattipati *et al.* [20]	F100 linear	Linear	SPRT	None	Digital simulation	SLS	5	Theoretical study of model uncertainty
Emami-Naeini *et al.* [21]	F100 linear	Linear	Hypothesis	Frequency-weighted filter estimates	Digital simulation	SLS	5	Theoretical study of model uncertainty
Brown *et al.* [22]	F110 nonlinear	Linear	Threshold checks	Substitution	Digital simulation	SLS	9	Application to multiple engines
Horak [23]	HYTESS	Linear	RMI	None	Digital simulation	SIS	1	

[a] Key: PS, parameter synthesis; SC, simplified component; PL, pseudolinear; SLS, sea-level static; SPRT, sequential probability ratio test; GLR, generalized likelihood ratio; and RMI, reachable measurement interval.

minicomputer. The implementation was executed in a 15-msec time frame, which allowed real-time interaction with the control. Testing in a sea-level static test stand compared idle to full-power step responses of rotor speed and thrust. For single failures, steady-state speed was held to within 1% of its final value and 92% of maximum thrust was achieved. For two sensor failures, steady-state speed was approximately 99% of its final unfailed value and thrust was 87% of maximum. Time to accelerate, however, had to be increased from 3 to 30 sec. Failures were induced at 50% power during a transient. Detection was reliable. The system also allows self-healing. An interesting feature of the DIA logic was its ability to learn, on line, all the data necessary to function. In a companion paper, Hrach *et al.* [5] used a real-time nonlinear hybrid computer simulation of a two-spool turbofan, the TF30-P-3 engine, to demonstrate the DIA logic of [4] over a wide operating range. For sensed variables: high-pressure rotor speed, high- and low-pressure compressor discharge static pressures, and nozzle total pressure; and five inputs: main fuel flow, nozzle area, afterburner fuel flow, and two compressor stage bleeds were considered. Again hard failure detection and isolation were obtained by individual rate checks.

Accommodation was achieved by replacement with averaged synthesized variables, which were a function of the remaining good sensors (1, 2, or 3). Synthesized variables were obtained from tabulations. However, the data were now stored as corrected values to allow a wide operating range. Data for the tables were collected at two operating points.

A real-time implementation of this DIA logic was programmed using assembly language in a minicomputer using a frame time of about 0.025 sec. Storage requirements include 4K bytes for the logic and 0.2K bytes for the tables. The logic was tested at five selected operating points (which include the two design points). Acceptable operation with no limit violations and approximately the same thrust was obtained for operation with one to three of the four sensors failed. For afterburning operating of the engine, acceptable control was possible for only a single failure and with a severe rate limit on accelerations. This logic also incorporated learning or adaptive logic.

Ellis [6] (January 1975) studied the use of AR techniques using a nonlinear digital simulation of a two-spool turbofan engine. The engine has five measured variables and two independent controlled variables. The DIA philosophy of this paper centers around estimates of the measured variables. First a multivariable linearized mapping (no explicit model dynamics) of corrected measurements to estimates is found. Since the engine has only two independent controls, it is assumed that only two measurements are required to generate an estimate. Taking unordered pairs of the five measured variables yields ten estimates of each measured variable. A weighted average estimate is obtained by combining these ten component estimates, each weighted by its relative accuracy. Detection and isolation are accomplished by a threshold check on both sides of each weighted average estimate. If a weighted estimate is outside of the threshold corridor then all weighting factors associated with this estimate are set to zero. Weighted estimates are used by the control at all times. Only the weightings change as failures occur. Thresholds for the weighted estimates are obtained from sensor error statistics assuming Gaussian distributions.

The next contribution to this area is documented in four reports [7–10] by de Silva and Wells. This series of reports applies Bayesian hypothesis testing to the detection of engine sensor failures. The engine studied is a simple turbojet with two outputs, speed and thrust, and one input, fuel flow. A second-order pseudolinear model of the engine was used on a mainframe computer to evaluate detection performance. A pseudolinear model consists of a dynamical, linear state-space structure where individual coefficients within the linear structure vary as a nonlinear function of the state. Bayesian hypothesis testing is implemented by (1) defining a risk function and (2) determining from measured data the hypothesis that minimizes this risk. This risk function defines the penalty associated with selecting a false hypothesis. Assuming Gaussian noise statistics, the lowest risk Bayesian hypothesis is also probabilistically most likely given the measured data. A "bank" of Kalman filters, one per hypothesis, uses measured data and an engine model to generate state estimates and filter residuals. The hypothesis associated with the most likely set of residuals, as determined by a likelihood ratio test, is taken as the true hypothesis. The mode of operation associated with this hypothesis (failed speed sensor, no failure, etc.) was assumed true. The approach worked well in simulation studies of this simple case. This work represents the first application of analytical redundancy to turbine engines based on modern control theory. Difficulties with this approach include the requirement of a different Kalman filter for each failure-mode hypothesis.

In June 1977 Spang and Corley [11] published an application of AR techniques to the quiet, clean, short-haul, experiment engine (QCSEE). This engine has seven measurements: fuel flow, compressor stator angle, fan speed, compressor speed, compressor discharge temperature and pressure, and turbine discharge temperature. Engine controls include fuel flow valve current and compressor stator vane blade torque motor current. In this study an extended Kalman filter approach is used to generate state estimates and residuals. A simplified nonlinear component model that is valid throughout the engine operating envelope and a simplified feedback gain matrix operating on engine measurements are used to update the filter estimates and residuals. Sensor failures were detected and isolated by a threshold comparison of the individual residual components. Thresholds are determined by sensor noise statistics. Only hard failures are considered. To accommodate failures, faulty measured values are replaced by sensor estimates from the filter. The approach was successfully demonstrated on a detailed, real-time, nonlinear hybrid computer simulation of the engine. The detection, accommodation, and control logic are implemented in a microprocessor-based control; also in real time. Successful operation for single hard sensor failures is demonstrated at sea-level static conditions for power chops and bursts in the idle to full take-off power range. This work, referred to as failure indication and corrective action (FICA), serves as the theoretical foundation for a significant portion of the work in the application of AR to turbine engines. Further applications based on FICA are given in a subsequent section.

Next, DeHoff and Hall [12] report a largely theoretical study that developed a unified framework to achieve engine performance monitoring, trending, and sensor fault DIA. This framework is based on maximum-likelihood state and parameter estimation methods. A simple turbojet example is used to illustrate the application of a maximum-likelihood-based, on-line, sequential-processing, parameter-estimation algorithm to the detection of sensor failures.

Sahgal and Miller [13] report on the design of a full-order observer that reconstructs fan turbine inlet temperature for an F100 engine. The observer is based on a fifth-order scheduled state-space model with four inputs: fuel flow, nozzle area, and compressor and fan variable geometries; and four outputs: fan and compressor speed, and compressor discharge temperature and pressure. Observer performance is compared with a full nonlinear digital simulation of the engine at sea-level static conditions. The reconstructed temperature tracks the actual temperature quite well. The analytical study proposes to use the reconstructed temperature to accommodate for fan turbine inlet sensor failures.

The next three papers [14–16] by Leininger and Behbehani report the application of the generalized likelihood ratio (GLR) technique to the QCSEE. The GLR technique is a hypothesis-based test with the time and type of failure unknown. Under linear Gaussian assumptions, if the Kalman–Bucy filter residuals are found to be nonwhite, a failure is declared. Next, various likelihood ratios are compared to determine the most probable failure time and type. The GLR method is used to detect and isolate hard and soft failures. Both single and multiple actuator and sensor failures were considered.

Detection and isolation studies are conducted by simplified simulation of the QCSEE. This simulation included six outputs: fan and compressor speeds, engine inlet static pressure, fan inlet duct static pressure, combuster pressure, and compressor discharge pressure; and three inputs: fuel-metering valve position, fan nozzle actuator position, and fan pitch angle. A linearized, eight-state model was used in the Kalman–Bucy filter. Successful detection and isolation of multiple sensor and actuator failures with noisy sensors and imperfect modeling were demonstrated. Accommodation by control reconfiguration using nonsquare multivariable Nyquist array methods was proposed. Designs were obtained but not demonstrated by simulation.

A doctoral dissertation by Meserole [17] uses detection filter theory to design a filter that detects sensor failures in a F100 engine. Similar to the Kalman filter, the detection filter incorporates a dynamic process model and generates error residuals. However, unlike the Kalman filter, a detection filter is designed to respond to a component failure with a residual that has a fixed, usually unique, direction. Also, this direction is independent of failure mode. Thus, sensor failures can be detected and isolated by detecting the occurrence of these fixed-direction residuals. A sixth-order state-space linear model with scheduled coefficients is used in the detection filter. Filter operation and detection capability are demonstrated using a detailed nonlinear digital simulation of the F100 engine. Fifteen components are checked for failure: the inlet pressure and temperature sensors, the fan and compressor speed sensors, the burner and augmentor total pressure sensors, the fan outer-diameter discharge and turbine inlet total temperature sensors, the fuel system, the nozzle, bleed, fan guide vane, and compressor stator vane actuators, and the high- and low-pressure turbines. Five inputs are considered: fuel flow, nozzle area, fan guide vane and compressor stator vane positions, and bleed. Filter performance was studied for sensor failures and component changes (failures) at sea-level static conditions for bias and scale-factor changes. Failures were detected for 2 to 5% changes in one or more output measurements. Minimum failure size for successful isolation is summarized by component in Table II.

A paper by Leininger [18] examines the impact of an inaccurate model on innovations-based detection and isolation procedures. The paper demonstrates that model inaccuracies appear as biases in the innovations (residuals). These biases are identified by a Student's t test. The t test is then related to a

TABLE II. MINIMUM FAILURE SIZE FOR A DETECTION[a] FILTER DESIGNED FOR AN F100 ENGINE

	Minimum failure size (%)	
Isolation	Steady engine state	Unsteady engine state
Output sensors	2	5–10
Inlet sensors	2	5–10
Fuel system, exhaust nozzle	5–10	10–20
Compressor vanes, fan vanes	10–30	20–60
Rotor efficiencies	2	5–10

[a] 2 to 5% changes in one or more output measurements.

recursive GLR detector using a sequentially updated Kalman filter. Model bias error is removed from the innovations data to remove the effect of model degradation and to allow more accurate soft and hard failure detection. Also, a finite-width-window sequential t test is used to update the bias term and provide a means of sensor failure detection and isolation. The theory was applied to an eighth-order linear model of the QCSEE. Model eigenvalues were perturbed by 10% to simulate model error. The t test successfully removed the bias, tracked a sensor drift followed by a low-frequency sinusoidal sensor bias, and exhibited a fail–heal–fail detection pattern for the sinusoidal test.

The next three papers present basic research in robust detection, isolation, and accommodation of sensor failures. This research focuses on one fundamental question: How accurately must engine dynamics be modeled for successful DIA? A definitive answer to this question would establish the quantitative trade-offs between complexity, detection time, and detection performance. An alternative viewpoint would be to define the robustness of a DIA algorithm to model inaccuracies or uncertainty. Two different approaches have been identified to the solution of this problem.

The research of [19] and [20] is based on the concept of redundancy, or parity, relations. These relationships among the measured system variables incorporate all possible redundant information available. Modeling uncertainty affects the reliability of these parity relations. For a quantified level of uncertainty, all parity relations can be ranked from most to least reliable. This allows the more reliable parity relations to be used to generate DIA strategies that are as robust to uncertainty as possible. A three-step design process is presented. First, the parity relations are rank ordered using a robustness metric. That set of relationships with acceptable robustness is identified. Second, the coverage (probability of detection for all failures) for this set of relationships is assessed. Finally, the ability of the set of parity relations to distinguish each failure mode from the others is assessed, again using a metric-based analysis. Iterations through this process are possible in order to expand the original set of relationships and to improve coverage or distinguishability by incorporating decreasingly robust parity relations. The parity relations can be generated efficiently from either a time- or frequency-domain description of the average process. The average process is defined as

$$\bar{A} = \sum_{l=1}^{n} \rho_l A_l$$

where A_l represents lth set of model parameters and ρ_l the *a priori* probability that A_l is correct. The methodology has been applied to the preliminary design of a robust DIA system for an F100 engine.

The research of [21] is based on the extension of recent advances in robust control system design to sensor DIA and estimator design. Model uncertainty effects on DIA robustness are quantified using conic sector uncertainty properties. Here, uncertainty that is bounded in a conic sector in the frequency domain, and which then propagates through a system, remains bounded by a conic sector. These sectors determine quantitatively the performance–robustness trade-off. This frequency-domain description of uncertainty along with a frequency-shaped filter yields optimally robust innovations to model uncertainty. Thus, sensor failure detection based on these innovations will also be robust.

The design process makes use of a threshold selector. The threshold selector determines the minimum detectable failure size for a given noise level, failure type, false-alarm rate, and model uncertainty description. This threshold selector determines maximum achievable performance for the given set of constraints. Optimally robust (to modeling errors) residuals are generated using filters designed using the internal model principle and frequency shaping. The results of this methodology are applied to the preliminary design of sensor DIA logic for an F100 engine.

Reference [22] presents an investigation of a variation of hardware redundancy to improve soft-failure DIA capability. This feasibility study examines a multiengine approach (in this case two engines) to soft-failure DIA. The underlying principle is to use a like sensor measurement from one engine as redundant information to improve DIA capability on another engine. This approach incorporates a model of potential engine differences, an average engine model, and decision logic. By looking at the sum and differences of redundant sensed values for the two engines, measured average and differential performance is obtainable. These are compared to the average and difference engine models contained in the DIA logic. This additional information allows improved DIA performance over a single-engine concept. This concept is demonstrated using a digital nonlinear simulation of two F110 engines.

The final paper in this section [23] determines the theoretical limits of failure detectability of sensor failures in systems with modeling errors. A method, called the reachable measurement intervals (RMI) method, is derived, which is based on this determination and which performs at the limit of detectability. This method is based on a linear, state-space model of the system and bounds on the uncertainties of the model parameters. The RMI, which are the smallest possible thresholds, are computed using an optimization procedure based on the maximum principle. Measurements that lie outside this interval indicate a failure. The method was applied to a hypothetical turbofan engine simplified simulation (HYTESS) [24]. High-performance failure detection was demonstrated for a fan speed measurement at a single operating point, for small fuel perturbations.

III. ANALYTICAL REDUNDANCY TECHNOLOGY DEVELOPMENT

Based on the encouraging, but preliminary, results of the AR technology base, several technology development programs were begun. The overall objective of these programs is the full-scale engine demonstration of improved control system reliability using AR technology. These important AR development programs are (1) advanced detection, isolation, and accommodation (ADIA), (2) energy-efficient engine (E^3) FICA, (3) full-authority digital electronic control (FADEC) FICA, (4) digital electronic engine control (DEEC) sensor DIA, and (5) analytical redundancy technology for engine reliability improvement (ARTERI). Also included is a discussion of the sensor redundancy approach used on the PW2037 engine.

A. ADVANCED DETECTION, ISOLATION, AND ACCOMMODATION

The objective of the advanced detection, isolation, and accommodation (ADIA) program is to demonstrate a viable DIA concept based on advanced methodologies. The ADIA program consists of four parts: development, implementation, real-time evaluation, and demonstration.

The development of the ADIA algorithm is reported by Beattie *et al.* [25, 26]. Here advanced detection and filtering methodologies were compared to develop a viable ADIA concept. Comparisons were made on a F100 engine and F100 multivariable control (MVC) [27] test-bed system. The type and severity of sensor failures were carefully defined. Typical state-of-the-art transducers were selected. Failure characteristics were defined and quantified according to the predominant failure categories of out of range, drift, and noise. Next, a failure mode and effects analysis was conducted to classify the various failure modes as critical or noncritical. Critical failures were defined as those that resulted in surge, a 10% or larger thrust variation, or a rotor overspeed. This classification was accomplished over the full operating range of the F100 engine. Five competing DIA concepts were developed by combining available detection and filtering technologies. These five concepts were specifically formulated to span as many applicable technologies as possible.

Since competing technologies were to be compared, a scoring system was developed. The scoring system evaluated the concepts for (1) exceeding minimum transient and steady-state operation requirements, (2) detection and isolation effectiveness, and (3) the qualitative benefits of bettering the requirements of the first item. Using the scoring system and a simplified simulation of the test-bed system, the five concepts before mentioned were screened. Two concepts were selected for a more detailed comparison. Based on this second screening, one concept was selected for evaluation on a detailed nonlinear simulation of the test-bed system. This detailed evaluation included simulated sensor failures for both steady-state and transient operation throughout the entire engine operating range. This evaluation showed the ADIA approach to be (1) viable for gas turbine applications and (2) more systematic and straightforward when compared to a parameter synthesis approach.

An accurate model of the engine is required to achieve high-performance failure detection. The ADIA algorithm uses a simplified simulation of the engine. Scheduled functions of engine performance define the steady-state portion of the simplified simulation. A scheduled state–space system forms the basis of the dynamic portion of the simplified simulation. In total, linear state-space models at 119 different operating points were used that uniformly span the entire flight envelope. Each individual element of the state-space matrices was corrected to reduce data scatter and then scheduled by a nonlinear polynomial of selected model output variables over the flight envelope. This approach yields a model with linear structure that maintains the essential nonlinearities of the engine. A complete description of this modeling technology, as applied to the development of a hypothetical turbofan engine simplified simulation (HYTESS), is given by Merrill *et al.* [24]. A comparison of the response of the simplified with actual engine performance demonstrates the excellent estimation capability of the simplified simulation. The ADIA algorithm incorporates this simulation and Kalman filter logic to improve these estimates further (Fig. 2).

The test-bed system with ADIA and MVC logic is shown in Fig. 3. The ADIA algorithm consists of three elements: (1) hard-failure detection and isolation logic, (2) soft-failure detection and isolation logic, and (3) an accommodation filter. The algorithm detects two classes of sensor failures—hard and soft. Hard failures are out-of-range or large bias errors that occur instantaneously in the sensed values. Soft failures are small bias errors or drift errors that accumulate relatively slowly with time.

The algorithm consists of an extended steady-state Kalman filter, called the accommodation filter, that generates sensor estimates and residuals based on the previously described simplified engine simulation. These residuals are compared to thresholds for hard failure detection and isolation. Soft-failure detection and isolation is accomplished using a bank of six Kalman filters (one for each sensor failure and one for the no failure case) and a likelihood ratio test of the five different filter residuals (Fig. 4). The likelihood ratio test calculates a weighted sum of squared residuals (WSSR) statistic for each of the six filters. This statistic represents the log of the likelihood of the particular

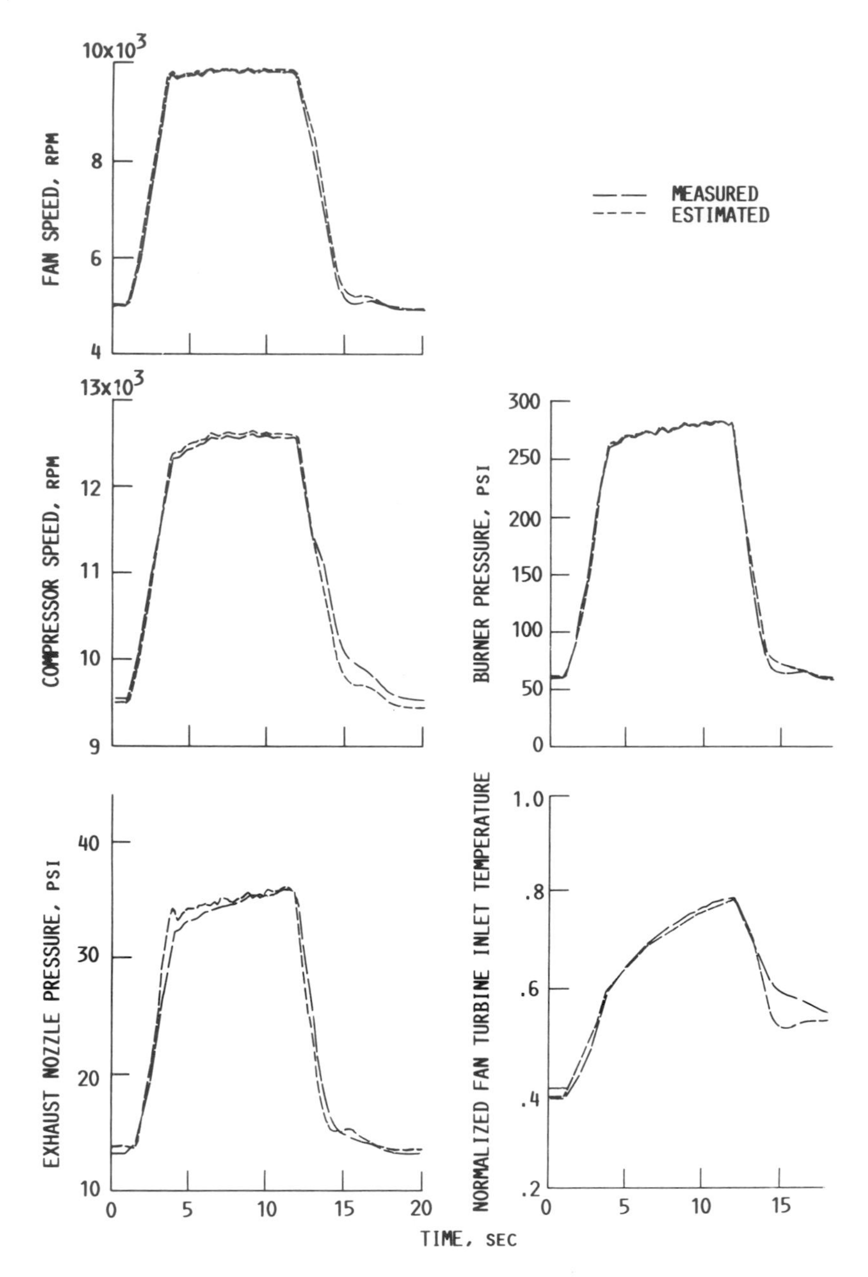

Fig. 2. A comparison of accommodation filter estimates with engine measurements for five engine outputs. Data show an acceleration and deceleration from idle to full power at 10,000-ft altitude and Mach number of 0.6.

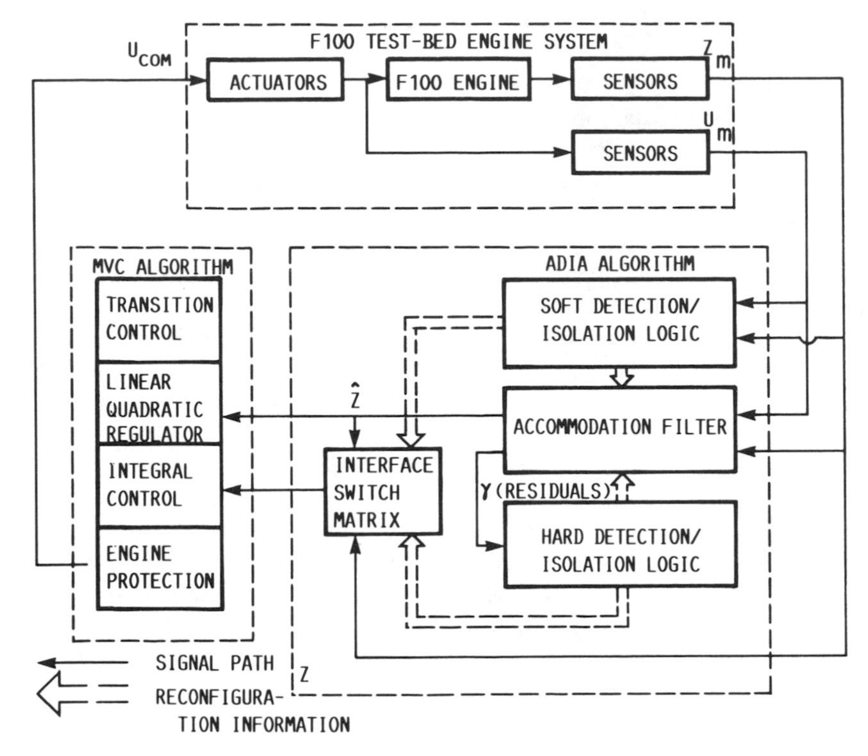

Fig. 3. F100 test-bed system with ADIA algorithm and MVC control.

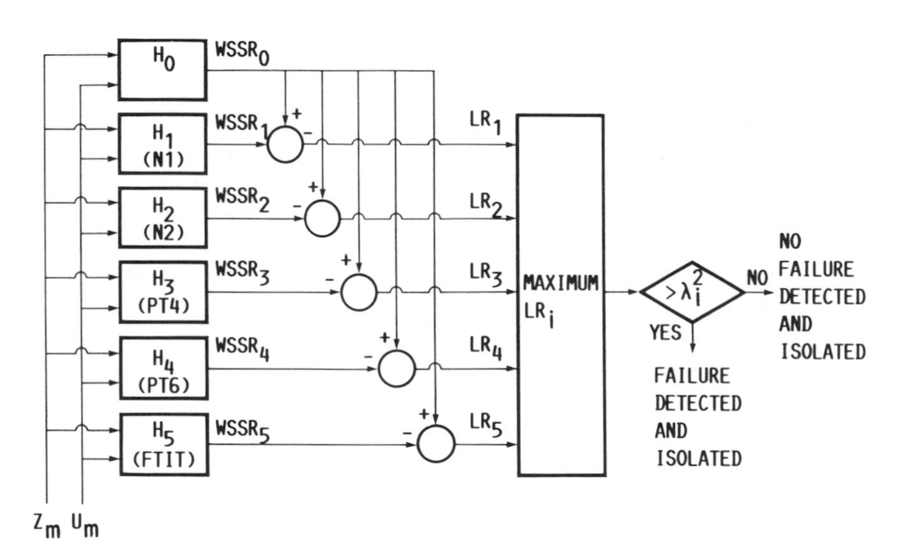

Fig. 4. Soft failure detection and isolation logic block diagram.

residuals being true. Subtracting the likelihoods of the five failure hypothesis filters from the normal-mode likelihood yields likelihhod ratios. The test then compares the maximum-likelihood ratio, which represents the maximum likelihood of a particular sensor failure hypothesis being true to a threshold. The threshold is adaptive and expands during transients to account for high-frequency modeling error. The adaptive threshold enables an 80% improvement in steady-state failure detection performance. After a failure is detected and isolated, the faulty information is removed from the accommodation filter by reconfiguration. Estimates of all sensor outputs are still produced, however, now they depend on the set of unfailed measurements. The ADIA algorithm interfaces with the MVC algorithm in two ways. First, it supplies the linear quadratic regulator (LQR) with estimates of the engine outputs at all times. Second, it supplies the integral control logic with actual sensed values in the normal mode. An individual sensed value is only replaced with an estimate when a failure occurs and is detected and isolated.

The second part of the ADIA program is the real-time microprocessor-based implementation of the MVC and ADIA algorithms. DeLaat and Merrill [28] describe a preliminary implementation of two 5-MHz. Intel-8086-based microprocessors operating in a parallel-processing environment. The first computer contains a fixed-point, assembly language, real-time implementation of the MVC that had been implemented and evaluated previously [29]. The second computer contains the detection and accommodation logic implemented in floating-point FORTRAN. Subsequent work has incorporated a third microprocessor into the implementation and replaced the 8086-based microprocessors with 8-MHz 80186-based microprocessors [30]. In this third computer the five isolation filters are implemented, again using floating-point FORTRAN.

The total control cycle time is 40 msec. Data are transferred between CPUs (central processing units) through dual-ported memory. Synchronization between CPUs is achieved through interrupts. The total memory requirement for the three CPUs is 54K bytes for the algorithm and 17K bytes for the real-time executive. In all cases the code and constants were about 65% and the data or variables about 35% of the total memory required.

In the evaluation phase [31, 32], an evaluation of algorithm performance was obtained using a real-time engine simulation running on a hybrid computer and microprocessor implementation described previously. The objectives of the evaluation were (1) validate the algorithm for sensor failure detection, isolation, and accommodation (DIA) effectiveness; (2) document algorithm performance; (3) validate the algorithm's real-time implementation; and (4) establish a data base for the demonstration phase of the ADIA program. All these objectives were successfully accomplished.

In the demonstration phase [33] the ADIA algorithm was tested on a full-scale F100 engine in the Lewis Research Center altitude test facility. The engine test successfully demonstrated the predicted performance of the ADIA algorithm on realistic hardware over a wide range of engine operating conditions. These conditions include altitude, Mach number, and power variations.

The criteria used to evaluate detection, isolation, and accommodation performance were (1) minimum detectable bias values and drift rates, (2) elapsed time between sensor failure and detection, (3) steady-state per-mance degradation after failure accommodation, and (4) transient response of the engine to the filter and control reconfiguration resulting from failure accommodation. Although the engine test demonstrated the capability to detect, isolate, and accommodate both hard and soft sensor failures, only soft-failure detection results are presented. This is because soft-failure detection is more difficult than hard-failure detection, and is therefore a more interesting problem. Two soft-failure modes, bias and drift, were studied.

The minimum detectable magnitudes of soft sensor bias failures for engine exhaust nozzle pressure (a variable closely related to engine thrust) demonstrated during testing are summarized in Fig. 5. Also shown in this figure, for comparison, are those minimum detectable magnitudes predicted by the real-time hybrid evaluation of the ADIA algorithm [32]. In general there is good agreement between predicted and observed detection magnitudes. This agreement demonstrates the excellent fidelity of the model used in the algorithm and the simulation used in the evaluation. Many of the values are the same. This is a result of the testing procedure. To minimize engine test time, the known evaluation predicted values were tested first. If the algorithm successfully detected the failure at the predicted magnitude, then that was the assumed minimum detectable value for the demonstration. If the detection was missed, the failure magnitude was increased until successful detection was demonstrated on the engine. Thus the demonstration values recorded in Fig. 5 are always equal to or greater than the predicted ones (except for N2 and PT4 at 55 K/2.2/72°). Although this was a conservative approach, it was clear from the test results that only minimal improvements over the predicted performance were possible at a limited number of operating points.

The times to detection for the soft bias failures were all less than 0.1 sec. The steady-state accommodation performance for this class of failure is shown in Fig. 6. Percentage changes in engine pressure ratio (EPR) are given by

$$\Delta \text{EPR} = 100[(\text{EPR}_{\text{T0}} - \text{EPR}_{\text{TF}})/\text{EPR}_{\text{T0}}]$$

$$\text{EPR} = \text{Exhaust nozzle pressure/engine inlet pressure},$$

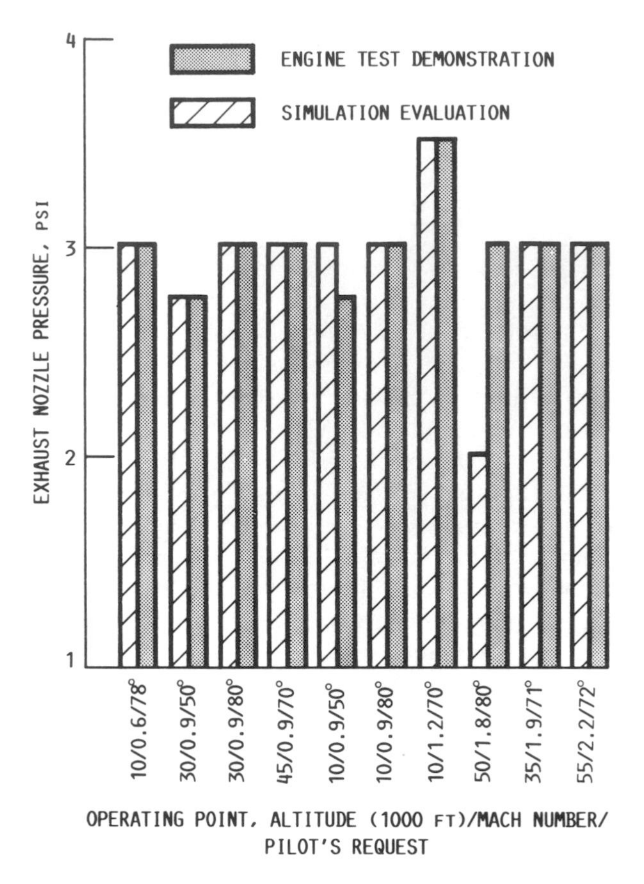

Fig. 5. Minimum detectable magnitudes of exhaust nozzle pressure sensor, soft bias failures at ten operating points (four supersonic and six subsonic).

where EPR_{T0} is the steady-state EPR before the failure and EPR_{TF} is the steady-state EPR after the failure. Performance is shown for several operating points demonstrating subsonic and supersonic operation at military and medium-power levels. Medium power is approximately half of military power. The parameter EPR is almost linearly related to engine thrust and is therefore a good measure of engine performance. All values are well below the 10% critical level except for operating condition 50 K/1.8 results (Fig. 6c), which show a 12% change in thrust for a PT6 sensor failure. This result is due to the low nominal value of PT6 at this condition (16.5 psi). The actual change in PT6 caused by the modeling error in the accommodation filter is only 2 psi and is considered relatively small. It appears large when compared with the low nominal value.

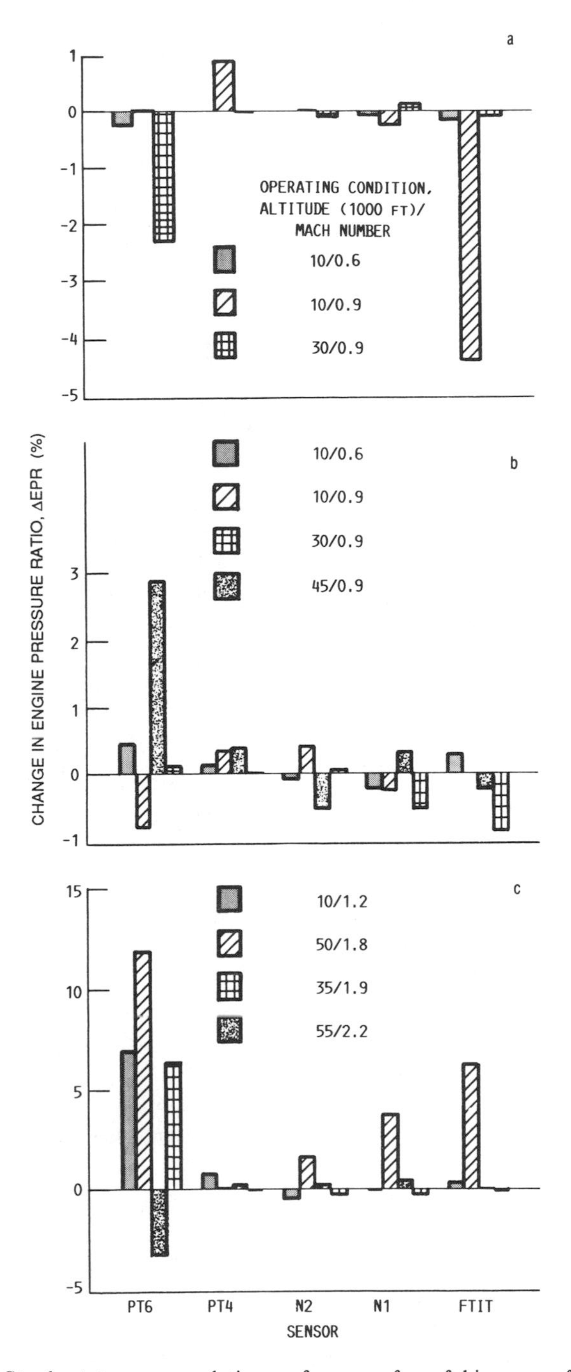

Fig. 6. Steady-state accommodation performance for soft bias sensor failures.

The minimum detectable drift magnitudes were determined by finding the smallest detectable drift failure such that a failure was detected approximately 5 sec after failure inception. Initial trial magnitudes were determined as before from predicted simulation values. The results are given in Fig. 7. Again shown in this figure for comparison are those minimum detectable magnitudes predicted by the real-time hybrid evaluation of the ADIA algorithm. As expected from the bias failure results, there is, in general, good agreement between predicted and observed detection magnitudes. Steady-state accommodation performance for subsonic and supersonic conditions at full and medium power for sensor drift failures was very good with most thrust changes being small and with none larger than the 10% level.

Additionally, detection performance for sequential failures was demonstrated. At condition 10 K/0.6 six different sequences of soft failures were injected into the test-bed system at medium power and one sequence was demonstrated at intermediate power. One example of a failure sequence was to fail N1, then 4 sec later fail N2, then PT4, and then PT6. In each case the algorithm successfully detected and accommodated each sensor failure in the correct order. Steady-state thrust changes are all close to the critical 10% level except for the intermediate power case. In each case these changes were experienced well into the transient when only two of the five sensors remained unfailed. These tests demonstrate the ability of the algorithm to continue to successfully perform even after most of the sensors have failed.

Finally, a simultaneous soft failure of PT4 and PT6 (both failed at the same instant of time) was injected into the engine system. The algorithm, although not designed for this extremely low probability event, successfully detected and accommodated this failure scenario. The change in EPR is about 1.5 psi or less than 7%.

All of these failures were electronically generated using special purpose hardware to give timed and repeatable results. The generated failures represent realistic sensor failures, which were injected into the engine control system. However, during engine testing an unplanned failure of actual sensor hardware was detected by the ADIA logic. Additionally, there were no missed detections of sensor hardware failures by the ADIA logic. The sensor hardware failure was associated with the fan turbine inlet pressure (FTIT) measurement. About 13 sec after the start of a nozzle pressure sensor failure detection experiment, a FTIT soft failure was detected. From the sensed FTIT signal shown in Fig. 8a it is clear that some transient anomaly occurred. The likelihood ratio for FTIT given in Fig. 8b shows the detection taking place at 13 sec (note that the detection threshold for this case is twice the normal size). The failure mode is indicative of a momentary “singing” of a signal conditioning amplifier. Although conclusive proof of a hardware failure was not obtained because of its nonrepeatability, this failure mode was not

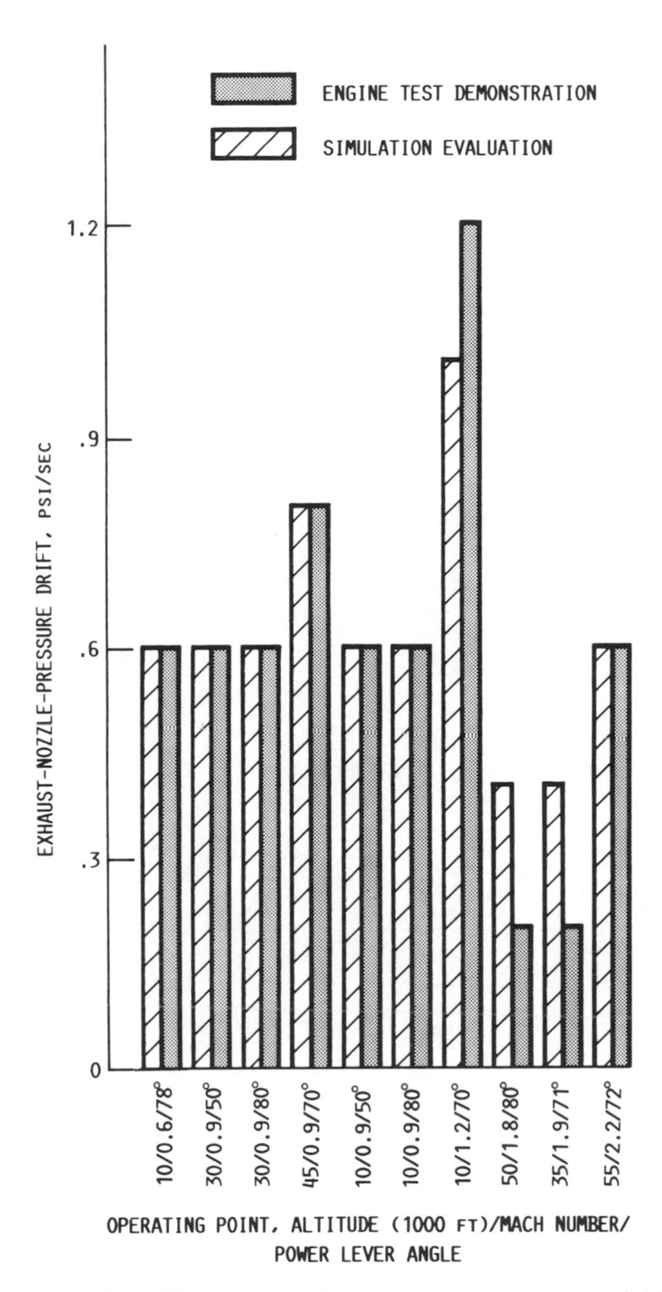

Fig. 7. Minimum detectable magnitudes of exhaust nozzle pressure sensor soft drift failures at ten operating points (four supersonic and six subsonic).

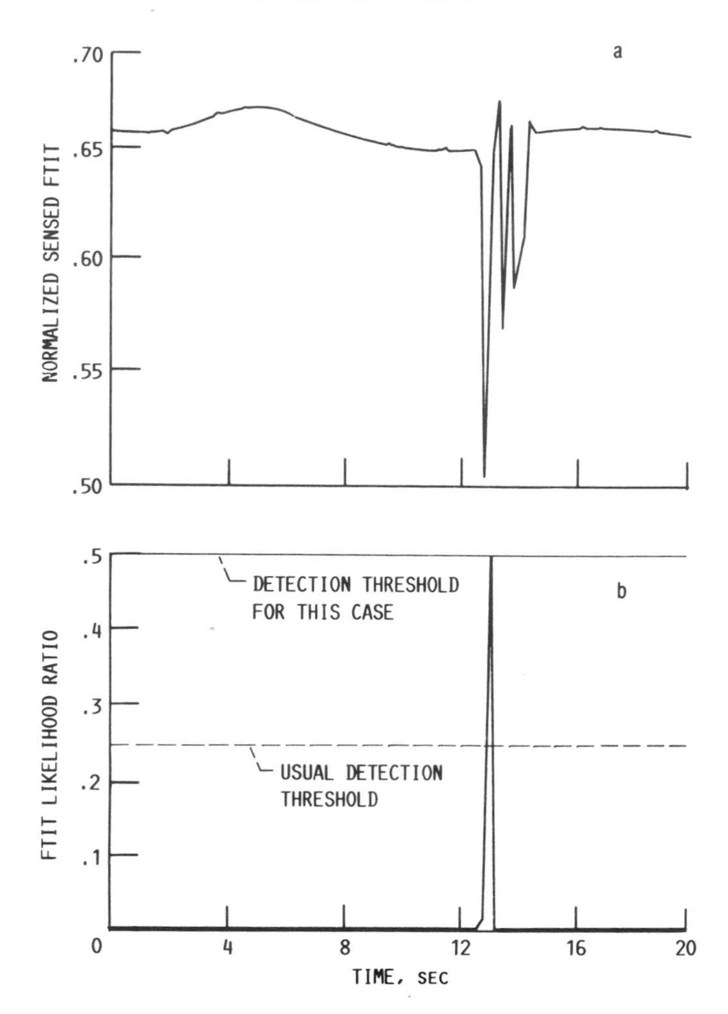

Fig. 8. Unplanned FTIT measurement hardware soft failure: (a) sensed and (b) likelihood ratio.

observed again once the suspected signal conditioning amplifier was replaced.

Two experiments were used to demonstrate the successful accommodation, or post-failure performance, of sensor failures. The first experiment consisted of injecting, detecting, and accommodating a single sensor failure and then commanding an acceleration–deceleration pulse transient. Engine performance, measured by the average absolute value of control error over the transient, with this accommodated failed sensor is compared to normal-mode engine performance. Eighteen of these single failure pulse tests were performed at five different operating points.

In general the change in performance is small for each experiment. The largest fan speed error change of about 160 rpm is, in fact, quite small when compared to the typical operating range of fan speed (5000 to 10,000 rpm). Results for an engine acceleration–deceleration pulse response with a single exhaust nozzle pressure sensor failure and its baseline response are shown in Fig. 9. Results are shown for both fan speed and exhaust nozzle pressure. In general, performance was good since the desired or request values were closely maintained. A slight drop in actual nozzle pressure can be seen but this is acceptable. In all other cases the accommodated single failure transient performance was good. The fluctuations evident in nozzle pressure at the high power level are caused by an airflow interaction between the altitude test cell and the engine.

The second accommodation performance experiment demonstrated the excellent accuracy of the engine model. In this experiment all the engine output sensors were failed and accommodated. Then, the engine was commanded to respond to a PLA (power lever angle) pulse transient. Two all-sensors-failed pulse transient experiments were performed at different conditions. At the first condition (altitude 10,000 ft, Mach number = 0.6) the transient was from idle to about 75% of full power. As confidence in the ability to safely control the engine without engine output sensors increased, a second test went to full power. Fan speed and exhaust nozzle pressure results for the first condition are shown in Fig. 10. Here excellent performance was demonstrated. Little or no overshoot was observed and engine steady-state performance was good. This demonstrates the capability of safe, predictable engine operation without any engine feedback information over a slightly restricted power range. Again the fluctuations in nozzle pressure at high power were caused by an airflow interaction between the facility and the engine. Performance for the second condition was similar.

Based on the results of the engine test several conclusions have been reached. First, the ADIA failure detection algorithm works and works quite well. Sensor failure detection and accommodation were demonstrated at 11 different operating points, which included subsonic and supersonic conditions and medium- and high-power operation. The minimum detectable failure magnitudes represent excellent algorithm performance and compare favorably to values predicted by simulation. Accommodation performance was excellent. Transient engine operation over the full power range with single sensors failed and accommodated was successfully demonstrated. Open-loop engine operation (all sensors failed and accommodated) over at least 75% of the power range was also demonstrated at two different operating conditions.

Second, the algorithm is implementable in a realistic environment and in an update interval consistent with stable engine operation. Off-the shelf microprocessor-based hardware and straightforward programming

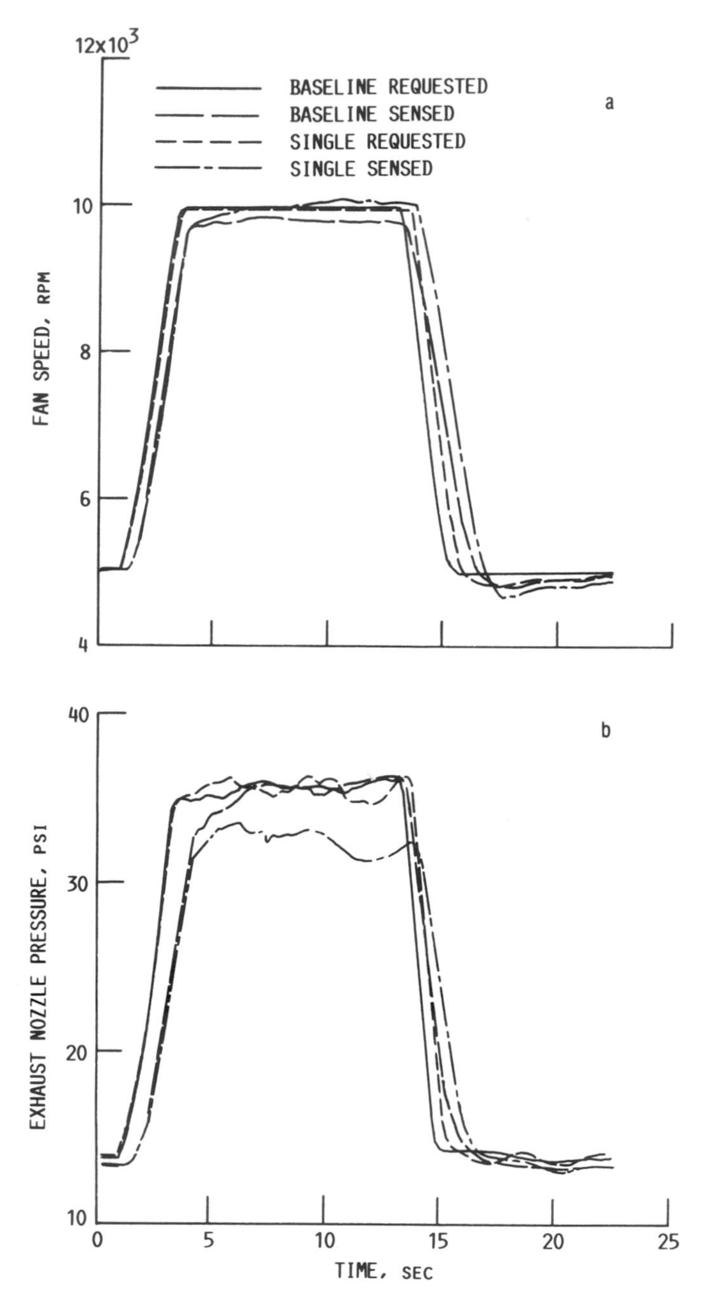

Fig. 9. Engine acceleration–deceleration response with exhaust nozzle pressure failure: (a) fan speed and (b) exhaust nozzle pressure.

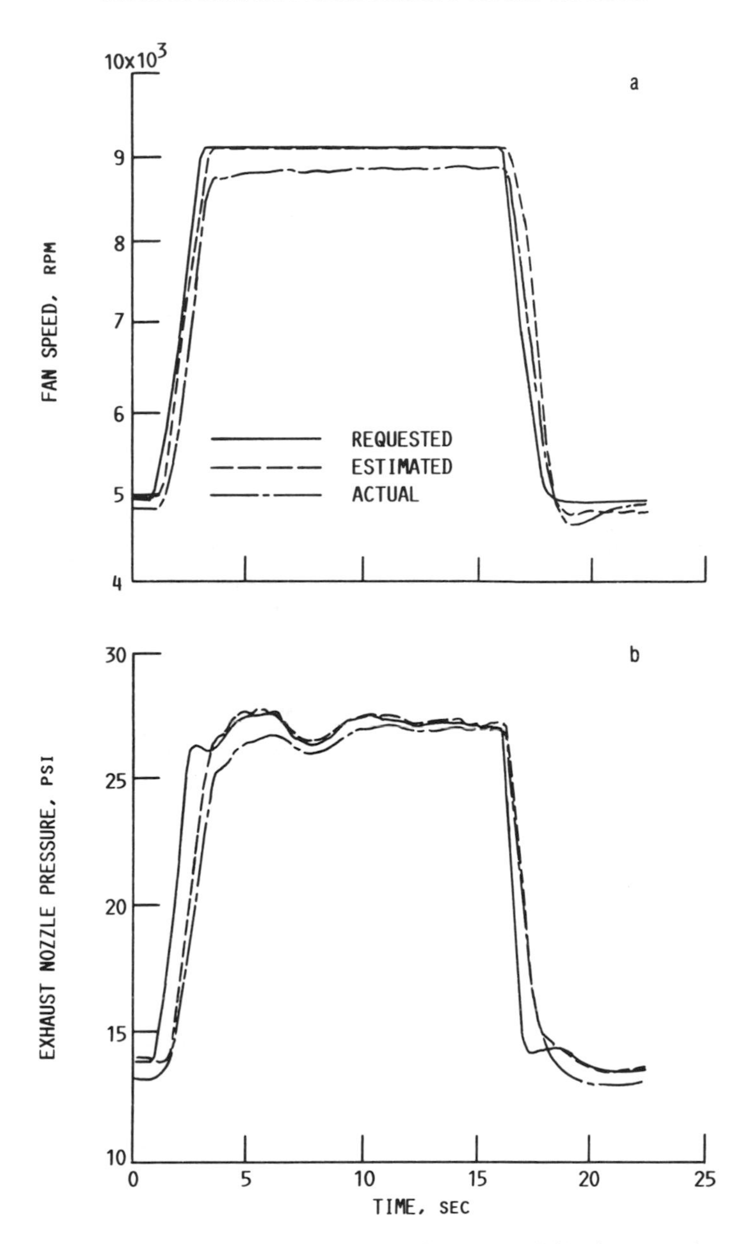

Fig. 10. Open-loop acceleration–deceleration: (a) fan speed and (b) exhaust nozzle pressure.

procedures, including FORTRAN and floating-point arithmetic, were used. Parallel processing was also used and shown to be an effective approach to achieving a real-time implementation using off-the-shelf (cost-effective) computer resources.

B. ENERGY-EFFICIENT ENGINE (E^3) FAILURE INDICATION AND CORRECTIVE ACTION

The E^3 program [34] is developing technology to improve the energy efficiency of future commercial transport aircraft engines. A FADEC based on the bit-slice AMD 2901 microprocessor is used to implement the control and FICA logic for the engine developed under this program. The FICA logic is based on the concept of Spang and Corley [11]. Here, a sixth-order extended Kalman filter is used to generate seven sensor estimates: fan and core speed, compressor inlet and discharge temperatures, turbine discharge temperature, fuel-metering value position, and compressor discharge static pressure. The Kalman filter uses a dynamic model of simplified engine aerothermodynamics and rotor dynamics. Actuator and sensor dynamic models are also included. This model accurately describes the engine over the full-power range and flight envelope using simplified component modeling. The Kalman gain matrix is computed at a key operating point using a linearized engine model. Sensor failures are detected when the sensed versus estimated difference is greater than a prespecified tolerance. Out-of-range failures are also detected. The tolerance is estimated by statistical analysis and adjusted during simulation trials. Accommodation of failures is accomplished by replacement of sensed values with estimated values. A nonlinear real-time simulation evaluation of the FICA logic showed that the filter estimate tracked the sensed values within the specified tolerance and successfully detected, isolated, and accommodated all hard sensor failures except fuel-metering valve position. The E^3 FICA logic does not detect slow-drift (i.e., soft) sensor failures.

C. FULL AUTHORITY DIGITAL ELECTRONIC CONTROL FAILURE INDICATION AND CORRECTIVE ACTION

Under the FADEC program [35] AR techniques (in particular, FICA) were applied to two engines, a joint technology demonstrator engine (JTDE) and the F404 afterburning turbofan engine. Each of these applications is discussed in the following.

The JTDE FICA was designed for a variable-cycle engine with seven manipulated variables and nine sensed variables. The engine model used in the

JTDE FICA is a second-order, dynamic pseudolinear model valid throughout the flight envelope. The model is updated by an observer. Observer gains were chosen as the reciprocals of corresponding engine model steady-state gains at a high-power condition. Gains were then adjusted to achieve adequate stability margins. For failure detection, sensor model errors were compared to a preset threshold. Substitution of estimated variables was demonstrated using a simulation and, subsequently, a full-scale engine. The engine demonstration was limited to sea-level static conditions and single substitutions. Single substitutions for fan speed, compressor discharge static pressure, and compressor inlet temperature were performed successfully. Also demonstrated by simulation in this program was the application of FICA techniques to actuator sensor failures. In particular, fuel flow and nozzle area actuator hard out-of-range sensor failures were detected and accommodated.

A second application of the FICA technology was to the F404 engine. The F404 is an afterburning turbofan engine with a rear variable-area bypass injector to permit selective cycle rematch. The rear injector adjusts the bypass-to-core-air ratio to match cycle demands. The engine includes five inputs and five outputs. A simplified, fourth-order, component-level model [36] is used in the FICA system. The model is accurate throughout the flight envelope and was implemented in FADEC microprocessor hardware in a 0.01-sec update time increment. The model along with the FICA update logic was checked against actual engine operation during full-scale engine tests at sea-level static and altitude conditions [37]. Steady-state and transient model accuracies were judged to be excellent. Single, double, and triple substitutions of FICA-generated estimates were performed successfully during the engine tests. These combinations are summarized in Fig. 11. Actuator FICA was also demonstrated successfully for exhaust nozzle hard open and closed failures. Thrust level in these cases was maintained by adjusting the gas generator speed reference schedule.

D. DIGITAL ELECTRONIC ENGINE CONTROL DETECTION, ISOLATION, AND ACCOMMODATION

The DEEC system [38] is a digital full-authority engine control containing selectively redundant components and fault-detection logic. The system also contains a hydromechanical backup control. Most of the sensors in the control are hardware redundant. However, failures of the inlet static pressure (PS2), burner pressure (PB), and fan turbine inlet pressure (FTIT) are covered using a form of AR called parameter synthesis.

In parameter synthesis an estimate of one measured variable is synthesized from an algebraic function of one or more different measured variables. This

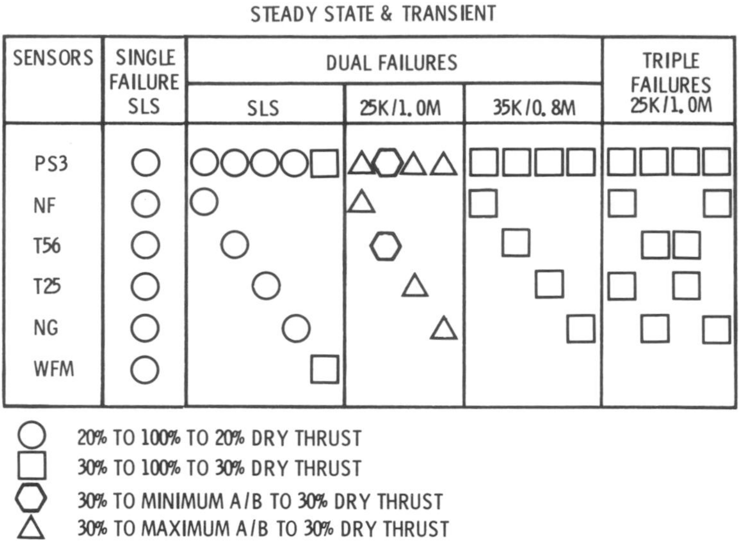

STEADY STATE & TRANSIENT

SENSORS	SINGLE FAILURE SLS	DUAL FAILURES SLS	DUAL FAILURES 25K/1.0M	DUAL FAILURES 35K/0.8M	TRIPLE FAILURES 25K/1.0M
PS3	○	○○○○□	△⬡△△	□□□□	□□□□
NF	○	○	△	□	□ □
T56	○	○	⬡	□	□□
T25	○	○	△	□	□ □
NG	○	○	△	□	□ □
WFM	○	□			

○ 20% TO 100% TO 20% DRY THRUST
□ 30% TO 100% TO 30% DRY THRUST
⬡ 30% TO MINIMUM A/B TO 30% DRY THRUST
△ 30% TO MAXIMUM A/B TO 30% DRY THRUST

Fig. 11. F404 FICA sensor substitution results.

relationship is static (i.e., no explicit dynamics are included). If PS2 fails a range check, a synthesized PS2 is determined from PB, compressor speed N2, and inlet total temperature TT2. IF PB fails, a synthesized PB is calculated from inlet total pressure, PT2, N2, and TT2. Fault detection of PB failures is based on a comparison of measured and synthesized values. A comparison tolerance of $\pm 25\%$ determines failures. This large tolerance precludes detection of soft failures. Both PS2 and PB failures are accommodated by substitution.

There are two groups of FTIT sensors. This allows hardware redundancy. However, if both FTIT sensor groups fail a range check, synthesized FTIT is substituted into the control. Synthesized FTIT is a function of PB and PT2.

The DEEC DIA logic was verified by closed-loop bench testing. Simulated sea-level and altitude engine transients were performed. Faults were intentionally produced to evaluate effectiveness. Subsequent sea-level and altitude full-scale engine tests uncovered no new problems with the DIA logic. A series of flight tests of a F15 aircraft with an F100 engine and DEEC control further demonstrated the DEEC logic [39]. During the flight program, the DEEC DIA logic did not detect any false alarms and did not cause any reversions to backup hydromechanical control. Two sensor failures occured during the flight program. One, inlet temperature, was covered by redundant hardware. The second, exhaust nozzle pressure, failed to a high-scale sensor limit. Appropriate accommodation action was taken by the logic in each case.

Neither of the two sensor failures encountered in the flight-test program demonstrated the AR-based logic of the DEEC DIA. In a subsequent flight test program [40] the objective was to induce selected hard sensor faults and evaluate the resulting actions of the control. The test program included both an extensive ground-thrust stand evaluation and a flight test. In flight failures were introduced at steady-state conditions and during throttle transients. Throttle transients were performed by inducing the failures before and during the throttle movement. The sensors failed during the flight test included the compressor inlet variable geometry sensor, PS2, PB, and FTIT. Most failures were detected and accommodated. However, a recreation of a broken line (hard) PB failure went undetected. Pilot response to aircraft performance after accommodation was favorable.

E. ANALYTICAL REDUNDANCY TECHNOLOGY FOR ENGINE RELIABILITY IMPROVEMENT

ARTERI [41] is a program to develop AR techniques based on FICA to the point where they may be employed in a full-scale engine development program. Both hard and soft failures must be covered over the full range of engine power and flight conditions. A component-tracking module, which tunes the engine model to match the actual engine by updating engine model dynamic states, inputs, outputs, and component performance parameters, is used to extend FICA to include a soft failure DIA capability. Nonlinear simulation results have demonstrated the ability of the logic to discriminate among sensor, and actuator hard and soft failures.

The real-time implementation and demonstration of the ARTERI logic on an actual engine remain to be accomplished. Also, the component tracking filter adapts the engine model at a selected operating point. Its global capabilities need to be improved to allow soft-failure detection during large excursions in power or operating condition.

F. PW2037 ENGINE

The PW2037 engine is a modern, high-bypass-ratio turbofan engine and is the first to incorporate a completely digital, full-authority electronic control system [2]. The control is engine mounted and dual channeled to meet reliability requirements. As part of the control's redundancy management strategy, a combination of hardware and software sensor redundancy is used to ensure engine operability whenever the capability is available. Dual hardware is used for seven sensors (two speeds, two pressures, two temperatures, and thrust level angle). All of these sensors are covered by channel-to-channel comparisons, as well as software range and rate checks, to detect failures. In the case of the two engine speeds and the two pressures, sensor failures are further covered by comparisons to synthesized estimates. In the case of a dual-channel failure (both low-spool-speed sensors, for example), operation continues using the synthesized estimate of low spool speed. The two pressures and high spool speed are synthesized from low spool speed using a parameter synthesis method. Low spool speed is synthesized from high spool speed.

IV. ANALYTICAL REDUNDANCY TECHNOLOGY ASSESSMENT

From the preceding survey an assessment of the relative state of the art of applied AR can be obtained. The results presented in the technology base, and summarized in Table I, demonstrate the feasibility of AR-based DIA. In particular, straightforward range or rate checks have provided successful detection of hard sensor failures. Further, advanced DIA approaches based on advanced statistical decision theory and optimal filtering have demonstrated soft failure DIA feasibility. However, this soft-failure DIA capability is obtained at the cost of increased computational complexity. This additional complexity consists of two parts: the filtering and decision-making logic, and a more accurate, and therefore more detailed, model. These results also demonstrate a trade-off between ability to accurately detect and time to detect. Where hard failures can be detected almost instantly, soft failures are reliably detected only after some finite amount of time. This time to detect is a function of threshold level, which determines detection reliability, required model accuracy, and logic complexity. Usually for soft failures, more time is available before accumulated error is damaging.

Further results presented in the technology development section demonstrate AR-based DIA capability for hard and soft sensor failures on full-scale engines over a wide range of power and flight conditions. State-of-the-art operational systems, such as the DEEC and the PW2037 control, use only limited AR in combination with more extensive hardware redundancy.

The work presented in this article clearly emphasized the fundamental importance of modeling in successful DIA. A model detailed enough for accurate DIA throughout the flight envelope is a significant technical challenge. Expectantly, when faced with a difficult technical problem, different approaches are pursued. Three different modeling approaches have been used: (1) parameter synthesis, (2) pseudolinear, and (3) simplified component. Both the parameter synthesis and simplified component modeling approaches have been used in successful hard-failure DIA on full-scale engines. The pseudolinear method has been demonstrated for both hard and soft-failure DIA on a full-scale engine. Each approach has its own advantages and disadvantages.

The parameter synthesis approach, which was used in the DEEC DIA and the PW2037, is simple to understand and straightforward to implement. Explicit dynamics normally are not included. However, this simplicity implies a less accurate model. Also, the most accurate interrelationships between measured and synthesized variables can not be identified easily or systematically. Model modifications are made easily.

The simplified component approach, which was used in the FADEC FICA, results in more accurate models than the parameter-synthesis approach. Simplified component models are based on detailed nonlinear engine simulations. Detail is selectively removed from the detailed simulation to maximize simplicity while maintaining accuracy. This process requires a great deal of judgment and is not straightforward or systematic. In addition, simplified model performance is not easily predicted. A simplified component model relates naturally to the physics of the actual engine and, therefore, is readily understandable. However, modification of a simplified component model is not straightforward since changes in component performance can have unpredictable effects on model performance.

The pseudolinear modeling method used in the ADIA algorithm, is a very organized, systematic approach. However, to achieve accuracy through a wide range of conditions requires a large amount of stored data. The relationship of a pseudolinear model with engine physics is not as straightforward as for a simplified component model. However, steady-state and dynamic model performance can be separated and modified independently. Due to the linear structure of the model equations, analysis and performance prediction is much easier with a pseudolinear model than with parameter synthesis or simplified component models. In addition, the complexity/accuracy trade-off is defined more clearly for a pseudolinear model.

V. CONCLUDING REMARKS

This article has surveyed the technology base and technology applications for analytical redundancy (AR)-based sensor failure detection, isolation, and accommodation (DIA) strategies for gas turbine engines. Several observations and conclusions are made. Comparisons of PW2037 technology with that of the F8 digital fly-by-wire program, or the approach used in the ADIA program with that proposed by Montgomery and Caglayan [42], show that engine AR technology often builds or expands on technology developed for flight controls. Also, modeling is the key issue in the success of AR techniques. Three types of models are used. Each has its advantages and disadvantages and no clear preferred type emerges. Because of this strong dependence of performance on modeling accuracy, fundamental questions about detection performance and robustness have been posed and addressed in robust DIA programs. Finally, simulation or full-scale engine testing has conclusively shown the feasibility of AR-based DIA for hard and soft failures.

REFERENCES

1. L. E. BAKER, D. E. WARNER, and C. P. DISPARTE, "Design of Fault Tolerant Electronic Engine Controls," AIAA Paper 81-1496 (1981).
2. J. F. KUHLBERG, D. M. NEWIRTH, J. KNIAT, and W. H. ZIMMERMAN, "Integration of the PW2037 Engine Electronic Control System in the Boeing 757 Airplane," SAE Paper 841554 (1984).
3. J. C. DECKERT, M. N. DESAI, J. J. DEYST, and A. S. WILLSKY, "F-8 DFBW Sensor Failure Identification Using Analytical Redundancy," *IEEE Trans. Autom. Control* **AC-22**, 795–803 (1977).
4. R. E. WALLHAGEN, and D. J. ARPASI, "Self-Teaching Digital-Computer Program for Fail-Operational Control of a Turbojet Engine in a Sea-Level Test Stand," NASA TM X-3043 (1974).
5. F. J. HRACH, D. J. ARPASI, and W. M. BRUTON, "Design and Evaluation of a Sensor Fail-Operational Control System for a Digitally Controlled Turbofan Engine," NASA TM X-3260 (1975).
6. S. H. ELLIS, *In* Third International Symposium on Air Breathing Engines" (D. K. Hennecke and G. Winterfeld, eds.), pp. 171–186. Dtsch. Ges. Luft Raumfahrt, Cologne.
7. C. W. DE SILVA, "Sensor Failure Detection and Output Estimation for Engine Control Systems," M.S. Thesis, University of Cincinnati (1976).
8. C. W. DE SILVA, *Arabian J. Sci. Eng.* **7**, 45–53 (1982).
9. W. R. WELLS and C. W. DE SILVA, *in* "Failure State Detection of Aircraft Engine Output Sensors," "Proceedings of Joint Automatic Control Conference," Vol. 2, pp. 1493–1497. IEEE, Piscataway, New Jersey, 1977.
10. W. R. WELLS, "Detection of Sensor Failure and Output Reconstruction for Aircraft Engine Controls," AIAA Paper 78-4 (1978).
11. H. A. SPANG, III and R. C. CORLEY "Failure Detection and Correction for Turbofan Engines," Rep. No. 77CRD159, General Electric Co., Schenectady, New York, 1977.
12. R. L. DEHOFF and W. E. HALL, JR., "Advanced Fault Detection and Isolation Methods for Aircraft Turbine Engines," Systems Control, Inc., ONR-CR-215-245-1 (1978). (Avail. NTIS AD-A588991.)
13. R. K. SAHGAL and R. J. MILLER, *in* "Proceedings of Joint Automatic Control Conference," pp. 381–386. IEEE, Piscataway, New Jersey, 1979.
14. G. G. LEININGER, and K. BEHBEHANI, *in* "Proceedings of Joint Automatic Control Conference," Vol. 2, Paper TP4-B. IEEE, Piscataway, New Jersey, 1980.
15. K. BEHBEHANI, "Sensor Failure and Multivariable Control for Airbreathing Propulsion Systems," Ph. D. Dissertation, University of Toledo (1980). (Also NASA CR-159791.)
16. K. BEHBEHANI and G. G. LEININGER, *in* "Propulsion Controls 1979," NASA CP-2137, pp. 139–143. National Aeronautics and Space Administration, Washingon, D.C., 1980.
17. J. S. MESEROLE, JR., "Detection Filters for Fault-Tolerant Control of Turbofan Engines," Ph. D. Dissertation, Massachusetts Institute of Technology (1981).
18. G. G. LEININGER, *in* "Proceedings of Joint Automatic Control Conference," Vol. 2, Paper FP-3A. IEEE, Piscataway, New Jersey, 1981.

19. J. L. WEISS, K. R. PATTIPATI, A. S. WILLSKY, J. S. ETERNO, and J. T. CRAWFORD, "Robust Detection/Isolation/Accommodation for Sensor Failures," Alphatec, Inc., Burlington, Mssachusetts, 1985. TR-213, (Also NASA CR-174797.)
20. K. R. PATTIPATI, A. S. WILLSKY, J. C. DECKERT, J. S. ETERNO, and J. S. WEISS, *in* "Proceedings of the 1984 American Control Conference," pp. 1755–1762. IEEE, Piscataway, New Jersey, 1984.
21. A. EMMANI-NAEINI, M. M. AKHTER, and M. M. ROCK, "Robust Detection, Isolation, and Accommodation for Sensor Failure," NASA CR-174825 (1985).
22. H. BROWN, R. C. CORLEY, J. A. ELGIN, and H. A. SPANG, "Multi-Engine Detection, Isolation, and Accommodation of Sensor Failures," R84AEB359, General Electric Co., Cincinnati, Ohio, 1984. (Also NASA CR-174846.)
23. D. T. HORAK, "Failure Detection in Dynamic Systems with Modeling Errors," *J. Guidance, Control, Dyn.* **11**, No. 6 (1988).
24. W. C. MERRILL, E. C. BEATTIE, R. F. LAPRAD, S. M. ROCK, and M. M. AKHTER, "HYTES: A Hypothetical Turbofan Engine Simplified Simulation," NASA TM-83561 (1984).
25. E. C. BEATTIE, R. F. LAPRAD, M. E. MCGLONE, S. M. ROCK, and M. M. AKHTER,, "Sensor Failure Detection System," PWA 5736-17, Pratt & Whitney Aircraft Group, East Hartford, Connecticut, 1983. (Also NASA CR-165515.)
26. E. C. BEATTIE, R. F. LAPRAD, M. M. AKHTER, and S. M. ROCK, "Sensor Failure Detection for Jet Engines," PWA-5891-18, Pratt & Whitney Aircraft Group, East Hartford, Connecticut, 1983. (Also NASA CR-168190.)
27. B. LEHTINEN, W. G. COSTAKIS, J. F. SOEDER, and K. SELDNER, "F100 Multivariable Control Synthesis Program—Results of Engine Altitude Tests," NASA TM S-83367 (1983).
28. J. C. DELAAT and W. C. MERRILL, "A Real-Time Implementation of an Advanced Sensor Failure Detection, Isolation, and Accommodation Algorithm," NASA TM-83553 (1983).
29. J. C. DELAAT and J. F. SOEDER, "Evaluation of a Microprocessor Implementation of the F100 Multivariable Control," NASA TM-87130 (1985).
30. J. C. DELAAT, and W. C. MERRILL, "A Real-Time Microprocessor Based Implementation of the ADIA Algorithm," NASA E-4391 (1989).
31. W. C. MERRILL and J. C. DELAAT, "A Real-Time Simulation Evaluation of an Advanced Detection, Isolation and Accommodation Algorithm for Sensor Failures in Turbine Engines," NASA TM-87289 (1986).
32. W. C. MERRILL, J. C. DELAAT, and W. M. BRUTON, "Advanced Detection, Isolation, and Accommodation of Sensor Failures: Real-Time Evaluation," *AIAA J. Guidance, Control Dyn.* **11**, No. 6 (1988). (Also NASA TP-2740).
33. W. C. MERRILL, J. C. DELAAT, S. M. KROSZKEWICZ, and M. ABDELWAHAB, "Advanced Detection, Isolation, and Accommodation of Sensor Failures—Engine Demonstration Results," NASA TP-2836 (1988).
34. R. S. BEITLER and J. P. LAVASH, "Energy Efficient Engine (E^3): Controls and Accessories Detail Design," R82AEB400, General Electric Co., Cincinnati, Ohio, 1982. (Also NASA CR-168017).
35. T. M. KREITINGER *et al.*, "Full Authority Digital Electronic Control, Phase II, Final Report—Industry Version," R82AEB435, General Electric Company, Cincinnati, Ohio, 1983.
36. M. W. FRENCH, "Development of a Compact Real-Time Turbofan Engine Dynamic Simulation," SAE Paper 821401 (1982).
37. K. L. LINEBRINK and R. W. VIZZINI, "Full Authority Digital Electronic Control (FADEC)—Augmented Fighter Engine Demonstration," SAE Paper 821371 (1982).
38. L. P. MYERS, *in* "Digital Electronic Engine Control (DEEC) Flight Evaluation in an F-15 Airplane," NASA CP-2298, pp. 33–54. National Aeronautics and Space Administration, Wasington, D.C., 1984.

39. L. P. MYERS, K. G. MACKALL, F. W. BURCHAM, JR., and W. A. WALTER, "Flight Evaluation of a Digital Electronic Engine Control System in an F-15 Airplane," AIAA Paper 82-1080 (1982).
40. L. P. MYERS, J. L. BAER-RIEDHART, and M. D. MAXWELL, "Fault Detection and Accommodation Testing on an F100 Engine in an F-15 Airplane," NASA TM-86735 (1985).
41. H. BROWN and J. A. SWANN, "Analytical Redundancy Technology for Engine Reliability Improvement," Naval Air Propulsion Test Center, NAPC-PE-171C (1987).
42. R. C. MONTGOMERY and A. K. CAGLAYAN, "Failure Accommodation in Digital Flight Control Systems by Bayesian Decision Theory," *J. Aircr.* **13**, 69–75 (1976).

HUMAN RESPONSE MODELS FOR INTERPRETIVE TASKS

D. D. SWORDER
KARYN S. HAALAND

Department of Applied Mechanics and Engineering Sciences
University of California—San Diego
La Jolla, California 92093

I. INTRODUCTION

Human decision makers play a ubiquitous but controversial role in the architectures of many complex systems. A trained individual has a unique talent for recognizing changes in situation and for allocating resources appropriately in a dynamically varying encounter that is subject to significant uncertainty. Such skills are utilized, for example, in an application in which a person must distinguish the highest priority target in a cluster of like objects in an environment permeated with a high level of structured clutter. In this case the decision maker is able to bring some semblance of order to observations of the motion of interacting groups of hostile and friendly vehicles, and, on the basis of his or her understanding of the implications of the perceived events, can make rational judgments on suitable strategies. Indeed, the ability of a human to employ powers of analogical reasoning and problem restructing in response to sudden and unexpected events contrasts sharply with algorithmic approaches to decision-making problem [1].

A full exploitation of this human capability is sometimes difficult to achieve. The physical support for a human presence is frequently lacking in precisely those locations in which such a presence would be of most value. One way to obviate these environmentally imposed constraints would be to assign some of the more primitive human roles to an autonomous system. For example, a self-navigating vehicle with a sophisticated sensor suite could, in principle, direct itself to a proper position for monitoring an encounter, and an

Copyright © 1990 by Academic Press, Inc.
All rights of reproduction in any form reserved.

algorithmic surrogate could then perform the observation and decision-making functions in lieu of direct human presence. Such autonomous systems have not proven to be as effective as was initially hoped. Expeditious navigation over broken terrain has proven to be an obdurate task. The implications of cluttered observations of the motions of a multiple-vehicle, dynamic encounter are unfathomable by algorithms derived on the basis of current technology. At present, only the most modest and precisely focused tasks are capably dealt with autonomously.

The motivation for the work presented here arose from a study of the utility of a teleoperated vehicle (TOV). A TOV provides a platform for projecting the human essence into a desired but inhospitable location while maintaining the corporeal being in a more secure site. Improvements in fiber optic technology permit a high fidelity transfer of sensory information between a mobile platform and a human operator situated in a remote and sheltered location. A system of this type in which an anthropomorphic robot mimics the motion of a remote operator has been constructed by the Naval Oceans Systems Center (NOSC) (Fig. 1a). Stereoscopic visual and stereophonic auditory stimuli as "seen and heard" by the robot as it moves (Fig. 1b) are transferred back in a form that mirrors the operator's direct perception. The operator can then change the direction and magnification of his or her visual field of view, or control the vehicle by making the appropriate motions in the control room (Fig. 1c).

There are some relevant differences between the milieu as it appears to the remote operator, and that which would be experienced by an *in situ* operator. Thus, the motion cues so useful in rapid transit over irregular terrain are not available to the operator of the TOV. Similarly, there are some limitations on the acuity of the sensory information conveyed from the TOV because of constraints induced by the sensorial hardware and the transmission mechanism. Therefore, it is important to study the interpretive capability of a remote operator as a function of the nominal descriptors of the decision-making environment.

To provide a description of operator behavior that will aid in defining the appropriate human role in the system architecture, it is advantageous to have a behavioral model that is both relatively simple and compatible with the models of the other components of the encounter. Analytical models of human response have a long history, with careful development beginning in the 1940s. More recently, as the tasks assigned to the human have become more multifarious, increasingly sophisticated models have been required. In [2] Johannsen and Rouse proposed a framework within which human activities could be organized. Their hierarchical perspective is amenable to a quantitative computerlike interpretation of human functions, but at the same time accounts for higher level psychological and intellectual activities

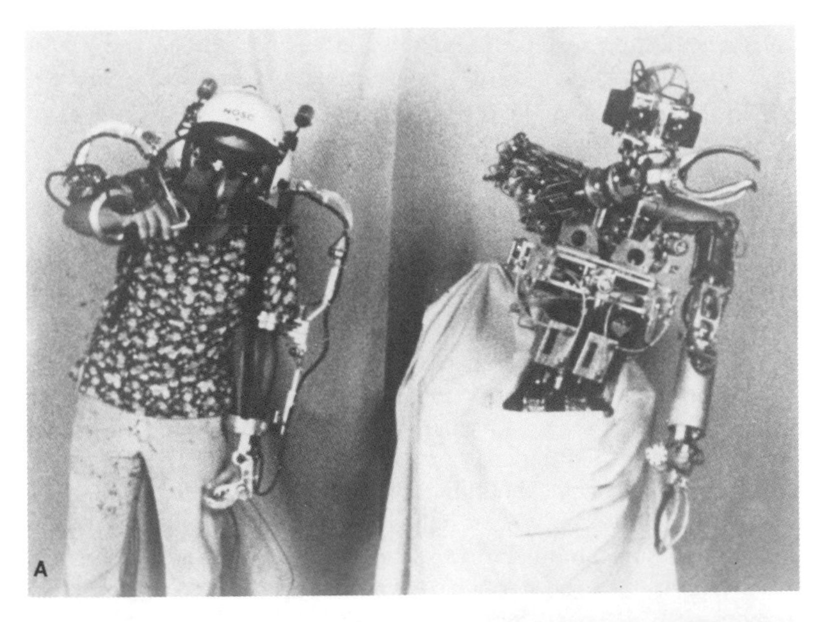

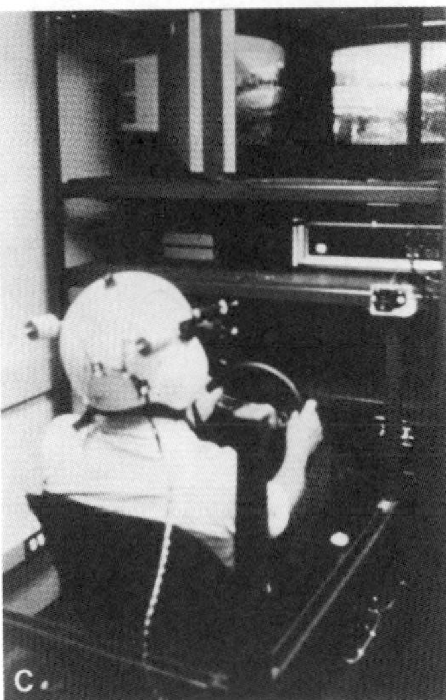

Fig. 1. (A) NOSC teleoperated robot, (B) NOSC teleoperated vehicle, and (C) remote operator in a control room.

such as reflection and planning. At the lowest level, the operator behaves in an essentially automatic way. Indeed, in highly trained operators, proper behaviors, once learned, become reflexive and are probably performed at the level of the cerebellum. Johannsen and Rouse point out that the events that elicit these activities tend to occur relatively frequently and the response becomes instinctive. Rasmussen [3] continued the trend toward a hierarchical representation of human control and decision-making behaviors and provided a taxonomy of human behavior patterns.

When concern centers on such composite problems as tracking of a dynamic target in clutter, the optimal control model (OCM) of Baron, Kleinman, and Levison has proven quite useful (see [4] for a clear description of this approach along with numerous references). Normative–descriptive models of which the OCM is a notable example have as their rationale the fact that "the motivated expert decision makers strive for optimality but are constrained from achieving it by inherent human perceptual limitations and cognitive biases" [5]. When applied to situations in which actions must be taken in response to an evolving encounter, the OCM and its more recent counterparts have been phrased within the stimulus/hypothesis/options/response (SHOR) paradigm of human decision making (see [6] and a recent example of its application [7]).

In [8] Wohl *et al.* discuss some of the fundamental issues that must be addressed in the formulation of the stimulus and hypothesis evaluation portion of the operator model. Paraphrasing these themes in the current context, it is evident that the operator will see a variety of things as an area is reconnoitered. There may be important targets of various types—such as decoys—or targetlike objects of little significance and open areas containing nothing of interest. The operator classifies these alternatives as different hypotheses, and makes a choice between them on the basis of what is observed. If this decision is made contingent on the motion patterns of the observed objects, then the hypothesis evaluation block in the operator model becomes a bank of Kalman filters tuned to the various dynamic hypotheses, along with a suitable combination of the outputs to generate the conditional likelihoods of the various hypotheses.

In [9] an alternative model, the decision directed model (DDM), of human response is proposed for the TOV application. The hypotheses that give structure to the encounter are more clearly distinguished by their panoramic features than they are by their local motion attributes. The human acts as an explicator of ambiguous observations. For example, a tank (high priority) may maneuver in concert with other vehicles of less worth. The remote operator will tend to identify the relevant object in the field of view on the basis of its visual signature rather than by its motion pattern. Clearly, motion cues and visual aspect are complementary stimuli. In this application, however, attention is centered on ranges at which the extended properties of an object provide an indication of the realized hypothesis superior to that of the motion.

For these types of applications, a recognition model having peculiar properties is required. It has been observed that the human explicator has two noticeable cognitive proclivities [10]:

1. Recency: Subjects consistently "overweigh" recent information with respect to conventional normative models.
2. Anchoring: Subjects consistently anchor on prior knowledge.

Although the cursory descriptions of recency and anchoring make them appear to be mutually exclusive behaviors, more reflection indicates that they are two distinct aspects of a multifaceted behavior pattern. When a person is convinced that he or she has identified the status of a situation, there is a tendency to under appreciate the value of new data. This is an important trait because the brief occurrence of inconsistent information will cause little vacillation. An assured person is unwavering or "anchored" in his or her belief.

Alternatively, when a person is exposed to an ambiguous data steam of reasonable length, he or she is much more open to a modification in his or her view of the condition of the environment. Recency is manifested by high valuation placed on more recent observations when a person is unsure of the current status of his surroundings. Thus, recency and anchoring are not contradictory behavior patterns, but are instead descriptions of the human response characteristics in different regions of his knowledge space.

In this article, the relationship between stimulus and hypothesis evaluation is explored in some detail within the context of tasks that are encountered in the TOV application. The explicator portion of the DDM is a nonlinear stochastic differential equation, which quantifies the operator's uncertainty when attempting to distinguish the relevent features in a changing scene containing visual clutter. The proper understanding of the impact of this uncertainty on overall system performance is essential if the human operator is to be best utilized. In what follows, the operator's environment is characterized by three descriptors: the frequency of changes in condition, the distinguishability of these changes, and the level of clutter in the observation. The next sections provide an indication of the dependence of the explicator's performance on these factors. One distinctive human skill is the ability to recognize novel situations and adapt to them. The pliancy of the model in this hypothesis generation role is also explored. The degree to which the DDM can be anthropomorphized depends on how well its response mimics that of a human in a similar environment. A test comparing the response of the model with that of a human subject suggests that "human" indecisiveness is captured by the model.

II. TELEOPERATOR RESPONSE MODELING

Since one of the primary functions of a system incorporating the TOV is reconnaissance, it is important to model the operator's proficiency in situation assessment. The ability to infer the mode of evolution of an encounter requires a panoramic view, and a faculty for placing the observations within a well-defined pattern. The human role becomes preeminent when the encounter involves sudden and unpredictable changes in the operational environment. An appearance of a target, a change in aspect of that target in a manner that indicates a threat to the TOV, an abrupt change in the terrain over which the vehicle must maneuver are all possible events that the operator may experience and to which he or she must respond. These are represented by a distinct set of hypotheses in [8]. In contrast to the structure developed in [8], the currently realized hypothesis that describes the TOV operation may change in time; that is, the target may be obscured, the terrain may change again.

Suppose that the operative hypothesis is denoted by an integer valued "feature" or schema [11] of the encounter.[1] Indicate the feature process by $\{r_t\}$, where the state space of r_t is $S = \{1, \ldots, s\}$; that is, there are s different possible features and r_t is an indicator of the current one. A specific example of this structure, which will be discussed in more detail in the sequel, can be described as follows. Suppose that the operator is using the TOV optics to scan a region for high-priority targets. Within the field of view there is at any given time either (1) a target, (2) a targetlike decoy, or (3) open terrain. The appropriate options for the operator are contingent on the realized feature. The most important event is that a target is within the operator's field of view, and this is indicated by $r_t = 1$. The events $r_t = 2$ or 3 are interpreted in an analogous fashion and $S = \{1, 2, 3\}$.

For calculations, it is expedient to introduce an alternative notation for the feature process. If $r_t = i$, let $\phi_t = e_i$.[2] Then $\{\phi_t\}$ is a unit vector indicating the current feature. Assume that $\{r_t\}$ is a Markov process with transition matrix $Q = [q_{ij}]$

$$\Pr(r_{t+\delta} = j \mid r_t = i) = \begin{cases} 1 + q_{ii}\delta + o(\delta), & i = j \\ q_{ij}\delta + o(\delta), & i \neq j \end{cases}. \tag{1}$$

---

[1] A "schema is an abstract cognitive structure that specifies both the defining features and the relevant attributes of some stimulus domain as well as the interrelation among the features and attributes" [11].

[2] Denote the unit vector in the ith direction in E^s by e_i.

Then $\{\phi_t\}$ satisfies a stochastic differential equation

$$d\phi_t = Q'\phi_t\,dt + dm_t, \tag{2}$$

where $\{m_t\}$ is a purely discontinuous martingale with respect to the natural filtration.

Equation (2) is a simple model of the feature process, but it will suffice for this introductory study. Feature variation manifests itself in panoramic changes in sensory data, and the operator attempts to identify feature changes from stimuli provided by the optical and auditory sensors on the robot. Such observations may be quite cluttered, and it is difficult to eliminate the ambiguity that naturally surrounds the interpretation of a scene. To be specific, suppose that each feature has a distinct sensory signature. In keeping with the referenced analytical models of human response, this signature is only remotely connected with the actual physiological processes in the human, but is rather a pseudostimulus used as the input in the normative–descriptive model of operator response. Let the signature of feature i be given by the real number h_i. The operator receives a cluttered measurement $\{y_t\}$ of the current feature. This measurement will be represented by the stochastic differential equation

$$dy_t = h'\phi_t\,dt + dn_t, \tag{3}$$

where h is the indicated s vector, and $\{n_t\}$ is Brownian motion

$$(dn_t)^2 = dt. \tag{4}$$

Equation (3) relates the operator's observation to a signal $h'\phi_t$ generated by the feature. This modal indication is contaminated by an exogenous wide-band clutter. On the basis of $\{y_t\}$, the operator infers the likelihood of the various possible features. This structure fits well with the SHOR paradigm, but since the motions give little information, the structure of the hypothesis estimator is significantly different.

Let $\{\mathbf{Y}_t\}$ be the natural filtration associated with $\{y_t\}$, and denote the conditional expectation with respect to $\{\mathbf{Y}_t\}$ by a circumflex. Then the operator's conception of the relative probabilities of different features is given by the information state $\{\hat{\phi}_t\}$,

$$\{\hat{\phi}_t\} = E\{\phi_t \mid \mathbf{Y}_t\} = [\Pr(\phi_t = e_i \mid \mathbf{Y}_t)]. \tag{5}$$

The equations of evolution of $\{\hat{\phi}_t\}$ are given in [9], and they represent the situation assessment portion of the DDM used to describe the response of a remote operator engaged in directing the TOV in a multitask environment:

$$d\hat{\phi}_t = Q'\hat{\phi}_t\,dt + (\operatorname{diag} h - \hat{\phi}_t' h I)\hat{\phi}_t\,dv_t, \tag{6}$$

where $\{v_t\}$ is the innovations process

$$dv_t = dy_t - h'\hat{\phi}_t\,dt. \tag{7}$$

Equation (6) is a nonlinear stochastic differential equation. It quantifies the way in which the information state changes as the operator processes new observations. The components of $\{\hat{\phi}_t\}$ are the conditional likelihoods of the various hypotheses concerning the encounter. It is on the basis of these quantities that strategy is formulated and resources allocated.

Equation (6) is an Ito equation, and considerable care must be exercised in its interpretation. In Section III the flexibility of the DDM is explored. Different personalities are created as model parameters are changed. The effect that the rate of change in the encounter, the feature discernibility, and the level of exogenous clutter have on hypothesis evaluation are exhibited in the sample behavior of the DDM.

In Section IV, the measured response characteristics of an operator is compared with the simulated behavior of the model. A time-varying scene was presented to a subject who was asked to indicate a level of confidence that a target (a tank) was within the field of view. This response along with the actual locations of targets, decoys, and open regions was recorded. A sample function of the information state $\{\hat{\phi}_t\}$ generated from (6) for the same feature sequence was also recorded. For different sample functions of the observation noise, the operator's perception of the scenario features will change. As a consequence, $\{\hat{\phi}_t\}$ is a random process even for a predetermined scenario $\{\phi_t\}$. Thus one would not expect the model to reproduce a specific human response any more than a human would be unaffected by the realization of the exogenous visual clutter, or indeed that two humans would respond identically to the same stimuli. With this caveat, (6) does indicate an appropriate indecisiveness in identifying feature changes.

III. PLIANCY OF THE DECISION DIRECTED MODEL

The model given in (6) is intended to represent the behavior of a remote operator in a variety of different situations, and to do this it must posses the human quality of formability. Thus, when the scenario is slowly varying and the clutter level is low, the DDM should respond surely and expeditiously. Alternatively, when the features that distinguish different hypotheses lack contrast, and when there is a considerable clutter, the DDM may be confused and vacillating in its response.

The operator model has two multidimensional parameters, and each is related to a different aspect of the encounter. The rate at which the underlying features of the scenario change is quantified by Q. If hypothesis i has a short mean lifetime, then q_{ii} is large. If there are targets of different classes in close proximity, then the transition rates q_{ij} would be large for appropriate i, j and so on. The correct specification of Q is dependent on the dynamic structure of the encounter, and it parameterizes the drift term in (6); that is,

$$E\{d\hat{\phi}_t \mid \mathbf{Y}_t\} = Q'\hat{\phi}_t \, dt. \tag{8}$$

In the same way that Q quantifies the encounter dynamics, h quantifies both the sensory acuity and the clutter. Equation (3) is phrased in terms of a normalized observation signal $\{y_t\}$ with the noise term $\{n_t\}$ having unit intensity [see (4)]. The scale of h is then contingent on the actual intensity of the wide-band clutter in the sense that if the exogenous disturbance is increased by a power factor of four, this same effect can be induced into (3) by mapping $h \to h/2$.

To study the flexibility of the DDM it is expedient to generalize (3)–(4) slightly, even though this means over parameterizing the system. Thus, instead of (4), suppose that

$$(dn_t)^2 = r \, dt. \tag{9}$$

Then r is a direct indicator of the intensity of the observation noise. Although redundant, (9) permits the effects of varying clutter levels to be separated from variations in the distinguishability of the various modal hypotheses. The equation for $\{\hat{\phi}_t\}$ must be modified in an obvious way, and this has been done in what follows. It was further noted that the numerical stability of (6) was inadequate for the simulation task. A change of variables transforms (6) into the Zakai form, which is a linear stochastic differential equation for an unnormalized version of $\{\hat{\phi}_t\}$. This equation proved to be superior in this application, and the solutions were renormalized for presentation here.

To illustrate the behavior of the model as a function of the observation noise, consider a situation in which there are only two hypotheses ($S = \{1, 2\}$) and $h' = (-1, 1)$. Hence the observation structure is

$$dy_t = -1 \, dt + \text{noise} \qquad \text{if} \quad r_t = 1$$

and (10)

$$dy_t = 1 \, dt + \text{noise} \qquad \text{if} \quad r_t = 2.$$

The signature of hypothesis 1 is the constant -1, and if observed long enough the operator would deduce the correct response by averaging (i.e., $\hat{\phi}_t \to e_1$). The operator response must, however, be expeditious and further, the modal variable may change anyway. Suppose that the scenario dynamics are given by

$$Q = \begin{bmatrix} -0.1 & 0.1 \\ 0.9 & -0.9 \end{bmatrix}. \tag{11}$$

The mean sojourn in mode 1 is 10 sec (1/0.1) while the mean sojourn in mode 2 is 1.1 sec (1/0.9). Equation (11) implies that the system is reluctant to leave e_1, and the operator knows it.

The decision-maker's response is clearly dependent on the clutter. The sharp edges of modal transitions are masked by noise. The operator realizes that these changes are inevitable, but the DDM may have difficulty discerning actual transitions from anomalous variation in the observation noise. Figure 2a shows a sample of the output of the DDM when the scenario has

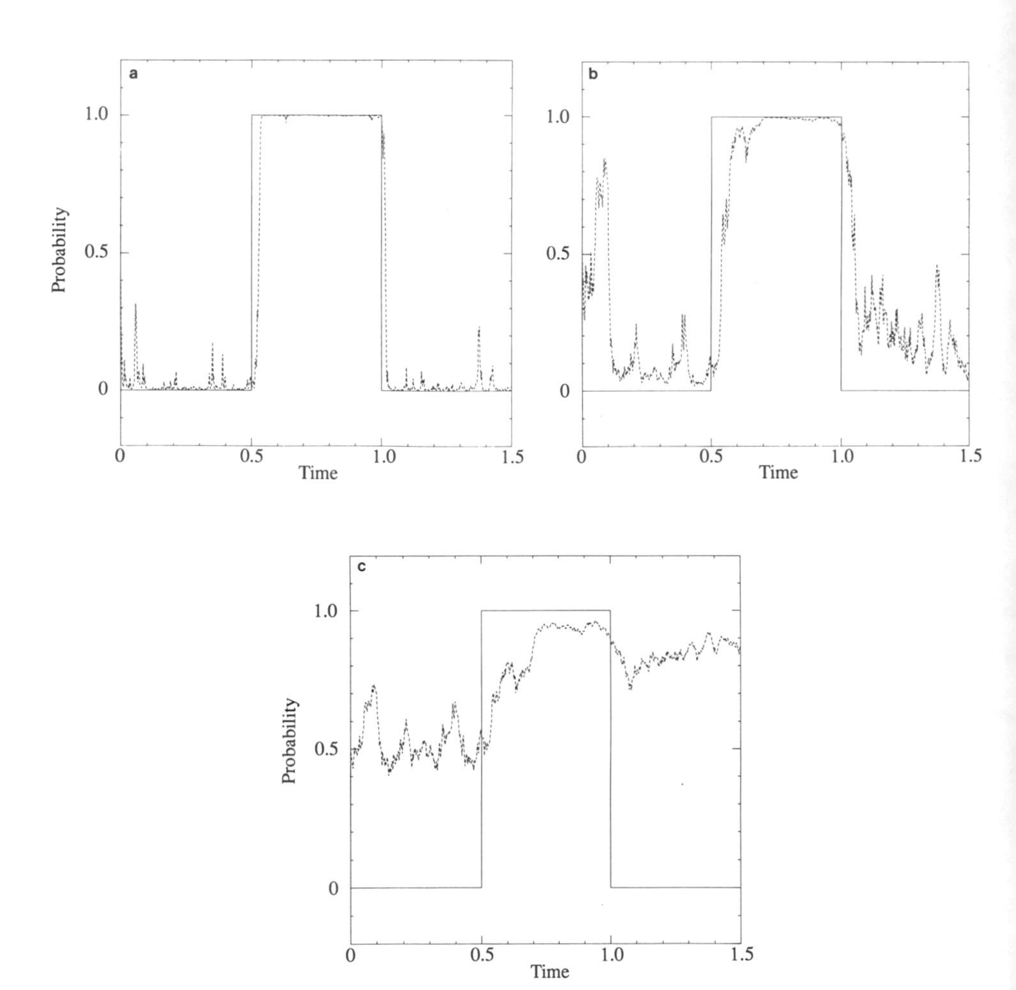

Fig. 2. Indicator of state 1 (solid) and simulated probability of state 1 (dashed) when noise intensity is (a) low, (b) moderate, and (c) high.

a single jump to e_1,

$$\phi_t = e_1 \quad \text{when} \quad t\varepsilon A \quad \text{and} \quad e_2 = e_2 \quad \text{when} \quad t\varepsilon A^c, \tag{12}$$

where $A = [0.5, 1.0]$. This interval is indicated in Fig. 2a along with the DDM response.[3] The noise level is quite small ($r = 0.01$), and the response is unequivocal. The delay in identifying an $e_2 \rightarrow e_1$ transition is negligible.[4] There are no significant false alarms, but the false indication of change are more evident during sojourns in e_2. This is a natural consequence of the fact that the mean lifetime in e_2 will be far shorter than that in e_1.

As the noise is increased ($r = 0.1$), hypothesis evaluation becomes more challenging (Fig. 2b). The delay in detecting a transition is now on the order of 0.1 sec. The volatility in more 2 increases significantly. If the amplitude of the noise is increased still further (see Fig. 2c), the hypothesis evaluation block no longer responds effectively to the stimuli at all, and tends toward the stationary distribution for $\{\hat{\phi}_t\}$, which would be obtained in the absence of observations.

Figure 2 gives a quantitative indication of how a increase in exogenous clutter manifests itself in increased indecisiveness in the DDM. The DDM moves in a continuous fashion from assured response to capriciousness as r increases. This figure also indicates that the volatility during the midst of a sojourn time is a direct function of the expected lifetime of the associated feature in all but the highest noise case.

To study the way in which encounter dynamics influence the personality of the DDM, consider Fig. 2a and 3. These figures show the response of the DDM to changes Q with a fixed value of r ($r = 0.1$) and h [$h' = (-1, 1)$]. In addition to the Q matrix given in (11), two other matrices were studied

$$Q = \begin{bmatrix} -0.5 & 0.5 \\ 0.9 & -0.9 \end{bmatrix} \tag{13}$$

for Fig. 3a and

$$Q = \begin{bmatrix} -0.5 & 0.5 \\ 5.0 & -5.0 \end{bmatrix} \tag{14}$$

for Fig. 3b. The behavior when $\phi_t = e_1$ does not vary radically with the changes in Q. In each case e_1 is the long-lived mode, and the detection of e_1 is

---

[3] A plot of $(\phi_t)_1$ is shown as the solid rectangle, and $(\hat{\phi}_t)_1$ is shown as a dashed line.

[4] A natural index for scaling clutter is the ratio of the size of the modal jump over the $\sqrt{r}$. In this example, this index would be $2/0.3 = 7$.

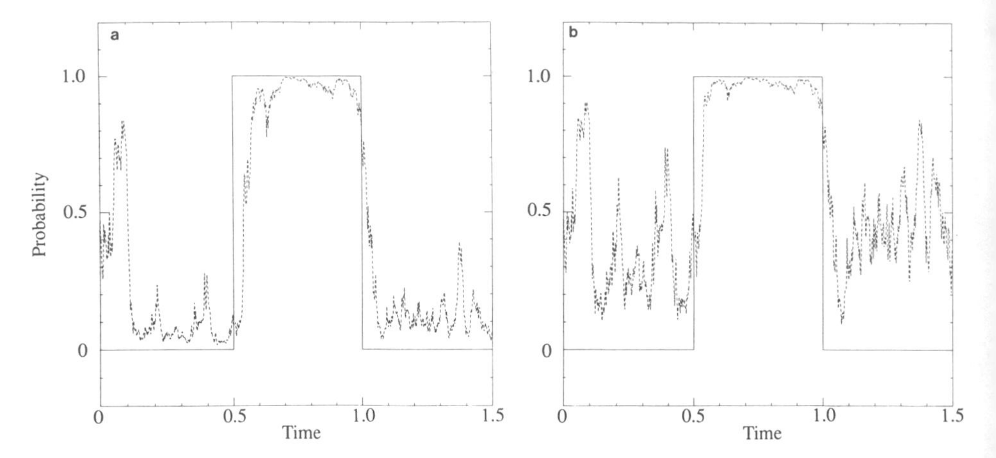

Fig. 3. Indicator of state 1 (solid) and simulated probability of state 1 (dashed) when (a) average staying time in state 1 is decreased ($t_1 = 2$, $t_2 = 1.1$) and (b) average staying time in state 2 is decreased ($t_1 = 2$, $t_2 = 0.2$).

made expeditiously. There is an increase in uncertainty in detecting e_2 when the time scale of the problem is changed (compare Fig. 2a with Fig. 3b for a time-scale variation of five). A short-lived mode is difficult to detect in wide-band noise.

The behavior of the DDM as image discernibility changes is of fundamental interest. Expensive operator enhancements are justified by an improved ability to differentiate hypotheses that lack definition in the image. Such sensor–processor modifications directly influence h, and through this intermediary they are reflected in the response of the model. To study hypothesis distinguishability as distinct from simple changes in exogenous clutter, at least three distinct hypotheses must be considered (i.e., the dimension of ϕ_t must be greater than two). With two hypotheses, changes in h can be mimicked by changes in r.

To investigate separability of the hypotheses, the feature state was three dimensional and the scenario dynamics were fixed;[5]

$$Q = \begin{bmatrix} -1.0 & 0.2 & 0.8 \\ 0.2 & -1.0 & 0.8 \\ 0.25 & 0.25 & -0.5 \end{bmatrix}. \tag{15}$$

[5] $S = \{1, 2, 3\}$ and e_1 is in R^3.

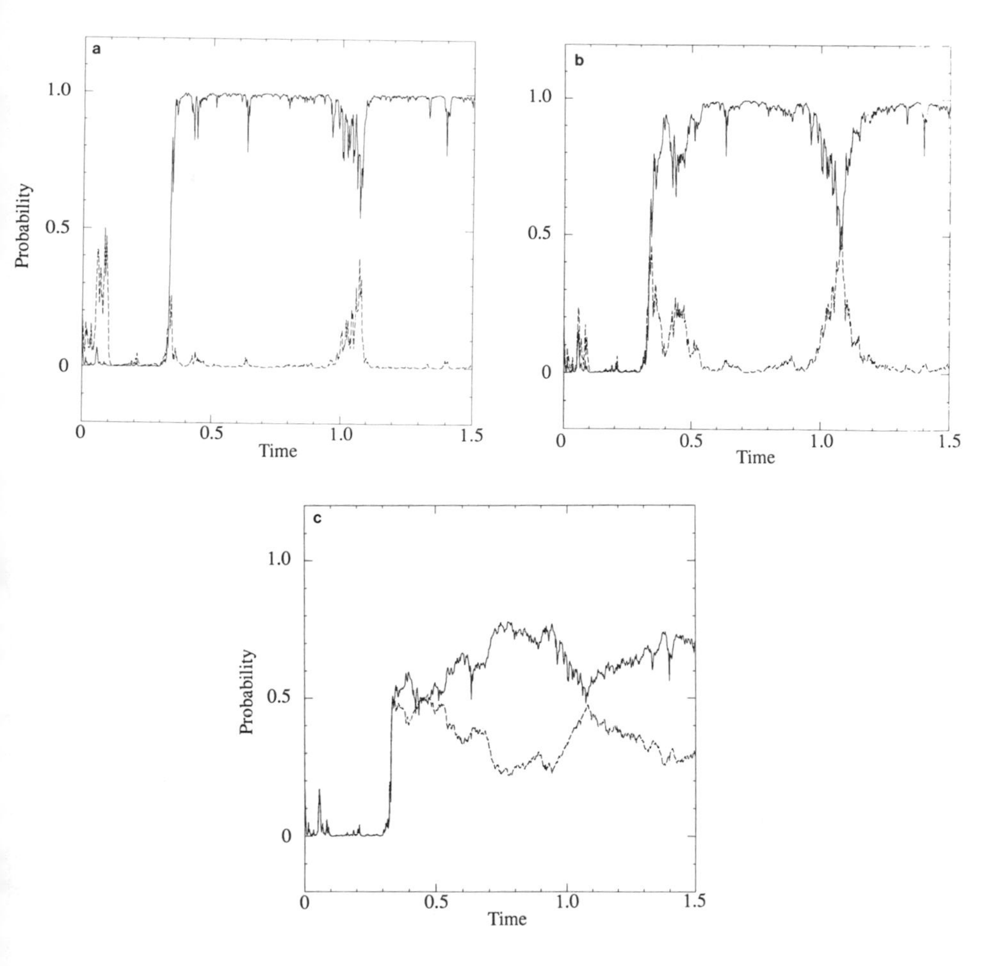

Fig. 4. Probability of state 1 (solid) and probability of state 2 (dashed) when states 1 and 2 are (a) quite distinguishable ($h^t = [-1.0 \;\; 0.0 \;\; 1.0]$), (b) less distinguishable ($h^t = [-1.0 \;\; -0.5 \;\; 1.0]$), and (c) indistinguishable ($h^t = [-1.0 \;\; -0.9 \;\; 1.0]$).

The noise had a constant intensity ($r = 0.01$). Operator acuity is delineated by h. Figure 4 show the decision-maker's response for different h vectors:

$$h = \begin{cases} (-1.0, 0.0, 1.0) & \text{in Fig. 4a} \\ (-1.0, -0.5, 1.0) & \text{in Fig. 4b} \\ (-1.0, -0.9, 1.0) & \text{in Fig. 4c.} \end{cases} \tag{16}$$

In all of the cases indicated in (16), the features that define hypotheses e_1 and e_3 are clearly marked ($h_1 = -1.0$ and $h_3 = 1.0$). Based on the results presented in Fig. 2b, these two alternatives are clearly distinguishable in clutter of the intensity indicated. It is the behavior of the DDM as a function of the signature of the second hypothesis e_2 (h_2) that is of most concern here. In Fig. 4a, each of the hypotheses has an easily perceived marker. The scenario is like that given by (12) with the renumbering of hypotheses $e_2 \to e_3$ and $A = [0.3, 1.5)$. When an $e_3 \to e_1$ transition takes place, both $\hat{\phi}_1$ and $\hat{\phi}_2$ increase.[6] The mean drift rate in (3) has changed from 1.0 to -1.0. The hypothesis evaluation block quickly recognizes that a change in drift has occurred (in ~ 0.025 sec), but can not tell which of the alternatives is the correct one since both are identified with an decreased drift. It takes another 25 msec for the hypothesis e_2 to be discarded. After identifying e_1, the quiescent interval is relatively uneventful until $t = 1.0$.

The period around $t = 1.0$ sec is interesting because the DDM becomes rather volatile. This would not be expected on the basis of (15) since e_1 is not a particularly short-lived state. The response is due to chance fluctuation in the exogenous noise. During the, interval leading up to this period of confusion, the DDM displayed the anchoring character noticed in human subjects. The innovations gain was small, and the DDM responded with a confidence that it had correctly identified the current condition of the system. The sample function of $\{n_t\}$ is such that $\hat{\phi}_1$ began to noticeably descend at $t = 0.96$. This decrease is reflected in a corresponding increase in the likelihood that e_3 is the operative hypothesis and also in a significant increase in the innovations gain. Once the information state leaves the region around $\hat{\phi}_t = e_1$, the recency behavior begins to manifest itself. The DDM becomes more uncertain of the true status of the encounter. Even though no change had occurred, the DDM began to move in the direction of hypothesis 3 and nearly gave a false indication of modal change.

In Fig. 4b, the distinguishability of e_1 and e_2 has been reduced. The same broad outlines of the DDM response again manifest themselves, with the confusion surrounding the $e_3 \to e_1$ transition increasing significantly. It now takes 0.2 sec for the DDM to moderate. The same sample function for $\{n_t\}$ was used in all of these tests. Hence, the anomalous behavior near $t = 1.0$ reoccurs throughout these simulations.

As the values of h_1 and h_2 become closer, the distinguishability of e_1 and e_2 decreases with easily predicted consequences. Figure 4c shows the change in the DDM brought about by this narrowing in the differentiation of the hypotheses. When $|h_1 - h_2| < \sqrt{r}$ as it is in Fig. 4c, the probability of correct detection of the transition to e_1 becomes too low to be acceptable even when the interval over which observation takes place is relatively long. Since the

[6] In Fig. 4, $\hat{\phi}_1$ is shown solid and $\hat{\phi}_2$ is shown dotted.

information state never develops a certainty about the true modal character, the DDM spends a prolonged period during which a vacillating behavior is clearly displayed.

Modification of the explicator block permits the inclusion of the an adaptive behavior pattern in the DDM. This is an important aspect of the human response because it is in adaptation that the human distinguishes himself most from an algorithmic alternative. The previous examples provide an indication of how the dialectical block responds to a narrowly posed situation; that is, the set of acknowledged hypotheses of the explicator is identical with the set S. This will not always be the true. Often, the set S is simply an attempted enumeration of the set of possible modal alternatives. A basic problem arises when an unanticipated modal event occurs. Thus, if the system is designed on the premise that there are two classes of objects, and a third class actually appears, the explicator faces an unreconcilable dilemma unless it is able to adapt. To illustrate this suppose that S has $s + 1$ elements. The state space of $\phi_t = \{e_0, \ldots, e_s\}$ where e_0 is unacknowledged by the explicator. Since the existence of e_0 is not recognized at the beginning of the control interval, $(\hat{\phi}_0)_0 = 0$. Further, the perceived rate of transitions into e_0 is also zero. The zeroth row of (6) can be written

$$d(\hat{\phi}_t)_0 = \sum_i Q_{i0}(\hat{\phi}_t)_i\, dt + (h_0 - \hat{h}_t)\hat{\phi}_0\, dv_t. \tag{17}$$

Each of the terms in the expression are zero, and consequently, $\{(\hat{\phi}_t)_0\}$ is identically zero without regard to the realization of $\{v_t\}$.

Although ignored in (17), the occurrence of exceptional hypotheses is recognized by sophisticated processors. This is particularly true when humans are used in the hypothesis evaluation role. They appear to be able to create new hypotheses from the similarities that exist between the expected and the realized observation signatures. While a human does this frequently and almost unconsciously, an algorithmic replacement with the same ability is quite difficult to implement. In any event, the failure to exhaustively enumerate the elements of the modal state space results in anomalous response characteristics in the model.

As an example of how unmodeled hypotheses influence explicator performance, reconsider the example presented earlier with the addition of a latent mode e_0. Then $\phi_t \varepsilon R^4$ and suppose that $h = (-0.75, -1, -0.5, 1)$. The augmented transition rate matrix Q as seen by the explicator has the block form

$$Q = \begin{bmatrix} 0 & 0 \\ 0 & Q_r \end{bmatrix}, \tag{18}$$

where Q_r is given by (15). In the nominal scenario the modal hypothesis e_0 is unexcited. Hypothesis e_1 is distinguishable from the alternatives as shown in Fig. 4b.

When a prohibited transition occurs, the behavior of (6) becomes more irresolute. In Fig. 5 the scenario depicted in Fig. 4b is modified by having a $\phi_t \rightarrow e_0$ transition at $t = 0.3$ sec. The information state $\{\hat{\phi}_t\}$ displays an interesting pathology not evident in Fig. 4b. The signature of e_0 is equidistant from modes 1 and 2, and the explicator spends a considerable time in a confused state. As pointed out earlier, this capriciousness is a result of the high innovations gain in this region. Then, a chance occurrence in $\{v_t\}$ leads to the conclusion that $\phi_t = e_1$. During the interval [0.6, 1.0], $\{\hat{\phi}_t\}$ indicates a conviction that the correct mode has been identified, and because of the low innovations gain in this interval, this response differs little from that presented in Fig. 4b. However, the false alarm that was corrected expeditiously in the scenario of Fig. 4b causes a sensed modal reversal in Fig. 5. The explicator finds the data intelligible, but misinterpreted it.

Figure 6 shows an adaptive response of the explicator to the same scenario. Adaptation takes place through adjustment in the $\{q_{0i}, q_{i0}\}$ elements in (18). The figure shows a slow recognition of the existence of the latent state followed by a rapid identification. The misclassification decays rapidly after about a second. This general character is typical of such adaptive algorithms.

In this section a simple parametric study of the hypothesis evaluation block of the DDM has been presented. An indication of the behavioral flexibility of the DDM is apparent. The next section gives an indication of how this flexibility can be used to describe the response of a test subject.

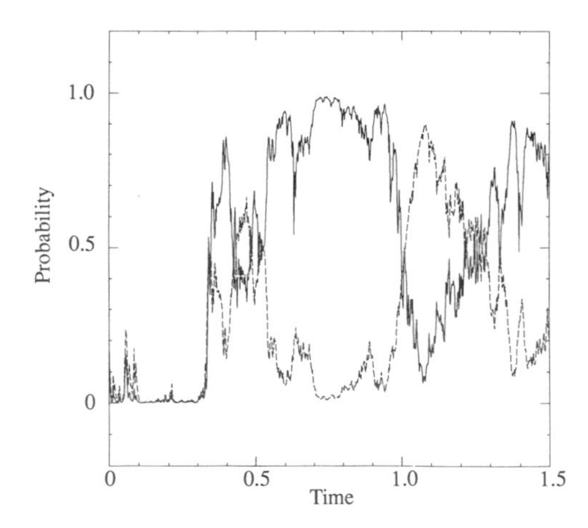

Fig. 5. Probability of state 1 (solid) and state 2 (dashed) in the presence of an unmodeled hypothesis.

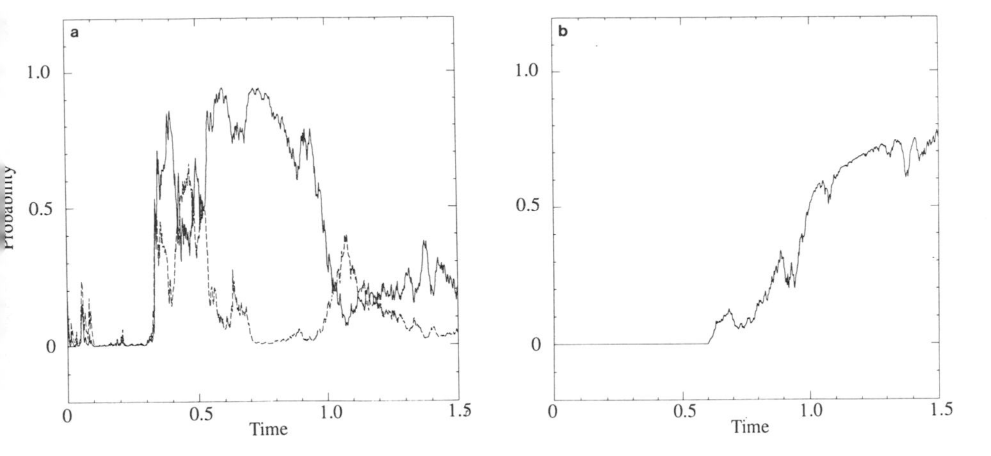

Fig. 6. Adaptive response of (a) state 1 (solid) and state 2 (dashed) of the DDM to an unmodeled hypothesis and (b) the DDM to an unmodeled hypothesis.

IV. OPERATOR RESPONSE CHARACTERISTICS

As stated earlier, the adequacy of the normative–descriptive model of a human decision maker must be established empirically. The previous section has indicated in some detail the performance attributes of the input–output model (3)–(6). In the SHOR paradigm the decision maker responds to new data by modifying the likelihoods of the various hypotheses that delineate the encounter. The DDM provides an algorithmic decription of the precise way in which this is done.

The hypothesis evaluation portion of the DDM is difficult to verify because this appraisal is an internal activity of the decision maker, and is normally reflected only indirectly by his or her actions. Nevertheless, to better understand the limitations of the DDM, a scene recognition task of the type to be encountered in the TOV application was given to a representative operator. The experimental facilities were rudimentary, and the operational environment of the TOV operator could not be reproduced. The experimental protocol can be described as follows. A set of photographs were taken at the Camp Pendleton Marine Base in California. They were appropriately juxtaposed to from a panoramic view of the terrain as seen from a fixed location. The scene consisted of rather open terrain containing scrub vegetation along with a scattering of discrete objects at various ranges. The actual object set contained jeeps, tracked land vehicles, (LVTs), outbuildings, tanks, and various natural structures similar in appearance to the foregoing. The vehicles had different aspect angles with respect to the viewer and had a variety of visual appearances.

To test the dynamic response of a decision maker, a single category of objects was made that of primary concern. In the experiment, the operator was asked to identify the presence of a tank in a changing scene, and to consider all other events as being inconsequential. At one level then the operator could be viewed as comparing a simple hypothesis (tank present; see Fig. 7a) with a single alternative (tank absent; see Fig. 7b). Actually both hypotheses are composite, with the latter containing many well-defined subsidiary hypotheses. Hence, this simple structure with $S = \{1, 2\}$ is not sufficiently rich to capture the scenario as the decision maker preceives it.

There was considerable visual ambiguity since other object classes share many features with a tank when viewed at a distance. Indeed, anything with an angular shape could be confused with the primary target class at first glance. Therefore, it was necessary to introduce a third operative hypothesis (secondary object present; see Fig. 7c) to account for the decoys. Surprisingly, given the intricacy of the scene, further additions to S were not found to be necessary. It should be noted that the DDM permits augmentation to the modal set in a direct manner if the actions of the decision maker warrant such an increase in dimension.

To inject scene dynamics into what is fundamentally a static encounter, a movable camera was made to pan a horizontal slice in the picture at a fixed angular rate. A local image was displayed on a monitor for the operator to view. Thus, static objects appeared in the monitor at random times with random clarity but with duration fixed by the linear dimension. The scan rate thus determined the time scale of changes in the events to which the operator responded. A frequent modal change is achieved at a high scan rate. At low scan rates, the operator has more time to contemplate and distinguish the relevant objects from clutter.

The subject was given a joystick with one degree of freedom, and was asked to position the rod to indicate her level of confidence that a target was contained in the image displayed on the monitor. The output voltage from the joystick was proportional to the angular position, with the lowest voltage representing certainty that there was not a target within the field of view, and maximum voltage representing certainty that a tank was displayed on the monitor. An intermediate position indicated that the scene was ambiguous.

Figure 8 displays a sample function of the subject's response. As mentioned earlier, the event sequence can be viewed as an alternation between distinct modal hypotheses. The actual time intervals during which a target is on the monitor are indicated in the figure by the dotted rectangles. Let the event that a tank is being observed be denoted by $r_t = 1$. Decoys are similarly distinguished by the dashed rectangles with the corresponding event denoted by $r_t = 2$. The remainder of the interval consisted of open fields and unstructured clutter, and this is denoted by $r_t = 3$.

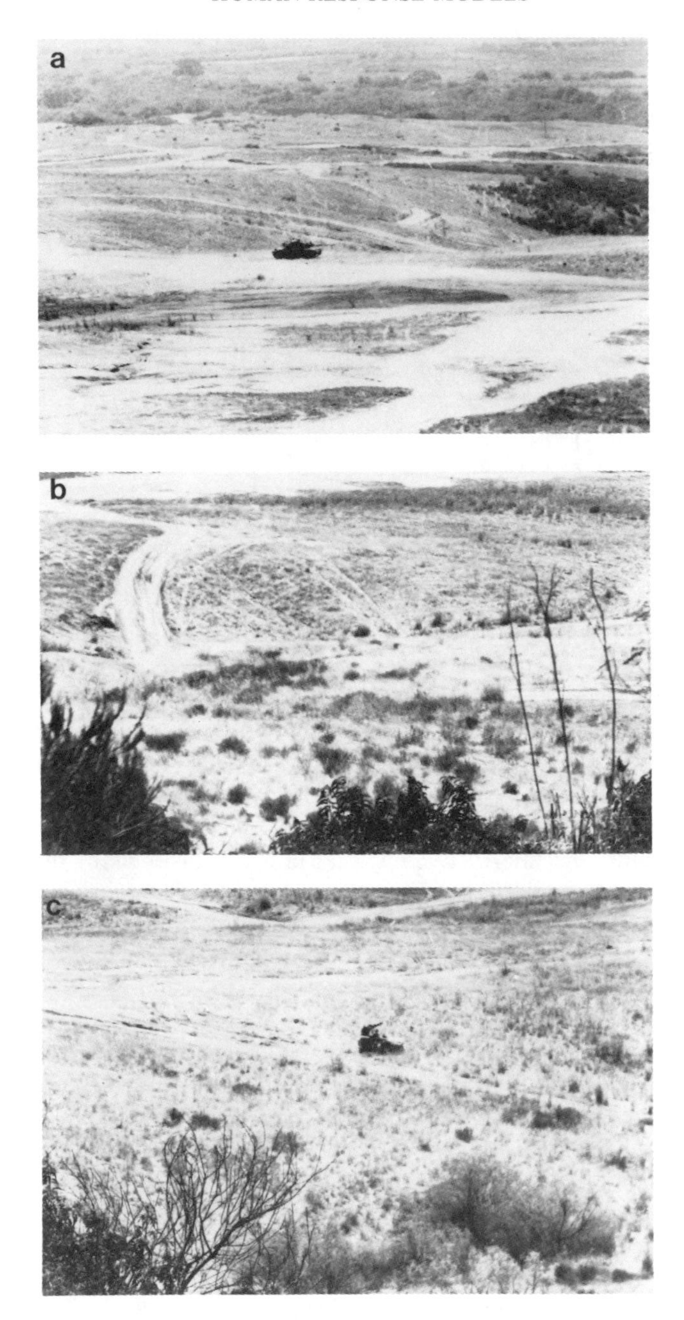

Fig. 7. Sample experimental test scene: (a) tank present, (b) tank absent, and (c) secondary object.

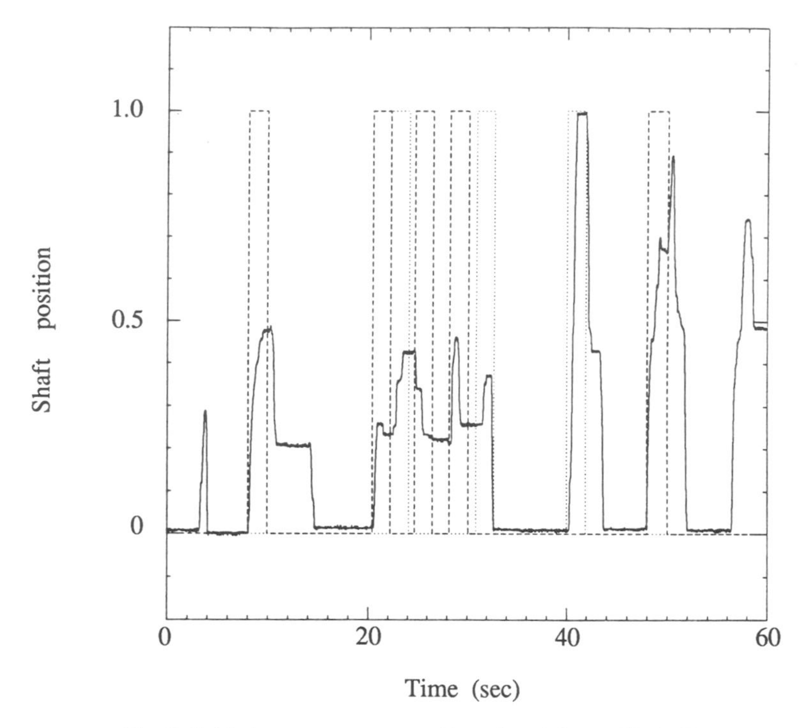

Fig. 8. Subject response to targets (dotted) and decoys (dashed).

From a cursory review of the picture, the broad outlines of the decision-maker's behavioral peculiarities, shown in Fig. 8, are easily predicted. During sojourns in regions of relatively open terrain, there is little operator confusion. The output of the joy stick should be essentially zero during such phases.

As the picture was scanned, there were regions in which isolated and distinguishable objects were encountered, and the subject attempted to identify them. The scanning direction is such that an object enters the subject's visual field from the right of the screen. When an object appears on the monitor, the operator will initally be uncertain as to the object classification. If the object has a distinct, angular shape that separates it from the more diffuse background, the operator will realize that a modal change has taken place but may delay making a confident identification. The subject indicates this confusion by moving the rod to an intermediate position, which is roughly proportional to the probability that the object is in the primary category. In an object-sparse region, the operator will concentrate on an unusual shape as it moves across the screen, and make more definitive judgments on the proper classification. Some of the things in the scene are quite deceptive. Careful study is required to distinguish these decoys from targets. The presence of these near targets produces significant false alarms.

In another region in the picture, a group of similar objects are in close proximity. While there is never a situation in which more than one object is within the field of view of the camera, when objects occur in rapid succession, the observer never has the time to focus clearly on the object at hand. Instead, the decision maker becomes preoccupied with the sequence itself, and lapses into an unsettled state in which the objects are not clearly differentiated.

In Fig. 8 the operator response is shown along with the regions of targets and decoys. The operator response is related to $\{\hat{\phi}_1\}$ in the notation of (6). The anticipated peculiarities of the decision-maker's response characteristics appear in the sample function. Open areas are clearly recognized as such for the most part, although some natural objects do cause confusion at certain light angles. What is more interesting is the response of the operator to targets and putative targets. Predictably, the operator has much more difficulty in differentiating the hypotheses when the objects appear more frequently than when they are separated by extended intervals.

A simulated response of the hypothesis evaluation portion of the DDM is shown in Fig. 9. The most notable difference between the simulated and the actual response is that the simulated sample function is locally much more volatile than is the human. This is not suprising. Equation (6) is an Ito

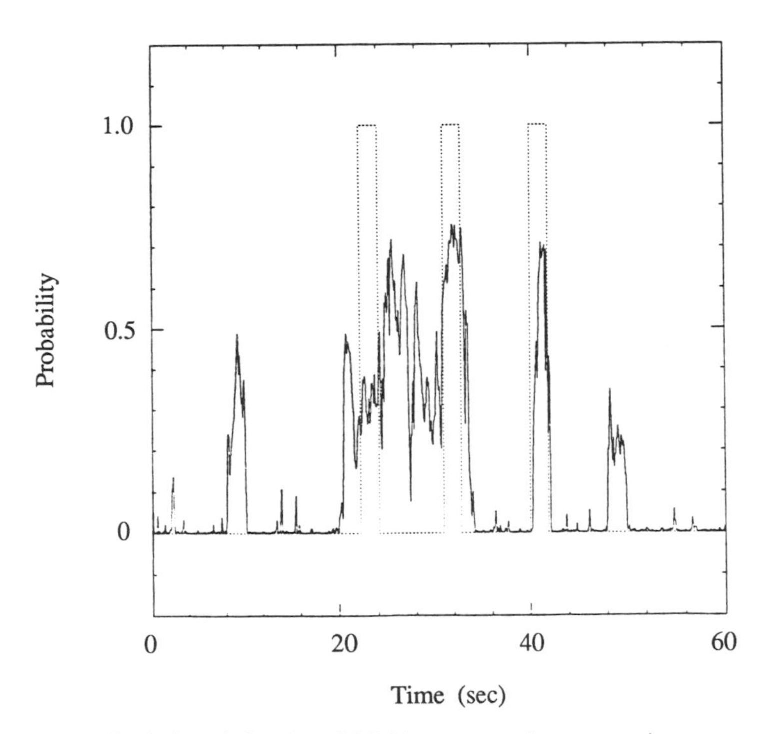

Fig. 9. Sample function of DDM response to the test scenario.

equation, and solutions to such equations are quite irregular on any short time scale. The decision-maker's evaluation of a situation is probably not nearly so changeable, and even if it were, the physiological lags that occur in translating a mental view of a scene into a shaft position would smooth the volatility and preclude its measurement. Furthermore, a pointing system under the direction of the operator has a low-pass character as well. Hence, the local volatility is not an important issue in operator modeling.

With this caveat, the response of the DDM has the human peculiarities that were predicted and measured. This similarity of response manifests itself clearly in the object-rich portion of the scan. The rapid modal variation is represented in the DDM by a Q matrix with high modal transition rates. As the modal variation becomes more frequent, the response of the DDM becomes more indecisive.

V. CONCLUSIONS

Human decision makers play a central role in many system architectures. Unfortunately it is difficult to delineate their characteristics in an analytically tractable manner. The decision maker is capable of so many different behaviors that it is hard to capture his or her persona in an analytical form. This article considers one aspect of his or her response characteristics, and indicates the capability and versatility of the hypothesis evaluation portion of the DDM. Some very human peculiarities have been observed in tests utilizing the model. The indecisiveness and uncertainty under which the decision maker must perform assigned tasks manifest themselves in the simulation study. As has been pointed out, the proper response of the DDM is dependent on the correct selection of the model parameters. The simulations indicate that considerable freedom exists in shaping the response characteristics. Indecisiveness in a rapidly changing scenario can be produced by increasing the size of the elements in the Q matrix. If two or more hypotheses are difficult to differentiate, the associated elements of the h vector should be made close.

Situation evaluation is only one component of the operator's function in the TOV application. The complete DDM requires a characterization of the options/response portion of the model. This block translates the vector of likelihoods provided by the hypothesis evaluation block into an appropriate action. This is a topic of current study.

ACKNOWLEDGMENTS

This research was partially supported by the National Science Foundation under Grant No. ECS-8607816, and by the Naval Ocean Systems Center under Contract No. N66001-87-D-0136.

REFERENCES

1. J. A. WELCH, "State of the Art of C2 Assessment," *Proc Quan. Assess. Utility Command Control Syst., National Defense Univ.* (1980).
2. G. JOHANNSEN and W. ROUSE, "Prospects on a Mathematical Theory of Human Behavior in Complex Man-Machine System Tasks," *14th Conf. Manual Control* (1978).
3. J. RASMUSSEN, "The Human as a Systems Component," *in* "Human Interaction With Computers" (H. T. SMITH and T. R. GREEN, eds.). Academic Press, New York, 1980.
4. S. BARON, "A Control Theoretic Approach to Modelling the Human Supervisory Control of Dynamic Systems," *in* "Advances in Man–Machine System" (W. B. ROUSE, ed.), Vol. 1. JAI Press, Greenwich, Connecticut, 1984.
5. D. SERFATY, L. G. BUSHNELL, and D. L. KLEINMAN, "Distributed Human Information Processing: A Normative–Descriptive Approach," *Proc. Symp. 5th Empirical Found. Inf. Software Sci., Roskilde, Denmark* (1987).
6. J. G. WOHL, "Force Management Requirements for Air Force Tactical Command and Control," *IEEE Trans. Syst., Man, Cybern.* **SMC-11**, 618–639 (1981).
7. J. G. WOHL ET AL., "Human Cognitive Performance in ASW Data Fusion," *Proc. 1st Tri-serv. Data Fusion Symp. Johns Hopkins Univ.* (1987).
8. J. G. WOHL, E. E. ENTRIN, and J. S. ETERNO, "Modeling Human Decision Processes in Command and Control," Alphatech Tech. Rep. TR-137, January 1983.
9. D. D. SWORDER and K. S. HAALAND, "Algorithms for the Design of Teleoperated Systems," *Control Dyn. Syst.* **30**, 167 (1988).
10. D. SERFATY, D. KLEINMAN, and E. ENTIN, "Information Fusion in a Team: Normative–Descriptive Modeling," *Proc. 3rd Annu. DTDM Coord. Meet., Newport, Rhode Island* (1987).
11. A. P. SAGE, "Human Information Processing Principles for Command and Control," *in* "AFCEA Magazine: Principles of Command and Control," Vol. 6, pp. 54—74. AFCEA Press, New York, 1987.

DECOUPLED FLIGHT CONTROL VIA A MODEL-FOLLOWING TECHNIQUE USING THE EULER OPERATOR

PETER N. NIKIFORUK

College of Engineering
University of Saskatchewan
Saskatoon, Saskatchewan S7N OWO, Canada

NORIYUKI HORI

Department of Mechanical Engineering
McGill University
Montreal, Quebec H3A 2K6, Canada

KIMIO KANAI

Department of Aerospace Engineering
National Defense Academy
Yokosuka City, Japan

I. INTRODUCTION

When conventional flight control is employed, the path of an aircraft is manipulated indirectly. The command inputs are applied to the aircraft through such control surfaces as the elevator, aileron, and rudder, which first generate angular accelerations that alter the aircraft's attitude, and then vertical and horizontal forces, which change its flight path. With this type of control, an aircraft can not change its flight path without first changing its attitude. For more than a decade, as advances have taken place in aircraft technology, considerable interest has been expressed in improving flight performance by directly controlling the vertical and horizontal forces [1]–[3], which would allow the linear and angular motions to be separated. Control

Copyright © 1990 by Academic Press, Inc.
All rights of reproduction in any form reserved.

configured vehicle (CCV) modes, some of which will be explained later, were introduced for this purpose. The design of flight control systems considered in this study is based on a decoupling technique using a model-following control method. If the inputs and outputs are selected properly, the CCV modes can be achieved using this control approach.

The purpose of model-following control, more specifically output model-following control, is to force the plant outputs to follow the model outputs. This problem has been investigated by a number of authors over the years. The characteristics of a model-following system were analyzed in [4] using an optimal control scheme. A design based on an algebraic condition imposed on the error equation was presented in [5] and the sufficient conditions for model following were established. Other authors studied the problem in terms of the transfer function matrix and called this the exact model-matching method [6, 7]. In this control technique, the controller structure must satisfy the condition that the transfer function matrix (TFM) of the plant–controller combination and that of the model coincide. When the numbers of input and output are identical, system decoupling can be achieved by selecting a TFM of a reference model that is diagonal. System decoupling using state feedback requires that a high-frequency gain matrix be nonsingular [8]. A method that overcomes this singularity problem is presented in [9] and a simpler method in [10].

Since modern aircraft are equipped with digital computers, it is desirable to design, from the outset, a control system in the discrete-time (DT) domain. To this end, the discretization of a continuous-time (CT) plant must be carried out to design a DT control system. There are various DT models from which one can select the best for a particular design purpose. These DT models include the invariant models (e.g., impulse and step), mapping models (e.g., backward difference and Tustin's), and the matched Z transform (MZT) model [11, 12]. It is shown in [13] that for a given CT system there exists an equivalence class of DT models whose response tends in some sense to that of a CT system when the DT interval (sampling interval) T approaches zero. In [13], a definition of the DT model of a CT system is given and various important theorems are derived. It is also shown that the operator $\varepsilon = (z - 1)/T$, where z is the Z operator, plays an important role in the analysis and synthesis of DT systems. In [14] it is shown that ε, called the delta operator, has better characteristics than the z operator. The operator ε is called the Euler operator in [15], where some of the properties of a DT system expressed in this operator are examined.

In this study the improved design of a model-following controller, which was proposed earlier in [9] and [10], is described for the linear, time-invariant, multi-input–multi-output (MIMO) case. In particular, a DT system which is expressed in the Euler operator and which has the same number of inputs and outputs is considered. An application of this model-following scheme to a CCV aircraft is considered with particular reference to the longitudinal motions. The lateral case can be treated in the same manner as described in this study for the longitudinal motions.

II. DISCRETE-TIME MODELS AND THEIR PROPERTIES

A number of techniques exist for obtaining a discrete-time model of a continuous-time system [11, 12]. They permit the selection of a model that is most suitable for a specific design purpose. Among these, the step invariant model, which is obtained by inserting a zero-order hold (ZOH) and a sampler into a CT system, has been widely used for the purpose of digital control [16]. The DT system representation, which is valid for all DT models that have been proposed in the literature, is used in the following, which permits other DT models to be used in the design. The definition of the DT model that follows is the extension of the one used in [13] for the single-variable case to the multivariable case.

Assume that a CT system S and a DT system S^* are initially at rest and that their respective input vectors $u(t)$ and $u^*(kT)$ are zero for $t\ (=kT) < 0$ and finite for a finite t. A DT system S^* whose output is $y^*(kT)$ is then said to be a DT model of a CT system S whose output is $y(t)$ if the following condition is satisfied:

$$\lim_{T\to 0} \|u(t) - u^*(kT)\| = 0 \qquad \text{for each fixed} \quad t \tag{1}$$

implies that

$$\lim_{T\to 0} \|y(t) - y^*(kT)\| = 0 \qquad \text{for each fixed} \quad t, \tag{2}$$

where k is an integer such that $kT \leq t < (k+1)T$ and $\|\cdot\|$ denotes a norm.

The Euler operator is defined as

$$\varepsilon = (z - 1)/T, \tag{3}$$

where z is used as the Z or the shift operator, depending on the context. The Euler operator plays an important role in the derivation and analysis of a DT model for a CT system. The operators ε and $1/\varepsilon$ are the Euler approximations, respectively, of a derivative and an integral; that is, the Euler operator can be considered as an approximation of the Laplace operator s. The basic and important calculation involving the Euler operator is

$$y(kT) = (1/\varepsilon)u(kT) = y((k-1)T) + Tu((k-1)T). \tag{4}$$

Many properties pertaining to DT systems expressed in the Euler operator approach those of CT systems as the sampling interval approaches zero, but this is not the case using the z operator [15]. For example, consider a system given by

$$\varepsilon x(kT) = Ax(kT). \tag{5}$$

The necessary and sufficient condition for the origin of this system to be asymptotically stable is that, for any positive-definite symmetrical matrix Q, there exists a unique positive-definite symmetrical matrix P that satisfies the equation

$$(TA^T + I)PA + A^T P = -Q. \tag{6}$$

The sufficiency can be proven by selecting a Lyapunov function candidate $V(x) = x^T(kT)Px(dT)$ and evaluating $\varepsilon V(x)$ using the relation

$$\varepsilon[x(kT)y(kT)] = x((k+1)T) \cdot \varepsilon y(kT) + \varepsilon x(kT) \cdot y(kT). \tag{7}$$

The necessity can be proven by choosing P as

$$P = \lim_{k \to \infty} T \sum_{i=0}^{k-1} (TA^T + I)^i Q (TA + I)^i. \tag{8}$$

It should be noted that as the sampling interval approaches zero, the stability condition of (6) approaches that of a continuous-time system which is

$$PA + A^T P = -Q. \tag{9}$$

Also, for a conventional discrete-time case

$$\mathbf{A}^T P \mathbf{A} - P = -TQ = -Q' \qquad (<0), \tag{10}$$

where $\mathbf{A} = T \cdot A + I$. The reason for using $\mathbf{A}$ is obvious in view of the relation $x((k+1)T) = [T \cdot A + I]x(kT)$, which is obtained from (5). Thus, the stability condition given by (6) includes that of the conventional discrete-time case and becomes that of the continuous-time case as T approaches zero.

Another example is the stability region in the ε-plane. The stability region in the z plane is inside the unit circle, which does not approach the stability region in the s plane even if $T \to 0$. The stability region in the ε plane is inside a circle of radius $1/T$ centered at $-1/T + j0$ (i.e., $|\varepsilon + 1/T| = 1/T$), which is called the stability circle in this study. As the sampling interval T approaches zero, the stability region in the ε plane approaches that in the s operator, as can be seen from Fig. 1.

Using the similarity of the ε operator to the s operator, a nonminimum phase problem can be solved. It is known that a minimum phase CT system whose relative degree is larger than two results in a nonminimum phase DT model if the sampling interval is sufficiently small [17]. However, using the

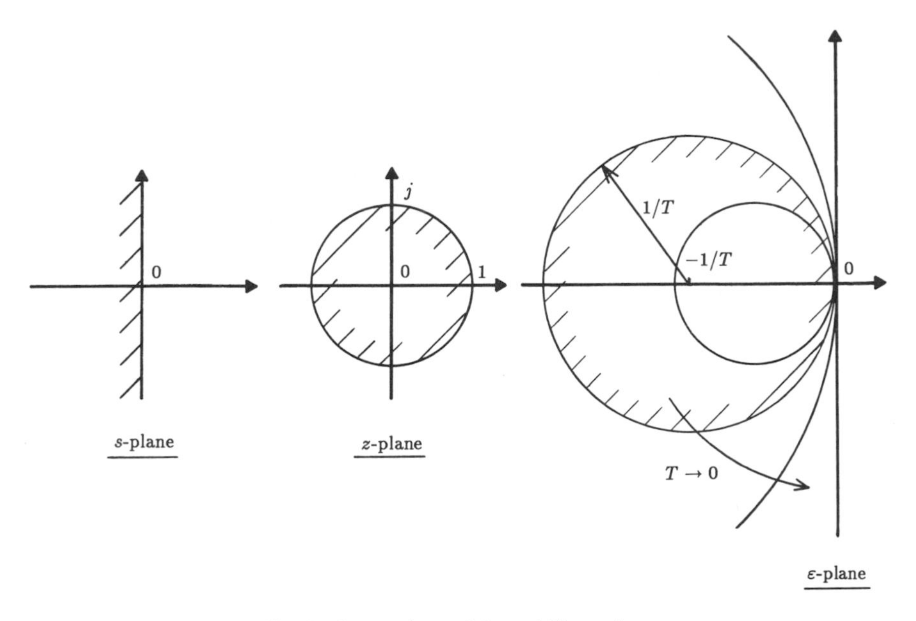

Fig. 1. Comparison of the stability regions.

Euler operator, the zeros of a DT model, which are introduced by the sampling process and become unstable as $T \to 0$, can be easily distinguished from the zeros that correspond to the CT system. The unstable zeros that are introduced by the sampling process can then be neglected so that the minimum phase property of a CT system can be preserved in a DT model, if T is selected to be sufficiently small [16]. Since the zeros are continuous functions of the CT system parameters, such a T always exists. Using the Euler operator, the following theorem is now in order.

Theorem 1. Let a CT system S be given by

$$\dot{x}(t) = Ax(t) + Bu(t) \tag{11}$$

$$y(t) = Cx(t) + Du(t). \tag{12}$$

Then, any DT system S^* that can be expressed as

$$\varepsilon x^*(kT) = (A + \Delta A)x^*(kT) + (B + \Delta B)u^*(kT) \tag{13}$$

$$y^*(kT) = (C + \Delta C)x^*(kT) + (D + \Delta D)u^*(kT) \tag{14}$$

is a DT model of a CT system S, where $\lim_{T\to 0} \Delta A = 0$, $\lim_{T\to 0} \Delta B = 0$, $\lim_{T\to 0} \Delta C = 0$, and $\lim_{T\to 0} \Delta D = 0$.

Proof. This theorem can be proven as in [13] by replacing the absolute values with norms. □

It should be noted that the Δ terms depend on A, B, C, D, T, and on the method of discretization, and not on ε. If the step invariant model is used, the DT model is expressed as

$$\varepsilon x^*(kT) = \left(\frac{e^{AT} - I}{T}\right) x^*(kT) + \left(\frac{1}{T}\int_0^T e^{A\tau}\, d\tau \cdot B\right) u^*(kT) \tag{15}$$

$$y^*(kT) = Cx^*(kT) + Du^*(kT), \tag{16}$$

which is a particular case of (13) and (14). This can be seen from the fact that $\lim_{T\to 0}(e^{AT} - I)/T = A$ and $\lim_{T\to 0}(\int_0^T e^{A\tau}\, d\tau)/T = I$. A computer subroutine that calculates this DT model is shown in the Appendix, together with a subroutine that computes the response using (4). Some interesting theorems follow:

Theorem 2. If S^* is a DT model of an asymptotically stable or unstable CT system S, then there exists a $\delta > 0$ such that $0 < T < \delta$ implies, respectively, an asymptotic stability or instability of S^*.

Theorem 3. Let S_1^* and S_2^* be the DT models of the CT systems S_1 and S_2, respectively. Then the series, parallel or feedback connections of S_1^* and S_2^* are, respectively, DT models of the series, parallel or feedback connected systems of S_1 and S_2.

Proof. The proofs of these theorems can be achieved in almost the same way as those given in [13]. □

III. MODEL-FOLLOWING CONTROLLER DESIGN

A. PROBLEM STATEMENT

Consider a minimum phase CT plant given by (11) with $D = 0$. Then, any DT model of the plant can be expressed as

$$\varepsilon x_P(kT) = A_P x_P(kT) + B_P u_P(kT) \tag{17}$$

$$y_P(kT) = C_P x_P(kT), \tag{18}$$

where the subscript P is added to and the asterisk is omitted from $x(kT)$, $u(kT)$, and $y(kT)$, and $x_P \in R^n$, u_P and $y_P \in R^r$. Also, A_P is defined as $A_P = A + \Delta A$ and the others in the same way. The transfer function matrix is given by

$$H_P(\varepsilon) = C_P(\varepsilon I + A_P)^{-1} B_P, \tag{19}$$

where its rank is assumed to be full and all of the zeros to be stable, that is, the zeros are located inside the stability circle. Since the CT plant is assumed to be of the minimum phase type, the unstable zeros introduced by discretization can be neglected to obtain a minimum phase DT model of the plant if T is selected to be sufficiently small [16].

A stable reference model is used to express the desired performance of a system and is given by

$$\varepsilon x_M(kT) = A_M x_M(kT) + B_M u_M(kT) \tag{20}$$

$$y_M(kT) = C_M x_M(kT), \tag{21}$$

where $x_M \in R^{\tilde{n}}$, u_M and $y_M \in R^r$. Its TFM is

$$H_M(\varepsilon) = C_M(\varepsilon I - A_M)^{-1} B_M. \tag{22}$$

In order for the resulting control system to be causal, the reference model must be selected such that each row-relative degree of this TFM is larger than or equal to the corresponding degree of the plant, or equivalently that

$$\tilde{n} - \partial_r[C_M \operatorname{adj}(\varepsilon I - A_M)B_M] \geq n - \partial_r[C_P \operatorname{adj}(\varepsilon I - A_P)B_P], \tag{23}$$

where ∂_r denotes the row degree.

The problem of concern here is to design a control system that makes the plant outputs follow the reference output in the sense of exact model matching. If the TFM of the reference model is chosen to be diagonal with desired pole–zero locations, the TFM of the plant–controller combination is forced to be decoupled and to have the desired characteristics. Therefore, this model-following scheme includes decoupling of the system, which implies that it can be used also for nondecoupled control, depending on the selection of the reference model.

B. OUTPUT ERROR EQUATION

From (17) and (18), it can be obtained for $i = 1, 2, \ldots, r$ that

$$\varepsilon^i y_{Pj}(kT) = C_{Pj} A_P^i x_P(kT), \qquad i = 0, 1, \ldots, m_j - 1 \tag{24}$$

$$\varepsilon^{m_j} y_{Pj}(kT) = C_{Pj} A_P^{m_j} x_P(kT) + C_{Pj} A_P^{m_j - 1} B_P u_P(kT), \tag{25}$$

where y_{Pj} denotes the jth element of y_P and m_j the jth row-relative degree, which is the difference in orders between the denominator and numerator polynomials of the plant TFM. For the reference model,

$$\varepsilon^i y_{Mj}(kT) = C_{Mj} A_M^i x_M(kT), \qquad i = 0, 1, \ldots, m_j - 1 \tag{26}$$

$$\varepsilon^{mj} y_{Mj}(kT) = C_{Mj} A_M^{mj} x_M(kT) + C_{Mj} A_M^{m_j - 1} B_M u_M(kT). \tag{27}$$

When the equality does not hold in (23) (i.e., when the reference model is strictly proper) $C_{Mj} A_M^{m_j - 1} B_M = 0$.

Selecting the coefficients $f_i^j (i = 1, 2, \ldots, m_j, j = 1, 2, \ldots, r)$ such that the polynomial

$$f^j(\varepsilon) = \varepsilon^{m_j} + f_1^j \varepsilon^{m_j - 1} + \cdots + f_{m_j}^j \tag{28}$$

is stable, (24)–(27) can be rewritten as

$$f^j(\varepsilon)y_{Pj}(kT) = (C_{Pj}A_P^{m_j} + f_1^j C_{Pj}A_P^{m_j-1} + \cdots + f_{m_j}^j C_{Pj})x_P(kT) + C_{Pj}A_P^{m_j-1}B_P u_P(kT) \tag{29}$$

$$f^j(\varepsilon)y_{Mj}(kT) = (C_{Mj}A_M^{mj} + f_1^j C_{Mj}A_M^{m_j-1} + \cdots + f_{m_j}^j C_{Mj})x_M(kT) + C_{Mj}A_M^{m_j-1}B_M u_M(kT). \tag{30}$$

From (29) and (30), the following output error equation is obtained:

$$\Phi(\varepsilon)e(kT) = F_P x_P(kT) + G_P u_P(kT) - F_M x_M(kT) - G_M u_M(kT) \tag{31}$$

where

$$e_j(kT) = y_{Pj}(kT) - y_{Mj}(kT) \qquad (j = 1, 2, \ldots, r) \tag{32}$$

$$e^T(kT) = [e_1(kT), e_2(kT), \ldots, e_r(kT)] \tag{33}$$

$$\Phi(\varepsilon) = \text{diag}[f^j(\varepsilon)] \qquad (j = 1, 2, \ldots, r) \tag{34}$$

$$G_P = [G_{Pj}] \qquad (j = 1, 2, \ldots, r), \qquad G_{Pj} = C_{Pj}A_P^{m_j-1}B_P \tag{35}$$

$$F_P = [F_{Pj}] \qquad (j = 1, 2, \ldots, r),$$

$$F_{Pj} = C_{Pj}A_P^{m_j} + f_1^j C_{Pj}A_P^{m_j-1} + \cdots + f_{m_j}^j C_{Pj}. \tag{36}$$

In these equations, G_P is called the gain matrix, and G_M and F_M are defined for the reference model in the same way as G_P and F_P for the plant.

C. NONSINGULAR CASE

It is assumed in this subsection that the gain matrix G_P is nonsingular and all the state variables are measurable. The determination of the control input is based on the algebraic condition of the output error such that the system governing the error has no input. To this end, define a control input such that the right-hand side of (31) is equal to zero; that is,

$$u_P(kT) = G_P^{-1}[-F_P x_P(kT) + q_P(kT)], \tag{37}$$

where

$$q_P(kT) = F_M(kT)x_M(kT) + G_M u_M(kT). \tag{38}$$

As $\Phi(\varepsilon)$ was chosen to be stable, the output error tends to zero with the input (37). In other words, the plant output follows the model output. Figure 2 shows the block diagram of this control system.

Consider this scheme from the point of view of the TFM (i.e., of the exact model matching problem). For the feedback portion the output $y_P(\varepsilon)$ can be

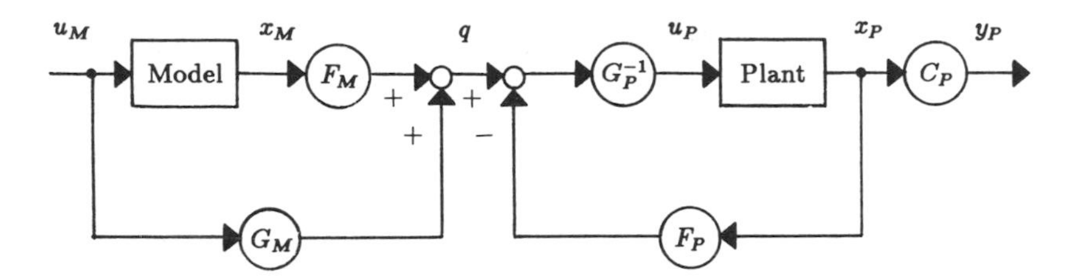

Fig. 2. Block diagram of the controller (nonsingular case).

expressed as

$$y_P(\varepsilon) = C_P(\varepsilon I - A_P + B_P G_P^{-1} F_P)^{-1} B_P G_P^{-1} q_P(\varepsilon) \tag{39}$$

$$= C_P(\varepsilon I - A_P)^{-1}[I + B_P G_P^{-1} F_P(\varepsilon I - A_P)^{-1}]^{-1} B_P G_P^{-1} q_P(\varepsilon). \tag{40}$$

The following equation can be proven by pre- and postmultiplying the inverted portions

$$[I + B_P G_P^{-1} F_P(\varepsilon I - A_P)^{-1}]^{-1} B_P = B_P[I + G_P^{-1} F_P(\varepsilon I - A_P)^{-1} B_P]^{-1}. \tag{41}$$

Using this relation, (40) becomes

$$y_P(\varepsilon) = H_P(\varepsilon)[G_P + F_P(\varepsilon I - A_P)^{-1} B_P]^{-1} q_P(\varepsilon). \tag{42}$$

The following equations can be verified easily

$$C_{Pj}\varepsilon(\varepsilon I - A_P)^{-1} - C_{Pj} = C_{Pj} A_P(\varepsilon I - A_P)^{-1} \tag{43}$$

$$A_P(\varepsilon I - A_P)^{-1} = (\varepsilon I - A_P)^{-1} A_P. \tag{44}$$

Using these relations, it can be shown that

$$C_{Pj} A_P^l(\varepsilon I - A_P)^{-1} = [C_{Pj}\varepsilon(\varepsilon I - A_P)^{-1} - C_{Pj}] A_P^{l-1} \tag{45}$$

$$= \varepsilon^l C_{Pj}(\varepsilon I - A_P)^{-1} - \sum_{i=1}^{l} \varepsilon^{l-i} C_{Pj} A_P^{i-1}. \tag{46}$$

Postmultiplying both sides of this equation by B_P, and using the condition

$$C_{Pj} A_P^{l-1} B_P = 0 \qquad (l = 1, 2, \ldots, m_j - 1) \qquad \text{and} \qquad C_{Pj} A_P^{m_j - 1} B_P \neq 0,$$

it can be proven that

$$C_{Pj} A_P^l(\varepsilon I - A_P)^{-1} B_P = \varepsilon^l C_{Pj}(\varepsilon I - A_P)^{-1} B_P \qquad (l = 1, 2, \ldots, m_j - 1) \tag{47}$$

$$C_{Pj} A_P^{m_j}(\varepsilon I - A_P)^{-1} B_P = \varepsilon^{m_j} C_{Pj}(\varepsilon I - A_P)^{-1} B_P - C_{Pj} A_P^{m_j - 1} B_P \tag{48}$$

for $(j = 1, 2, \ldots, r)$, which lead to

$$G_P + F_P(\varepsilon I - A_P)^{-1} B_P = \Phi(\varepsilon) H_P(\varepsilon). \tag{49}$$

Thus, from (42) and (49),

$$y_P(\varepsilon) = \Phi^{-1}(\varepsilon) q_P(\varepsilon). \tag{50}$$

It can be seen that all of the plant zeros are canceled by the state feedback, which is the reason for assuming the minimum phase property of the plant. The remaining portion of the TFM of this part is $\Phi^{-1}(\varepsilon)$, which is arbitrarily specified by the designer through the selection of the coefficients f_i^j.

For the feedforward path, $q_P(\varepsilon)$ can be written as

$$q_P(\varepsilon) = [F_M(\varepsilon I - A_M)^{-1} B_M + G_M] u_M(\varepsilon). \tag{51}$$

Using the same relation as (49) for the reference model, (51) becomes

$$q_P(\varepsilon) = \Phi(\varepsilon) H_M(\varepsilon) u_M(\varepsilon). \tag{52}$$

The overall TFM is obtained, therefore, as $H_M(\varepsilon)$ the desired reference model.

D. SINGULAR CASE

The requirement that the gain matrix G_P be nonsingular is identical to the necessary and sufficient condition for system decoupling using state feedback [8]. When the gain matrix G_P is singular, the control input $u_P(kT)$ cannot be synthesized using (37). One way to solve this singularity problem is to augment the system with the use of a unimodular polynomial matrix, which is a polynomial matrix whose determinant is a nonzero scalar, that is, whose inverse is also a polynomial matrix [18, 19]. In this regard, there are two ways to augment the system. The first is to augment both the plant and the reference model as described in [9]. The second is to augment only the plant as proposed in [10]. Here only the second approach is described because of its simplicity.

The philosophy of the second approach is to preaugment the plant using a unimodular polynomial matrix $U_R(\varepsilon)$. It should be recalled that (1) the orders of the plant and reference model can be different as long as condition (23) is satisfied, and (2) the pole–zero cancellation of the plant occurs in the feedback part, making the TFM of this portion $\Phi^{-1}(\varepsilon)$, which can be specified arbitrarily by the designer as long as it is stable. It is guaranteed by Theorem 2.5.14 in [19] that one can always find a suitable unimodular polynomial matrix $U_R(\varepsilon)$ such that

$$[C_P \cdot \operatorname{adj}(\varepsilon I - A_P) \cdot B_P] U_R(\varepsilon) \tag{53}$$

is row proper; that is, the coefficient matrix of the highest row degree terms has a full row rank. Define, then, the proper augmenting system

$$H_R(\varepsilon) = U_R(\varepsilon) \cdot L_R^{-1}(\varepsilon), \tag{54}$$

where

$$L_R(\varepsilon) = \text{diag}[L_{Ri}(\varepsilon);\ i = 1, 2, \ldots, r] \tag{55}$$

and $L_{Ri}(\varepsilon)$ is an arbitrary stable polynomial whose degree should be as low as possible. Its lowest possible degree is $\partial_{ci}[U_R(\varepsilon)]$, the ith column degree of $U_R(\varepsilon)$. At the least, $L_R(\varepsilon)$ can be selected as $L_{Ri}(\varepsilon)I_r$ where $\partial L_{Ri}(\varepsilon) = \partial_{\max} U_R(\varepsilon)$ and I_r is a $r \times r$ identity matrix. It should be noted that the above augmenting system has no zeros, since $U_R(\varepsilon)$ is a unimodular polynomial matrix, and thus always has a stable inverse. Perform now the minimal realization of (54) as

$$\varepsilon x_R(kT) = A_R x_R(kT) + B_R r(kT) \tag{56}$$

$$u_P(kT) = C_R x_R(kT) + D_R r(kT) \tag{57}$$

which must satisfy

$$H_R(\varepsilon) = C_R(\varepsilon I - A_R)^{-1} B_R + D_R, \tag{58}$$

where $r(kT) \in R^r$ is to be determined later. The augmented plant $H_P(\varepsilon)H_R(\varepsilon)$ can be expressed as

$$\varepsilon x_S(kT) = A_S x_S(kT) + B_S r(kT) \tag{59}$$

$$y_P(kT) = C_S x_S(kT), \tag{60}$$

where

$$A_S = \begin{bmatrix} A_R & 0 \\ B_P C_R & A_P \end{bmatrix}, \qquad B_S = \begin{bmatrix} B_R \\ B_P D_R \end{bmatrix}$$
$$C_S = [0, C_P], \qquad x_S(kT) = [x_R^T(kT), x_P^T(kT)]^T. \tag{61}$$

Let $r(kT)$ be defined for the feedback part, using the same technique as described for the nonsingular gain matrix case, as

$$r(kT) = G_S^{-1}[-F_S x_S(kT) + q_S(kT)], \tag{62}$$

where G_S and F_S are defined for the augmented plant in the same way as G_P and F_P for the original plant, and where $q_S(kT)$ will be determined still later. It should be remembered that G_S^{-1} always exists, since (53) holds. With this $r(kT)$, the TFM from $q_S(\varepsilon)$ through $y_P(\varepsilon)$ can be expressed as

$$\begin{aligned} C_S(\varepsilon I - A_S)^{-1} B_S[G_S + F_S(\varepsilon I - A_S)^{-1} B_S]^{-1} \\ = C_P(\varepsilon I - A_P)^{-1} B_P[C_R(\varepsilon I - A_R)^{-1} B_R + D_R] \\ \times [\Phi_S(\varepsilon) H_P(\varepsilon) H_R(\varepsilon)]^{-1} \\ = \Phi_S^{-1}(\varepsilon), \end{aligned} \tag{63}$$

where $\Phi_S(\varepsilon)$ is defined for the augmented plant as in $\Phi(\varepsilon)$ for the original plant.

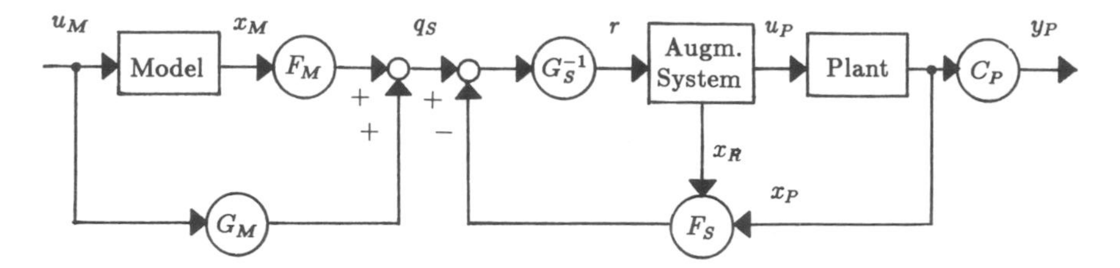

Fig. 3. Block diagram of the controller (singular case).

So that the TFM of the plant–controller combination can match that of the reference model, the precompensator part, from $u_M(kT)$ to $q_S(kT)$, must be $\Phi_S(\varepsilon)H_M(\varepsilon)$. Noting the row-relative degree observation (1) described earlier, it is possible to select among its family a reference model that satisfies the condition that each row degree of $H_M(\varepsilon)$ is larger than or equal to the corresponding degree of $H_P(\varepsilon)H_R(\varepsilon)$. It can be seen now that $q_S(kT)$ should be defined, with the same design parameters used in the feedback controller design, as

$$q_S(kT) = F_M x_M(kT) + G_M u_M(kT). \tag{64}$$

The TFM from $u_M(kT)$ *to* $q_S(kT)$ can be obtained as

$$F_M(\varepsilon I - A_M)^{-1}B_M + G_M = \Phi_S(\varepsilon)H_M(\varepsilon), \tag{65}$$

which makes the overall TFM the desired $H_M(\varepsilon)$. Equations (56), (57), (62), and (64) determine the controller structure that is shown in Fig. 3.

IV. MODEL-FOLLOWING CONTROL USING PLANT OUTPUT

It was assumed in the previous section that all of the state variables of the plant are available. The resulting control law requires no dynamic components, such as filters. When some of the state variables are not available through measurement, they must be estimated from the input and output of the plant. In [9] a state observer, which is a filter, was designed explicitly in the state-space formulation. The design, unfortunately, is relatively complex. A combined design of the state observer and state feedback controller using polynomial algebra is presented in [19], which is simpler than using the state-space equations. The design method presented in [20] and described in the following is based on this approach and achieves model matching; that is, the TFM of a plant–controller combination matches that of a reference model.

The resulting controller involves two extra filters, called the state variable filters, which correspond to a state observer. When all of the state variables are available, the design described in the previous section should be used because of its simplicity.

The plant to be controlled is given by (11) and (12), and any discrete-time model of one's choice can be used to design the controller. The discrete-time model in the Euler operator is then expressed as

$$y_P(kT) = H_P(\varepsilon)u_P(kT) = N_P(\varepsilon)D_P^{-1}(\varepsilon)u_P(kT), \tag{66}$$

where $D_P(\varepsilon)$ and stable $N_P(\varepsilon)$ are right-coprime polynomial matrices and each column degree of $N_P(\varepsilon)$ is less than or equal to the corresponding column degree $d_i (i = 1, 2, \ldots, r)$ of $D_P(\varepsilon)$ [18]. The stability of $N_P(\varepsilon)$ means that the roots of $\det[N_P(\varepsilon)] = 0$ lie inside the stability circle. Such $N_P(\varepsilon)$ can be obtained if T is made sufficiently small. The Hermite form [21] of $H_P(\varepsilon)$ is $\Upsilon(\varepsilon) = \text{diag}[\pi^{-q_i}(\varepsilon)]$ where $\pi(\varepsilon)$ is a stable polynomial of degree one and $q_i (i = 1, 2, \ldots, r)$ an integer. For a single-input–single-output system ($r = 1$), q_1 is the relative degree. The Hermite form is diagonal if and only if the gain matrix defined in the previous section is nonsingular [22]. If the Hermite form is not diagonal (i.e., lower triangular), a suitable precompensentor given by (54) can be used to diagonalize it. When the Hermite form is diagonal, so is K_P, where

$$K_P = \lim_{\varepsilon \to \infty} \Upsilon^{-1}(\varepsilon)H_P(\varepsilon) \tag{67}$$

and K_P^{-1} always exists. When $\pi(\varepsilon)$ is monic, K_P is the gain matrix of the plant (i.e., $K_P = G_P$).

The reference model should be selected such that the following dynamic equivalence [23] holds:

$$\Upsilon(\varepsilon) = H_P(\varepsilon)U_P(\varepsilon) = H_M(\varepsilon)U_M(\varepsilon), \tag{68}$$

where $U_P(\varepsilon)$ and $U_M(\varepsilon)$ are unimodular matrices over a set of all proper rational transfer functions [24]. In other words, $H_M(\varepsilon)$ must be an element of the dynamic equivalence class, which includes a canonical element $\Upsilon(\varepsilon)$ of the plant. This assures $\Upsilon^{-1}(\varepsilon)H_M(\varepsilon)$, which will appear in the control law later, to be causal.

To begin the design, the following theorem is used.

Theorem 4. Given $r \times r$ right-coprime polynomial matrices $D_P(\varepsilon)$ of column degrees d_i and $N_P(\varepsilon)$ of column degrees not greater than d_i, an arbitrary $r \times r$ polynomial matrix $\beta(\varepsilon)$ of column degrees $d_i + v - 1$, where v is the observability index of the minimal plant TFM, can be described by

$$\beta(\varepsilon) = \Gamma(\varepsilon)D_P(\varepsilon) + \Delta(\varepsilon)N_P(\varepsilon), \tag{69}$$

where the highest degree of $\Gamma(\varepsilon)$ and $\Delta(\varepsilon)$ is $v - 1$.

Proof. Refer to [18] or [19] for the proof. □

We are especially interested in selecting $\beta(\varepsilon)$ as

$$\beta(\varepsilon) = F(\varepsilon)[\Upsilon^{-1}(\varepsilon)N_P(\varepsilon) - K_P D_P(\varepsilon)], \tag{70}$$

where $F(\varepsilon) = I_r f(\varepsilon)$ and $f(\varepsilon)$ is an arbitrary stable polynomial of degree v. Equations (66), (69), and (70) lead to the following nonminimal expression of the plant

$$y_P(kT) = \Upsilon(\varepsilon)[K_P u_P(kT) + F^{-1}(\varepsilon)\Gamma(\varepsilon)u_P(kT) + F^{-1}(\varepsilon)\Delta(\varepsilon)y_P(kT)], \tag{71}$$

where the state variable filters $F^{-1}(\varepsilon)\Gamma(\varepsilon)$ and $F^{-1}(\varepsilon)\Delta(\varepsilon)$ are causal. It should be noted that a pole-zero cancellation of $N_P(\varepsilon)$ is involved in (71).

The control input is determined by equating (71) to $y_M(kT)$, which gives

$$u_P(kT) = K_P^{-1}[v(kT) - F^{-1}(\varepsilon)\Gamma(\varepsilon)u_P(kT) - F^{-1}(\varepsilon)\Delta(\varepsilon)y_P(kT)], \tag{72}$$

where

$$v(kT) = \Upsilon^{-1}(\varepsilon)H_M(\varepsilon)u_M(kT). \tag{73}$$

Figure 4 shows the block diagram of this control law.

To show that this control law achieves model matching, define the partial state vector $\omega(kT)$ such that

$$u_P(kT) = D_P(\varepsilon)\omega(kT) \tag{74}$$

$$y_P(kT) = N_P(\varepsilon)\omega(kT). \tag{75}$$

Substituting these relations into (72) gives

$$u_P(kT) = K_P^{-1}\{\Upsilon^{-1}(\varepsilon)H_M(\varepsilon)u_M(kT) - F^{-1}(\varepsilon)[\Gamma(\varepsilon)D_P(\varepsilon) + \Delta(\varepsilon)N_P(\varepsilon)]\omega(kT)\}. \tag{76}$$

Using Theorem 4 with $\beta(\varepsilon)$ selected as in (70), $\omega(kT)$ can be obtained as

$$\omega(kT) = N_P^{-1}(\varepsilon)H_M(\varepsilon)u_M(kT) \tag{77}$$

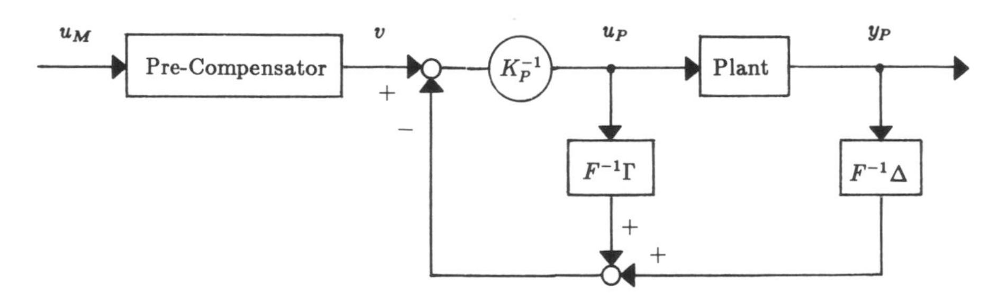

Fig. 4. Block diagram of the controller (using plant output).

which leads to

$$y_P(kT) = H_P(\varepsilon)u_P(kT) = N_P(\varepsilon)\omega(kT) = H_M(\varepsilon)u_M(kT). \tag{78}$$

It can also be shown that the TFM of the feedback part is $\Upsilon(\varepsilon)$ and that of the feed-forward part is $\Upsilon^{-1}(\varepsilon)H_M(\varepsilon)$. It should be noted that, as the plant is assumed to be minimum phase, $N_P(\varepsilon)$ has a stable inverse and $u_P(kT)$ is bounded.

To express the control law in a convenient and succinct form, a part of the state variable filters is rewritten as

$$\begin{aligned} \zeta_i(kT) &= F^{-1}(\varepsilon)\varepsilon^{i-1}u_P(kT) \qquad (i = 1, 2, \ldots, v) \\ \zeta_{v+i}(kT) &= F^{-1}(\varepsilon)\varepsilon^{i-1}y_P(kT) \qquad (i = 1, 2, \ldots, v) \\ \zeta^T(kT) &= [\zeta_1(kT), \zeta_2(kT), \ldots, \zeta_{2v}(kT)], \\ &\zeta_i(kT) \in R^r \qquad (i = 1, 2, \ldots, 2v) \end{aligned} \tag{79}$$

and

$$\xi^T(kT) = \begin{bmatrix} u_P(kT) \\ \zeta(kT) \end{bmatrix} \in R^{2v+1}. \tag{80}$$

With these notations, (71) can be written as

$$\Upsilon^{-1}(\varepsilon)y_P(kT) = \theta^T\xi(kT), \tag{81}$$

where

$$\begin{aligned} \theta^T &= [K_P, \alpha_0, \alpha_1, \ldots, \alpha_{v-1}, \alpha_v, \alpha_{v+1}, \ldots, \alpha_{2v-1}] \\ &= [K_P, \Psi^T] \in R^{r\times(2v+1)}. \end{aligned} \tag{82}$$

The control law can be rewritten as

$$u_P(kT) = K_P^{-1}[v(kT) - \Psi^T\zeta(kT)] = \Phi^T\rho(kT), \tag{83}$$

where

$$\Phi^T = [K_P^{-1}, -K_P^{-1}\Psi^T], \qquad \rho^T(kT) = [v^T(kT), \zeta^T(kT)]. \tag{84}$$

It can be seen that $u_P(kT)$ of (83) satisfies

$$\Upsilon^{-1}(\varepsilon)y_M(kT) = \theta^T\xi(kT). \tag{85}$$

When the plant parameters are unknown or uncertain, the parameters appearing in (83) are unknown. In this case, they can be estimated by one of the on-line identification schemes and used such that

$$\hat{y}_P(kT) = \Upsilon(\varepsilon)\hat{\theta}^T(kT)\xi(kT). \tag{86}$$

V. DECOUPLED CONTROL CONFIGURED VEHICLE FLIGHT CONTROL

An application of the model-following method described in the previous sections to a control configured vehicle (CCV) is considered with particular reference to the longitudinal motions; viz., pitch pointing, vertical translation, and direct lift modes. The lateral modes, such as the direct side force, yaw pointing, and lateral translation modes, can be treated in a manner similar to that described here. It is assumed in this section that all of the state variables are available through measurement and for this reason the model-following method based on the state equations is used. The model-following method using the plant output can be applied with equal ease to CCV modes using the same philosophy as described in the following.

A. CONTROL CONFIGURED VEHICLE FLIGHT MODES

During the past decade, the concept of the CCV has been developed to improve the flight performance of aircraft by the use of active control technologies (ACT) such as relaxed static stability (RSS), maneuver load control (MLC), and flutter mode control (FMC). When conventional control surfaces, the elevators, ailerons, and rudders, are employed, the path of an aircraft is controlled indirectly through three axes moments and thrust as shown in Fig. 5. With this type of control, called four-degrees-of-freedom

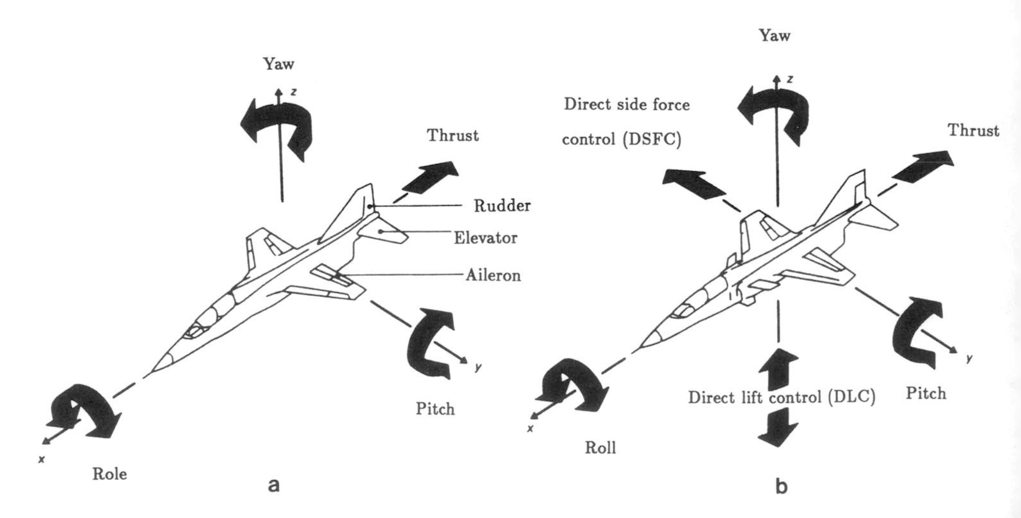

Fig. 5. Four- and six- degrees-of-freedom controls. (a) Conventional control (4 DOF). (b) CCV control (6 DOF).

control, the flight path of an aircraft cannot be changed without attitude change such as pitching, yawing, and rolling. By directly controlling the vertical and horizontal forces, the flight path of a CCV can be controlled directly without any attitude change and the attitude can be controlled directly without any flight path change. This is called six-degrees-of-freedom control and is shown in Fig. 5. It is said, therefore, that the flight path control is decoupled from attitude control. The newly added direct forces are called direct lift control (DLC) for the Z axis and direct side force control (DSFC) for the Y axis. Some canards or flapelons are required and are used to generate such direct forces. Figure 6 shows the typical CCV modes, and the possible advantages of CCV flight are compared with the conventional flight modes in Fig. 7.

B. EQUATIONS OF MOTION

The longitudinal dynamics of the aircraft considered are given by the following linearized time-invariant equations [9]:

$$\dot{u}(t) = X_u u(t) + X_w w(t) - g\theta(t) \tag{87}$$

$$\dot{w}(t) = Z_u u(t) + Z_w w(t) + U_0 q(t) + Z_{\delta f}\delta_f(t) + Z_{\delta e}\delta_e(t) \tag{88}$$

$$\begin{aligned}\dot{q}(t) = {} & (M_u + M_{\dot{w}} Z_u)u(t) + (M_w + M_{\dot{w}} Z_w)w(t) \\ & + (M_q + U_0 M_{\dot{w}})q(t) + (M_{\dot{w}} Z_{\delta f} + M_{\delta f})\delta_f(t) \\ & + (M_{\dot{w}} Z_{\delta e} + M_{\delta e})\delta_e(t)\end{aligned} \tag{89}$$

$$\dot{\theta}(t) = q(t), \tag{90}$$

whose terms are defined as:

$u(t)$	change in forward velocity
U_0	nominal forward velocity
$w(t)$	change in vertical velocity
$\theta(t)$	pitch angle
$q(t)$	pitch rate
M_i	stability and control derivatives of pitching moment
X_i	stability and control derivatives of forward force
Z_i	stability and control derivatives of vertical force
g	gravitational acceleration
$\delta_f(t)$	maneuvering flap deflection
$\delta_e(t)$	elevator deflection

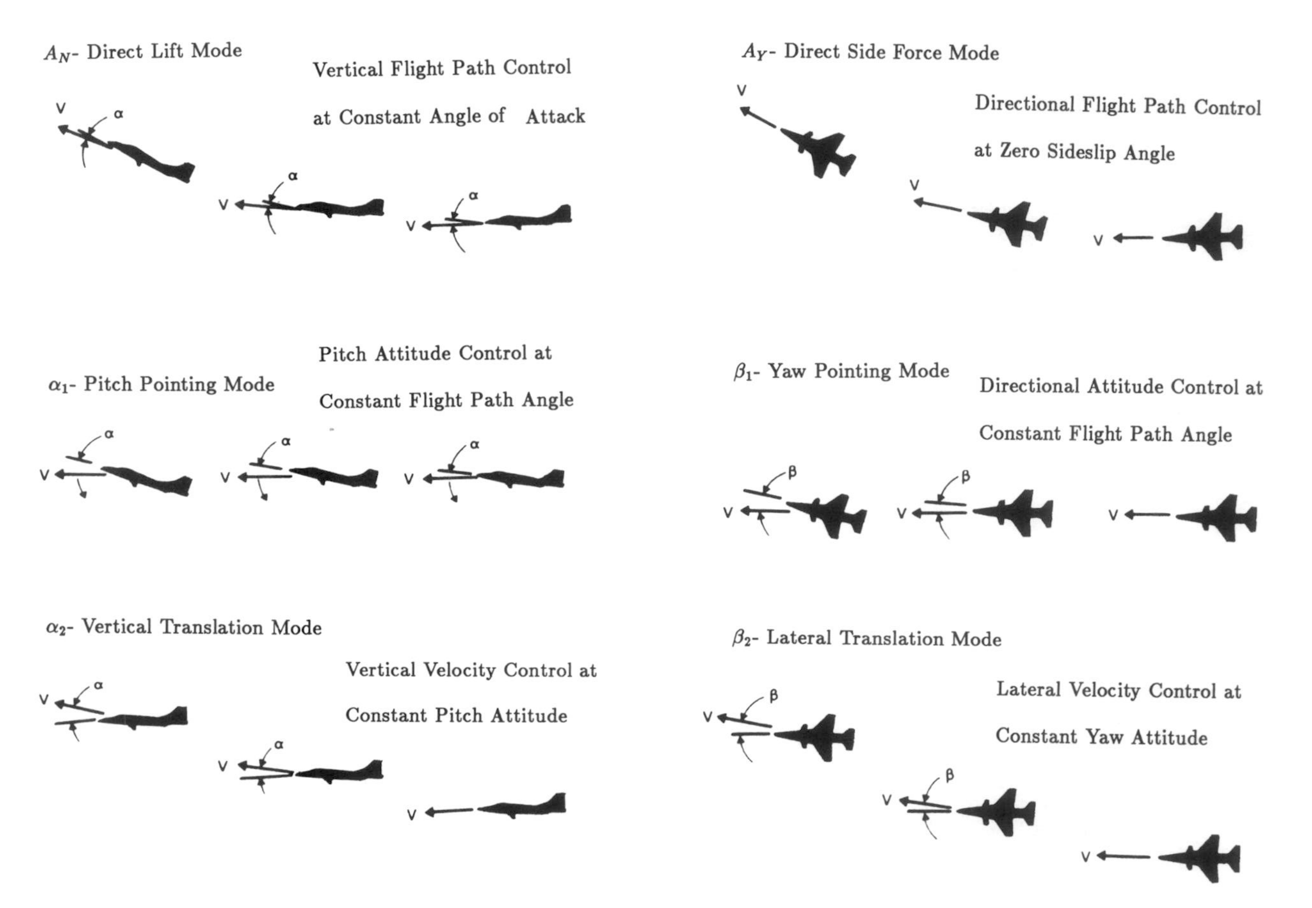

Fig. 6. CCV flight modes.

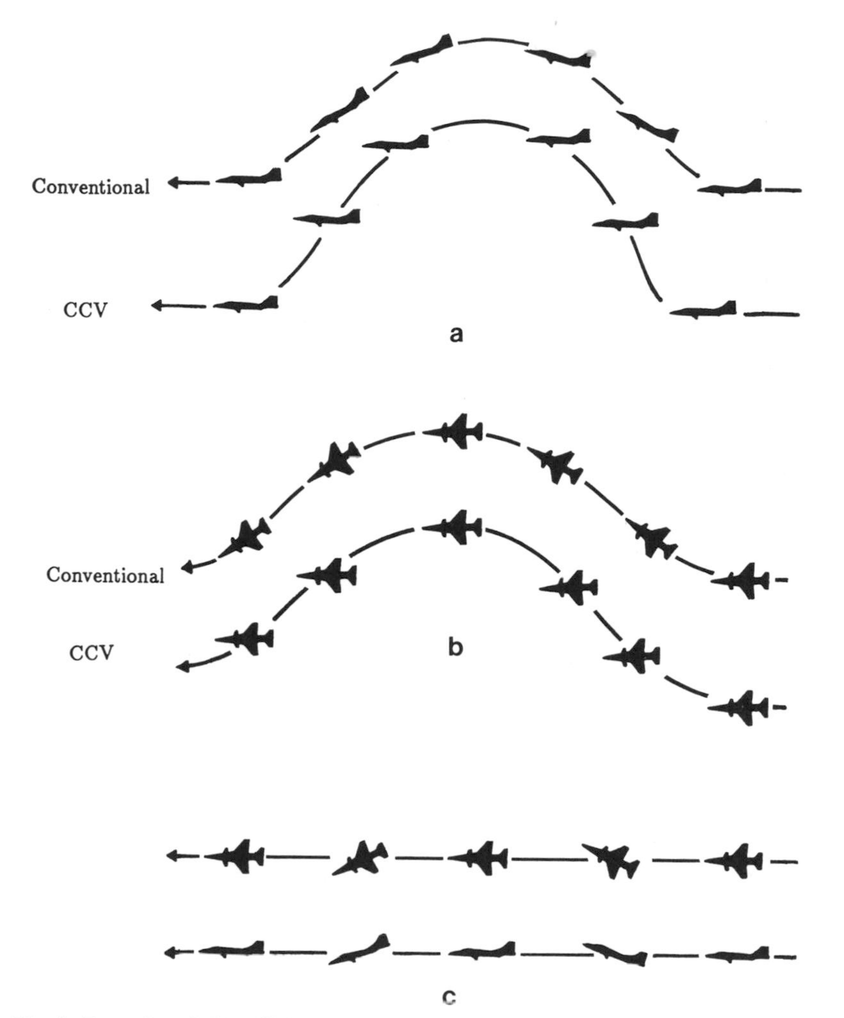

Fig. 7. Examples of CCV flight modes: (a) vertical translation, (b) lateral translation, and (c) attitude control without flight path change.

Other relations to be used are

$$a_Z(t) = \dot{w}(t) - U_0 q(t), \qquad n_Z(t) = -\frac{a_Z(t)}{g} \tag{91}$$

$$\alpha(t) = w(t)/U_0, \qquad \gamma(t) = \theta(t) - \alpha(t), \tag{92}$$

where $a_Z(t)$ is normal acceleration; $n_Z(t)$, normal acceleration in G (load factor); $\alpha(t)$ angle of attack (AOA); and $\gamma(t)$, flight path angle.

The state-space form of the aircraft is given by (11) and (12) with $D = 0$ and the input vector is given by

$$u_P(t) = [\delta_f(t), \delta_e(t)]^T. \tag{93}$$

Defining the state vector as

$$x^1(t) = [\gamma(t), \theta(t), q(t)]^T \tag{94}$$

the system matrices are obtained as

$$A = \begin{bmatrix} Z_w & -Z_w & 1 \\ 0 & 0 & 1 \\ -M'_w U_0 & M'_w U_0 & M'_q \end{bmatrix} \qquad B = \begin{bmatrix} \dfrac{-Z_{\delta f}}{U_0} & \dfrac{-Z_{\delta e}}{U_0} \\ 0 & 0 \\ M'_{\delta f} & M'_{\delta e} \end{bmatrix}, \tag{95}$$

where

$$M'_i = M_i + M_{\dot{w}} Z_i, \qquad M'_q = M_q + M_{\dot{w}} U_0 \tag{96}$$

and i denotes w, δ_f, or δ_e. Also, defining the state vector as

$$x^2(t) = [\alpha(t), q(t)]^T, \tag{97}$$

the system matrices are given by

$$A = \begin{bmatrix} Z_w & 1 \\ M'_w U_0 & M'_q \end{bmatrix} \qquad B = \begin{bmatrix} \dfrac{Z_{\delta f}}{U_0} & \dfrac{Z_{\delta e}}{U_0} \\ M'_{\delta f} & M'_{\delta e} \end{bmatrix}, \tag{98}$$

where $x^1(t)$ will be used for the α_1 mode and $x^2(t)$ for the α_2 and A_N modes.

C. PITCH POINTING (α_1) MODE

In the pitch pointing (α_1) mode, the pitch attitude is controlled at a constant flight path angle by varying the angle-of-attack (AOA) at a constant normal acceleration. This mode is particularly useful for quick and precise air-to-air tracking and air-to-ground missions.

The output is selected as

$$y_P(t) = [\gamma(t), \theta(t)]^T \qquad \text{or} \qquad C_P = \begin{bmatrix} 1 & 0 & 0 \\ 0 & 1 & 0 \end{bmatrix} \tag{99}$$

and the state vector as $x^1(t)$ to achieve this flight mode. The DT model of an aircraft can be obtained then by using one of the discretization schemes to express the CT aircraft in the DT form given by (17) and (18). The model-following method described earlier can be used such that the system is decoupled by choosing the reference model to be diagonal. When the decoupling is achieved, the reference input

$$u_M(t) = [\gamma^*, \theta^*]^T, \tag{100}$$

where the asterisk superscript denotes the command input, controls the pitch attitude with a specified flight path angle.

In all the simulation studies that were carried out the parameters of a hypothetical T-2 CCV at the condition of $H = 20{,}000$ ft and Mach number = 0.8 were used. Among the various DT models, the step invariant model [(15) and (16)] was used to obtain the discretized aircraft model at a sampling frequency of 40 Hz. The CT plant and its step invariant DT model are shown in Fig. 8a, which indicates the similarity of these two representations. The reference model was selected such that $\zeta = 1$ and $\omega_n = 3$ rad/sec. In the

Continuous-Time Form				Discrete-Time Form (40 Hz)			
A	-0.882	0.882	1	A_P	-0.847	0.847	0.985
	0	0	1		0.025	-0.025	0.985
	2.035	-2.035	-1.217		1.983	-1.983	-1.199
B	0.237	0.186		B_P	0.206	-0.089	
	0	0			-0.029	-0.273	
	-2.340	-22.060			-2.299	-21.723	

a

Continuous-Time Form					Discrete-Time Form (40 Hz)				
A_M^c	0	1	0	0	A_M	-0.107	0.928	0	0
	-9	-6	0	0		-8.350	-5.673	0	0
	0	0	0	1		0	0	-0.107	0.928
	0	0	-9	-6		0	0	-8.350	-5.673
B_M^c	0	0			B_M	0.012	0		
	1	0				0.928	0		
	0	0				0	0.012		
	0	1				0	0.928		

b

Fig. 8. (a) CT and DT forms of the aircraft for α_1 mode. (b) CT and DT forms of the reference model for α_1 and α_2 modes.

continuous-time domain, the system matrices are given by

$$A_M^c = \begin{bmatrix} 0 & 1 & 0 & 0 \\ -9 & -6 & 0 & 0 \\ 0 & 0 & 0 & 1 \\ 0 & 0 & -9 & -6 \end{bmatrix}, \qquad B_M^c = \begin{bmatrix} 0 & 0 \\ 1 & 0 \\ 0 & 0 \\ 0 & 1 \end{bmatrix},$$

$$C_M^c = \begin{bmatrix} 9 & 0 & 0 & 0 \\ 0 & 0 & 9 & 0 \end{bmatrix} \tag{101}$$

and then discretized to obtain the step invariant model (Fig. 8b). The gain matrix is obtained as

$$G_P = \begin{bmatrix} C_{P1} B_P \\ C_{P2} B_P \end{bmatrix} = \begin{bmatrix} 0.206 & -0.089 \\ -0.029 & -0.273 \end{bmatrix}, \tag{102}$$

where $m_1 = m_2 = 1$. As G_P is nonsingular and all the state variables are assumed to be available, the control input can be generated by (37) as

$$\begin{aligned} u_P(kT) = & \begin{bmatrix} 33.186 & -8.099 & 3.089 \\ -3.606 & -28.362 & -3.936 \end{bmatrix} x_P(kT) \\ & + \begin{bmatrix} 329.963 & 38.784 & -107.202 & -12.601 \\ -34.946 & -4.108 & -248.916 & -29.258 \end{bmatrix} x_M(kT) \\ & + \begin{bmatrix} 0.497 & -0.162 \\ -0.053 & -0.375 \end{bmatrix} u_M(kT), \end{aligned} \tag{103}$$

where $f^1(\varepsilon) = f^2(\varepsilon) = \varepsilon + 0.2/T = \varepsilon + 8$. In all the simulation studies, including other CCV modes that will be described in the following subsections, the same f_i^j were used. The simulation result for the reference input of $u_M(kT) = [0, 2°]^T$ is shown in Fig. 9, which indicates that the change in flight path angle is zero while the pitch angle is changed.

D. VERTICAL TRANSLATION (α_2) MODE

The vertical translation (α_2) mode controls the vertical velocity at a constant pitch attitude. For instance, the altitude can be changed at the same time as the fuselage is horizontal. This is ideal for small and precise altitude changes such as in aerial refueling, formation flight, and glide path control during approach and landing.

Using $x^2(t)$ as the state vector and selecting the output and input vectors as

$$y_P(t) = [\alpha(t), q(t)]^T \qquad \text{or} \qquad C_P = \begin{bmatrix} 1 & 0 \\ 0 & 1 \end{bmatrix} \tag{104}$$

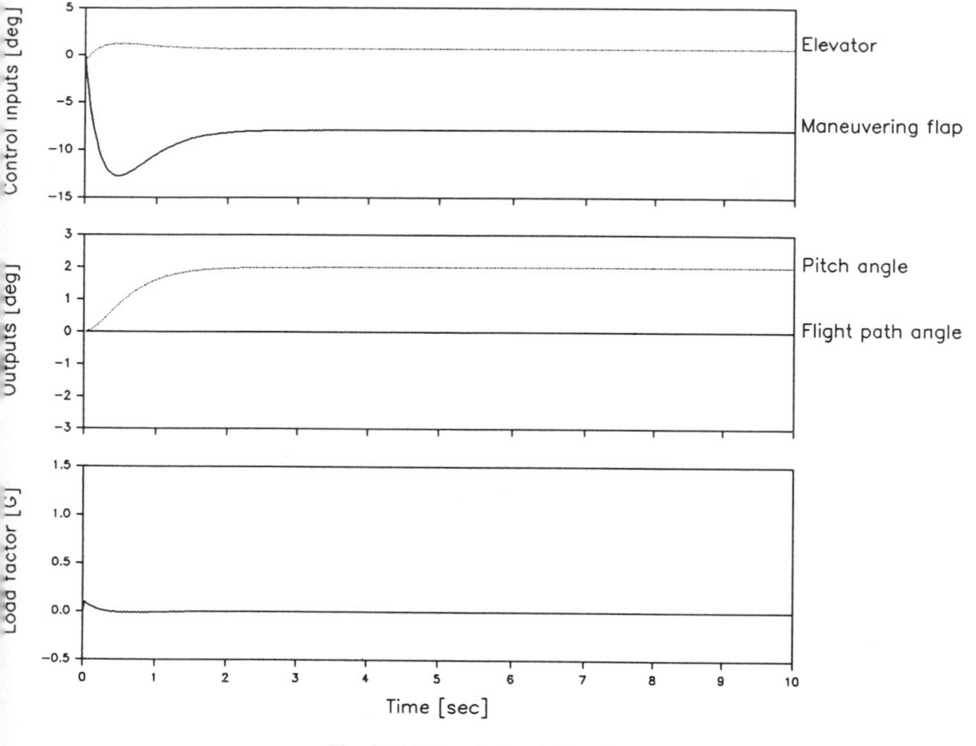

Fig. 9. Pitch pointing (α_1) mode.

the step invariant DT model of the aircraft can be obtained as shown in Fig. 10a. If the reference model used in the α_1 mode is chosen, the control law for the α_2 mode is given by

$$\begin{aligned} u_P(kT) &= \begin{bmatrix} -33.179 & -3.863 \\ 3.602 & 0.097 \end{bmatrix} x_P(kT) \\ &\quad + \begin{bmatrix} -329.904 & -38.777 & -0.094 & 0.813 \\ 34.912 & 4.104 & 0.054 & -0.470 \end{bmatrix} x_M(kT) \\ &\quad + \begin{bmatrix} -0.497 & 0.010 \\ 0.053 & -0.006 \end{bmatrix} u_M(kT), \end{aligned} \tag{105}$$

where $m_1 = m_2 = 1$.

The reference input

$$u_M(t) = [\alpha^*, 0]^T \tag{106}$$

controls the AOA without changing a pitch angle. Figure 11 is the simulation result for the reference input $u_M(kT) = [-2^\circ, 0]^T$, which shows that α_2 mode is achieved successfully.

Continuous-Time Form			Discrete-Time Form (40 Hz)		
	-0.882	1		-0.897	0.974
A	-2.035	-1.217	A_P	-1.982	-1.223
	-0.237	-0.186		-0.263	-0.455
B	-2.340	-22.060	B_P	-2.298	-21.719

a

Continuous-Time Form			Discrete-Time Form (40 Hz)		
	-2	0		-1.951	0
A_M^c	0	-2	A_M	0	-1.951
	1	0		0.975	0
B_M	0	1	B_M	0	0.975

b

Fig. 10. (a) CT and DT forms of the aircraft for α_2 and A_N modes. (b) CT and DT forms of the reference model for A_N mode.

E. DIRECT LIFT (A_N) MODE

The direct lift (A_N) mode controls the lift at a constant AOA so that the vertical flight path can be controlled. In this mode a recovery from dive can be achieved with small loss of altitude since the vertical velocity can be increased without changing the AOA.

When the AOA is kept constant, the normal acceleration can be controlled by the pitch rate alone, since in this case

$$n_Z(t) = U_0 q(t)/g. \tag{107}$$

Hence, the A_N mode can be achieved by using the same state and output vectors as those used in the α_2 mode and by selecting the reference input as

$$u_M(t) = [0, q^*]^T. \tag{108}$$

The reference model was selected to be

$$A_M^c = \begin{bmatrix} -2 & 0 \\ 0 & -2 \end{bmatrix}, \quad B_M^c = \begin{bmatrix} 1 & 0 \\ 0 & 1 \end{bmatrix}, \quad C_M^c = \begin{bmatrix} 2 & 0 \\ 0 & 2 \end{bmatrix}, \tag{109}$$

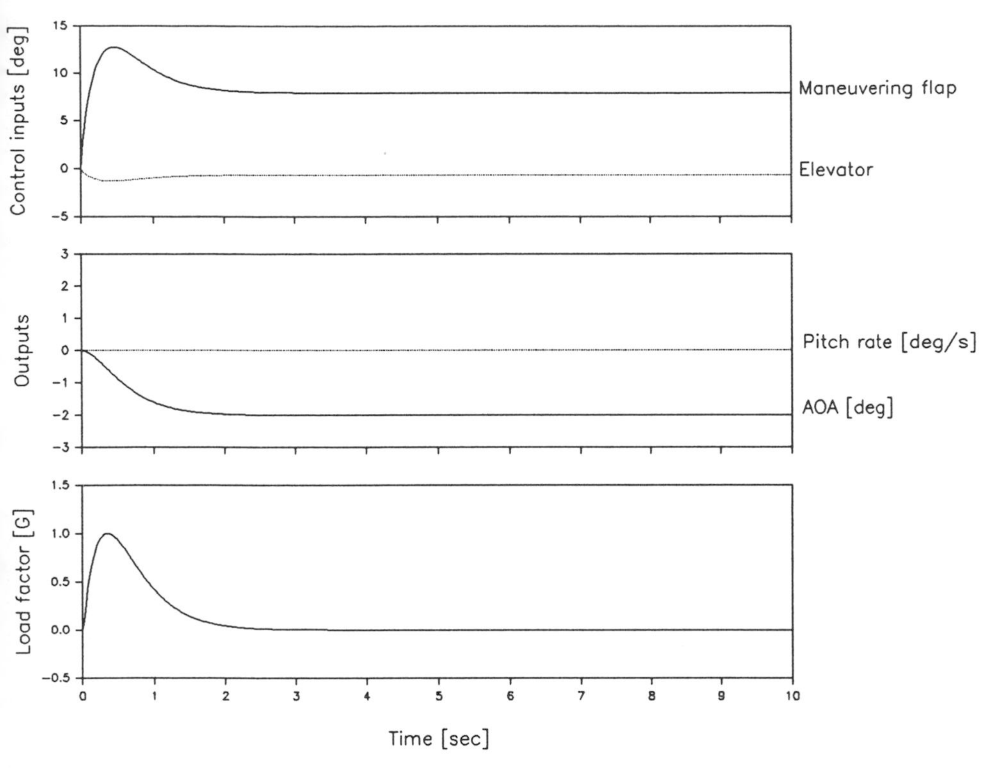

Fig. 11. Vertical translation (α_2) mode.

whose step invariant DT model is shown in Fig. 10b. The control law for the A_N mode is found to be

$$u_P(kT) = \begin{bmatrix} -33.179 & -3.863 \\ 3.602 & 0.097 \end{bmatrix} x_P(kT) + \begin{bmatrix} -56.186 & 1.178 \\ 5.946 & -0.682 \end{bmatrix} x_M(kT) + \begin{bmatrix} -9.060 & 0.190 \\ 0.959 & -0.110 \end{bmatrix} u_M(kT), \tag{110}$$

where $m_1 = m_2 = 1$. Figure 12 shows the simulation result using the reference input given by $u_M = [0, 2°/\text{sec}]^T$.

The three modes mentioned thus far are called the open-loop CCV modes and have a wide range of applications. Closed CCV modes also exist. In the maneuver enhancement (ME), or the automatic blended mode, for example, the error between pilot command G and actual aircraft G are kept zero by coordinating the maneuvering flap with the elevator. This results in a more rapid G response and a measure of gust alleviation [2].

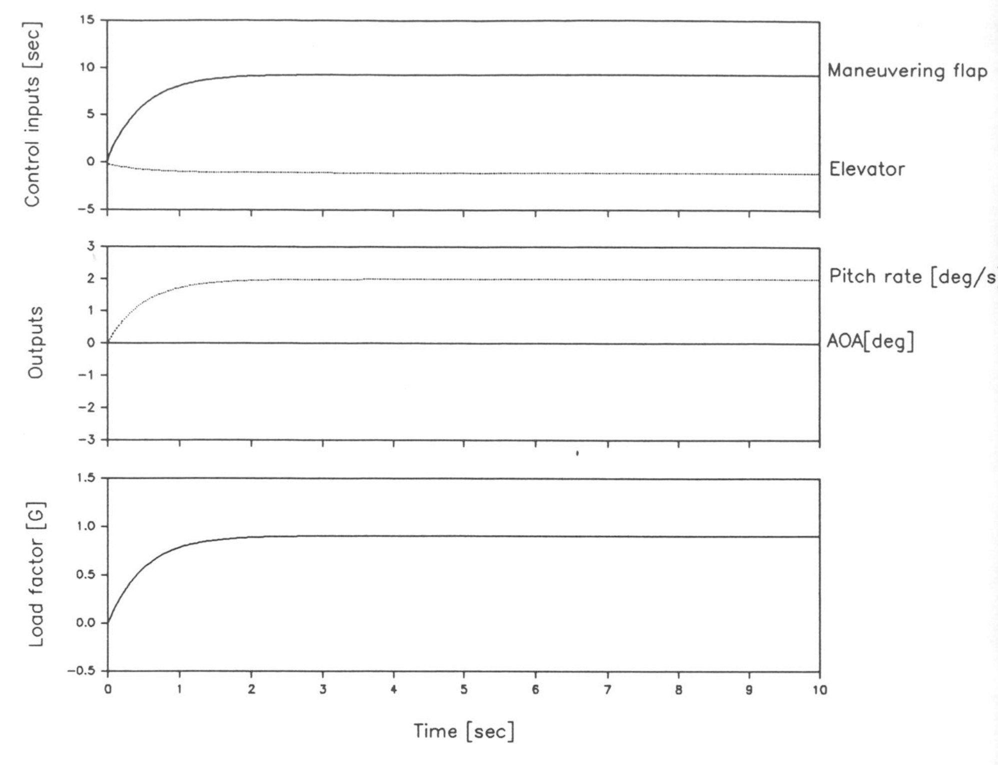

Fig. 12. Direct lift (A_N) mode.

VI. CONCLUSIONS

A discrete-time model-following controller, which is composed of a state feedback and input dynamics compensator, was described using the Euler operator. It was assumed in the design that all of the plant zeros are stable. Using the Euler operator, the unstable zeros, if any, that are introduced by the discretization process can be identified easily and be neglected by selecting the sampling interval to be sufficiently small. Thus, the minimum phase property of a continuous-time plant can be carried over to the discrete-time model. When all of the state variables are available through measurement, the state-

Fig. 13. FORTRAN subroutine programs DTMELR and RSPELR: N, order of the system; M, number of inputs; T, sampling interval; A, system matrix (CT); B, input matrix (CT); $E = I + AT/2! + (AT)^2/3! + \cdots + (AT)^{L-1}/L!$; L, number of terms in E required for convergence; F, system matrix (DT) $F = AE = EA$; G, input matrix (DT) $G = EB$.

```
      SUBROUTINE DTMELR(N1,N2,N,M,T,A,B,F,G,ITER)
      DIMENSION A(N1,N1),B(N1,N2),F(N1,N1),G(N1,N2)
     1,AA(50,50),BB(50,50),CC(50,50)
C— CALCULATION OF E —
      DO 400 I=1,N
      DO 400 J=1,N
      IF(I–J) 200,100,200
100   CC(I,J)=1.0
      GO TO 300
200   CC(I,J)=0.0
300   AA(I,J)=CC(I,J)
400   CONTINUE
      ITER=1
500   DO 600 I=1,N
      DO 600 J=1,N
      BB(I,J)=0.0
      DO 600 K=1,N
      BB(I,J)=AA(I,K)*A(K,J)*T+BB(I,J)
600   CONTINUE
      DO 700 I=1,N
      DO 700 J=1,N
      AA(I,J)=BB(I,J)/FLOAT(ITER+1)
      CC(I,J)=AA(I,J)+CC(I,J)
700   CONTINUE
      XMAX=0.0
      DO 900 I=1,N
      DO 900 J=1,N
      IF(ABS(A(I,J)–XMAX)) 900,800,800
800   XMAX=ABS(AA(I,J))
900   CONTINUE
      IF(XMAX–1.0E–10) 1100,1000,1000
1000  ITER=ITER+1
      GO TO 500
C— CALCULATION OF F & G —
1100  DO 1300I=1,N
      DO 1200 J=1,N
      F(I,J)=0.0
      DO 1200 K=1,N
      F(I,J)=CC(I,K)*A(K,J)+F(I,J)
1200  CONTINUE
      DO 1300 J=1,M
      G(I,J)=0.0
      DO 1300 K=1,N
      G(I,J)=CC(I,K)*B(K,J)+G(I,J)
1300  CONTINUE
      RETURN
      END

      SUBROUTINE RSPELR(N1,N2,N,M,T,F,G,X,U)
      DIMENSION F(N1,N1),G(N1,N2),X(N1),U(N2),XX(50)
      DO 110 I=1,N
      XX(I)=0.0
      DO 100 J=1,N
      XX(I)=F(I,J)*X(J)+XX(I)
100   CONTINUE
      DO 110 J=1,M
      XX(I)=G(I,J)*U(J)+XX(I)
110   CONTINUE
      DO 120 I=1,N
      X(I)=X(I)+T*XX(I)
120   CONTINUE
      RETURN
      END
```

space equations are used to design a simple controller whose feedback portion is nondynamic. This design is applicable to systems that have a singular gain matrix by suitably augmenting the plant using a unimodular polynomial matrix. When some of the state variables are not available, a polynomial method can be used to design a combined state observer and controller.

It was shown how the model-following method can be applied to the design of longitudinal CCV flight controllers. The model-following control system was found to be effective in achieving the various CCV modes by choosing a diagonal reference model with the proper choice of inputs, outputs, and state variables. The proposed scheme can be applied to the other modes, including lateral CCV modes, in a similar manner. The model-following method can be used also for nondecoupled control such as in-flight simulation.

Aircraft that operate at conditions not much different from their nominal flight conditions normally use fixed gain controllers. However, when an aircraft operates over widely different flight conditions, the performance degradation using a fixed gain controller often becomes unacceptable. In such cases, gain scheduling or some adaptive control method can be considered [25].

APPENDIX. FORTRAN SUBROUTINES

FORTRAN subroutines DTMELR, which computes the step invariant discrete-time model, and RSPELR, which computes the state response, are given in Fig. 13 in the Euler operator form.

REFERENCES

1. R. A. WHITMOYER and J. K. RAMAGE, "The Fighter Control Configured Vehicle (CCV) Program Development and Flight Test Summary," *Proc. 8th Annu. Symp., Soc. Flight Test Eng., Washington,* D.C. (1977).
2. F. R. SWORTZEL and J. D. McALLISTER, "Design Guidance from Fighter CCV Flight Evaluations," AGARD-CP-260, pp. 18.1–18.21 (1978).
3. S. L. GRUNWALD and R. F. STENGEL, "Design and Flight Testing of Digital Direct Side-Force Control Laws," *AIAA J. Guidance, Control Dyn.* **8**, 188—193 (1985).
4. J. S. TYLER, "The Characteristics of Model-Following Systems as Synthesized by Optimal Control," *IEEE Trans. Autom. Control* **AC-9**, 485–498 (1964).
5. H. ERZBERGER, "Analysis and Design of Model Following Control Systems by State Space Techniques," *Proc. J. Autom. Control Conf., Ann Arbor, Michigan*, pp. 572–581 (1968).
6. B. C. MOORE, "Model Matching by State Feedback and Dynamic Compensation," *IEEE Trans. Autom. Control* **AC-17**, 491–497 (1972).
7. W. A. WOLOVICH, "The Application of State Feedback Invariants to Exact Model Matching," *Proc. 5th Princeton Conf. Inf. Syst., Princeton, New Jersey* (1971).
8. P. L. FALB and W. A. WOLOVICH, "Decoupling in the Design and Synthesis of Multivariable Control Systems," *IEEE Trans. Autom. Control* **AC-12**, 651–659 (1967).
9. K. KANAI, S. UCHIKADO, P. N. NIKIFORUK, and N. HORI, "Application of a New Multivariable Model-Following Method to Decoupled Flight Control," *AIAA J. Guidance, Control Dyn.* **8**, 637–643 (1985).

10. K. KANAI, N. HORI, and P. N. NIKIFORUK, "A Discrete-Time Multivariable Model-Following Method Applied to Decoupled Flight Control," *AIAA J. Guidance, Control Dyn.* **9**, 403–407 (1985).
11. G. F. FRANKLIN and J. D. POWELL, "*Digital Control of Dynamic Systems.*" Addison-Wesley, Reading, Massachusetts, 1980.
12. P. KATZ, "*Digital Control Using Microprocessors.*" Prentice-Hall Int., London, 1981.
13. T. MORI, P. N. NIKIFORUK, M. M. GUPTA, and N. HORI, "A Class of Discrete Time Models for a Continuous Time System," *Proc. Am. Control Conf., Minneapolis Minnesota*, pp. 953–957 (1987).
14. R. H. MIDDLETON and G. C. GOODWIN, "Improved Finite Word Length Characteristics in Digital Control using Delta Operators," *IEEE Trans. Autom. Control* **AC-31**, 1015–1021 (1986).
15. N. HORI, P. N. NIKIFORUK, and K. KANAI, "On a Discrete-Time System Expressed in the Euler Operator," *Proc. Am. Control Conf., Atlanta, Georgia*, pp. 873–878 (1988).
16. G. C. GOODWIN, R. LOZANO-LEAL, D. Q. MAYNE, and R. H. MIDDLETON, "Rapproachement between Continuous and Discrete Model Reference Adaptive Control," *Automatica* **22**, 199–207 (1986).
17. K. J. ASTROM, P. HAGANDER, and J. STERNBY, "Zeros of Sampled Systems," *Automatica* **20**, 31–38 (1984).
18. T. KAILATH, "*Linear Systems.*" Prentice-Hall, Englewood Cliffs, New Jersey, 1980.
19. W. A. WOLOVICH, "*Linear Multivariable Systems.*" Springer-Verlag, New York, 1974.
20. N. HORI, P. A. PANNALA, P. R. UKRAINETZ, and P. N. NIKIFORUK, "Design of an Electrohydraulic Positioning System using a Novel Model Reference Control Scheme," *ASME J. Dyn. Syst. Meas. Control* **111**, 292–298 (1989).
21. A. S. MORSE, "Parameterizations for Multivariable Adaptive Control," *Proc. 20th IEEE CDC*, pp. 970–972 (1981).
22. R. P. SINGH and K. S. NARENDRA, "Prior Information in the Design of Multivariable Adaptive Controllers," *Tech. Rep. No. 8304*, Center for Systems Science, Yale University, New Haven, Connecticut, 1983.
23. A. S. MORSE, "System Invariants under Feedback and Cascade Control," *in* "*Mathematical System Theory*," pp. 61–74, Springer-Verlag, New York, 1976.
24. N. T. HUNG and B. D. O. ANDERSON, "Triangularization Technique for the Design of Multivariable Control Systems," *IEEE Trans. Autom. Control* **AC-24**, 455–460 (1979).
25. N. HORI, P. N. NIKIFORUK, and K. KANAI, "Robust Adaptive Flight Control in the Presence of Unmodeled Dynamics and Disturbances," *Proc. 25th Aircr. Symp. Tokyo*, pp. 176–179 (1987).

ADVANCES IN DIGITAL SIGNAL PROCESSING

GREGORY W. MEDLIN

Department of Electrical and Computer Engineering
University of South Carolina
Columbia, South Carolina 29208

I. INTRODUCTION

Sampling rate conversion is an important operation in modern digital signal processors. It is accomplished by either the interpolation or the decimation of a digital signal. Systems that implement sampling rate conversions require a filtering operation to limit the effects of aliasing. Symmetric linear phase finite impulse response (FIR) filter coefficients are commonly used for this application. These filters are usually designed using the Chebyshev criterion, which minimizes the maximum error in the frequency response. References [1–7] are examples of publications that discuss this approach. The actual filter coefficients are typically calculated by the popular McClellan–Parks–Rabiner computer program [8].

This article presents the method of Lagrange multipliers for the design of optimal linear phase FIR digital filters. In particular, the problem of optimizing these filters for multirate applications is emphasized. The filters are designed to minimize the total aliasing error rather than the maximum error contributed by any particular aliased signal. The advantages of this approach are demonstrated and a closed-form solution for obtaining the filter coefficients is presented. Several design examples with maximally flat passbands are included.

Copyright © 1990 by Academic Press, Inc.
All rights of reproduction in any form reserved.

II. MULTIRATE DIGITAL FILTER SPECIFICATIONS

Multirate digital filters are frequently used to implement a bank of equally spaced bandpass filters with identically shaped passbands. Banks of filters are used extensively in systems that rely on frequency division multiplexing. The technique of subdividing a wide-band signal into multiple narrow-band channels is important for another reason. It permits the parallel processing of a high data rate input signal with more than one computer. This increases the number of applications for which real-time digital signal processing is possible.

For the purpose of discussing the design of the FIR coefficients in the prototype filter for these systems, we can consider the implementation suggested in Fig. 1. Here, the center frequency of the ith channel, F_i, is translated to DC by the linear phase modulation, $e^{-j2\pi F_i n}$. The modulated signal is convolved with the impulse response of a prototype low-pass filter $h(n)$ with frequency response $H(e^{j2\pi f})$. The band-limited output is then decimated by a factor of $M:1$ to reduce the data rate. The filter coefficients $h(n)$ should be designed to minimize the effects of aliasing. This chapter presents a new approach to designing the filter coefficients.

Given a channel bandwidth of $2F_p$ cycles per sample and a decimation ratio of M, the low-pass prototype filter is required to have a passband edge frequency of F_p cycles per sample. If aliasing into the transition band is permitted, the maximum initial stopband edge frequency is given by$(1/M) - F_p$ cycles per sample. However, if it is desired to prevent significant aliasing into the transition band, the lower stopband frequency should be only $1/(2M)$.

For applications where aliasing into the transition band can be permitted, and where the set of filter coefficients is only used for a single decimation ratio, the multiple stopband specification may be used. In this case, the ith stopband

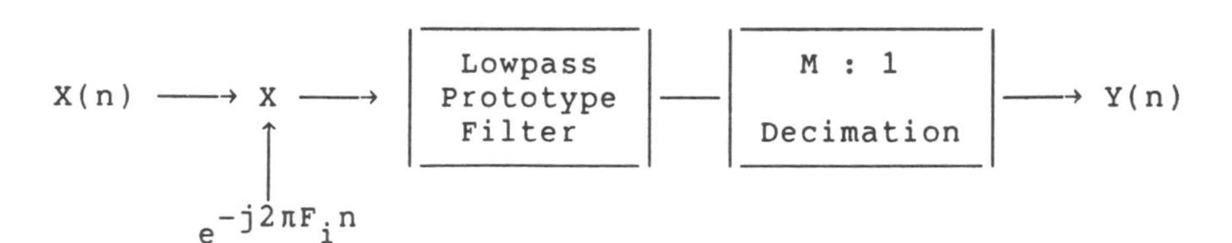

Fig. 1. Bandpass filter centered at the frequency F_i and with $M:1$ sampling rate decimation. The structure is a conceptually simple example used only for the purpose of discussing the specifications of the prototype filter. (This simple structure is inefficient compared to channel-bank filters, which employ the FFT.)

has edge frequencies given by

$$\text{lower edge frequency} = i/M - F_{\mathrm{p}} \quad \text{(cycles/sample)}$$

$$\text{upper edge frequency} = i/M + F_{\mathrm{p}} \quad \text{(cycles/sample).}$$

The frequency intervals between these stopbands are "don't care" regions. The derivation of the band-edge frequencies is well known and is discussed in [1, 2].

The filter specifications for interpolation to higher sampling rates are essentially in the same form as for decimation to lower rates [1]. Depending on the specific application, the stopband for interpolating filters occupies either a single-frequency interval or multiple bands with "don't care" intervals between them. Also, as discussed in [1], the basic form of the filter specifications for sampling rate conversions by integer factors is the same as for resampling by rational factors.

The proposed filter design method can be used for all of the varieties of multirate filtering problems (i.e., interpolation or decimation by integer or rational factors). However, since the nature of the filter design is so similar for the various categories, the focus will be on only one representative case. In the following discussions the important problem of designing optimal filters for decimation by integer factors is emphasized.

III. CRITERIA FOR DESIGN OF MULTIRATE FILTERS

The minimax criterion is typically used to determine the prototype filter frequency response in the stopband [1–7]. The resulting filter has approximately equiripple side lobes over the stopband frequency intervals. This minimizes the maximum error from any aliased frequency. However, it is more important to minimize the total integrated aliasing error. The minimax criterion is appropriate if one is concerned about suppressing a high-amplitude single-tone signal at an arbitrary frequency in the stopband. However, this situation does not usually occur in practice. It is usually reasonable to assume that the frequency content of the wide-band channel is distributed approximately uniformly. In this case it is important to minimize the integrated aliasing error from all of the channels.

Reference [1] provides a comprehensive survey of the commonly accepted methods for the design of FIR multirate filters. Here, the method of Lagrange multipliers is presented for the design of multirate FIR digital filters. We impose a maximally flat passband constraint and minimize the total aliasing error. The importance of minimizing the total aliasing error instead of the maximum error is demonstrated with several examples in a later section. The advantages of filters with maximally flat passbands over their equiripple counterparts are discussed in [9] and [10].

In addition to the advantages discussed in [9] and [10], filters with maximally flat passbands are generally much easier to equalize than filters with passband ripples. In a long-distance communications system where several repeater stations are involved, the passband error increases after each station. It is usually easier to compensate for the simple passband droop, which accumulates from several maximally flat filters in cascade than for the multiple ripples encountered with minimax filters. Also, channel bank filters, and multirate filters in general, typically require narrow-band specifications. Filters with maximally flat passbands have virtually the same efficiency as filters with equiripple passbands for narrow-band specifications.

IV. MULTIRATE FILTER DESIGN

The integrated aliasing error is denoted as E_s. It is simply the total stopband energy for the prototype filter. In the multiple-stopband filter where aliasing into the transition band is permitted, this error is given by

$$E_s = \sum_{i=1}^{M-1} \int_{(i/M)-F_p}^{(i/M)+F_p} |H(e^{j2\pi f})|^2 \, df. \tag{1}$$

In the case of a single-stopband filter where aliasing into the transition band is not allowed, we have

$$E_s = 2 \int_{F_s}^{0.5} |H(e^{j2\pi f})|^2 \, df, \qquad \text{where} \quad F_s = \frac{1}{M} - F_p. \tag{2}$$

In the following sections the optimal filter design problem is stated, the closed-form solution is presented, a design algorithm is provided, and, finally, filter design examples are discussed.

V. STATEMENT OF THE PROBLEM

Given a FIR filter with impulse response coefficients $h(n)$, the frequency response is

$$H(e^{j\omega}) = \sum_{n=0}^{L-1} h(n) e^{-j\omega n}, \tag{3}$$

where ω is the frequency in units of radians per sample and the filter length is L samples. Requiring the filter to have an exactly linear phase response, the impulse response falls into four cases [11]. Two of these apply to low-pass filters–Case 1 and Case 2 filters. Case 1 filters have an odd length and Case 2 filters are of even length. The zero-phase response of such filters is given by is given by

$$H(e^{j\omega}) = \begin{cases} \sum_{n=0}^{(L-1)/2} a(n) \cos n\omega; & \text{Case 1} \\ \sum_{n=1}^{L/2} b(n) \cos(n - 1/2)\omega; & \text{Case 2.} \end{cases} \tag{4}$$

For Case 1 filters,

$$a(n) = \begin{cases} h[(L-1)/2] & n = 0 \\ 2h[(L-1)/2 - n] & n = 1, \ldots, (L-1)/2 \end{cases} \tag{5}$$

and for Case 2 filters,

$$b(n) = 2h[(L/2) - n] \qquad n = 1, \ldots, L/2. \tag{6}$$

Vector–matrix notation lends itself particularly well to this problem. Defining a filter coefficient vector,

$$x = \begin{cases} [a(0) \quad a(1) \quad \cdots \quad a[(L-1)/2]]^T; & \text{Case 1} \\ [b(1) \quad b(2) \quad \cdots \quad b(L/2)]^T; & \text{Case 2} \end{cases} \tag{7}$$

and a Fourier transform kernel vector,

$$s(\omega) = \begin{cases} [1 \cos\omega \quad \cdots \quad \cos[(L-1)/2]\omega]^T; & \text{Case 1} \\ [\cos(1/2)\omega \quad \cdots \quad \cos[(L/2) - (1/2)]\omega]^T; & \text{Case 2} \end{cases} \tag{8}$$

the zero-phase response can be written as a vector inner product

$$H(e^{j\omega}) = x^T s(\omega). \tag{9}$$

The filter energy over a frequency interval $[\alpha, \beta]$ is given by

$$E(\alpha, \beta) = \frac{1}{2\pi} \int_\alpha^\beta [H(e^{j\omega})]^2 \, d\omega = \frac{1}{4\pi} x^T Q x, \tag{10}$$

where the energy matrix

$$Q = 2 \int_\alpha^\beta s(\omega) s^T(\omega) \, d\omega. \tag{11}$$

Q is obviously a symmetric, positive definite matrix.

For the single stopband energy discussed in Section IV,

$$E_S = E(\Omega_s, \pi) \tag{12}$$

$$Q = 2\int_{\Omega_s}^{\pi} s(\omega)s^T(\omega)\,d\omega = [q_{km}].$$

For Case 1 filters; $k, m = 0, 1, \ldots, (L-1)/2$, and

$$q_{km} = \begin{cases} 2\displaystyle\int_{\Omega_s}^{\pi} d\omega & k = m = 0, \\ 2\displaystyle\int_{\Omega_s}^{\pi} \cos k\omega \cos m\omega\,d\omega, & \text{otherwise.} \end{cases} \tag{13}$$

Using the trigonometric identities,

$$\cos^2 k\omega = \tfrac{1}{2}[\cos 2k\omega + 1]$$

and

$$\cos k\omega \cos m\omega = \tfrac{1}{2}[\cos(k+m)\omega + \cos(k-m)\omega], \tag{14}$$

and integrating, we find

$$q_{km} = \begin{cases} 2(\pi - \Omega_s) & k=m=0 \\ \pi - \Omega_s - (\sin 2k\Omega_s/2k) & k=m\neq 0 \\ \dfrac{-\sin[(k+m)\Omega_s]}{k+m} - \dfrac{\sin[(k-m)\Omega_s]}{k-m} \quad k \neq m. & k\neq m. \end{cases} \tag{15}$$

Similarly, for case 2 filters; $k, m = 1, 2, \ldots, L/2$,

$$q_{km} = \begin{cases} \pi - \Omega_s - \dfrac{\sin[(2k-1)\Omega_s]}{(2k-1)} & k = m \\ -\dfrac{\sin[(k+m-1)\Omega_s]}{(k+m-1)} - \dfrac{\sin[(k-m)\Omega_s]}{(k-m)} & k \neq m. \end{cases} \tag{16}$$

In the multiple-stopband formulation, the total stopband energy becomes the sum of energy from the individual stopbands. Writing the energy for the nth stopband,

$$E_n = E\left[\frac{2\pi}{M}n - \Omega_p, \frac{2\pi}{M}n + \Omega_p\right] = \frac{1}{4\pi}x^T Q_n x, \tag{17}$$

where

$$Q_n = 2\int_{(2\pi/M)n - \Omega_p}^{(2\pi/M)n + \Omega_p} s(\omega)s^T(\omega)\,d\omega = [q_{km}]. \tag{18}$$

For Case 1 filters; $k, m = 0, 1, \ldots, (L-1)/2$, and

$$q_{km} = \begin{cases} 2\displaystyle\int_{(2\pi/M)n-\Omega_p}^{(2\pi/M)n+\Omega_p} d\omega, & k = m = 0 \\ 2\displaystyle\int_{(2\pi/M)n-\Omega_p}^{(2\pi/M)n+\Omega_p} \cos k\omega \cos m\omega \, d\omega, & \text{otherwise.} \end{cases} \tag{19}$$

Using trignometric identities and integrating, it follows that

$$q_{km} = \begin{cases} 4\Omega_p & k = m = 0 \\ 2\Omega_p + \cos\left(\dfrac{4\pi}{M}nk\right)\dfrac{\sin(2\Omega_p m)}{m} & k = m \neq 0 \\ 2\cos\left[\dfrac{2\pi}{M}n(k+m)\right]\dfrac{\sin[(k+m)\Omega_p]}{(k+m)} & \\ \quad + 2\cos\left[\dfrac{2\pi}{M}n(k-m)\right]\dfrac{\sin[(k-m)\Omega_p]}{(k-m)} & k \neq m. \end{cases} \tag{20}$$

Similarly, for Case 2 filters; $k, m = 1, 2, \ldots, L/2$, and

$$q_{km} = \begin{cases} 2\Omega_p + 2\cos\left[(2k-1)\dfrac{2\pi}{M}n\right]\dfrac{\sin[(2k-1)\Omega_p]}{(2k-1)}, & k = m \\ 2\cos\left[\dfrac{2\pi}{M}n(k+m-1)\right]\dfrac{\sin[(k+m-1)\Omega_p]}{(k+m-1)} & \\ \quad + 2\cos\left[\dfrac{2\pi}{M}n(k-m)\right]\dfrac{\sin[(k-m)\Omega_p]}{(k-m)}; & k \neq m. \end{cases} \tag{21}$$

The total stopband energy is then

$$E_s = \sum_{n=1}^{\text{NB}} E_n = \frac{1}{4\pi} x^T \sum_{n=1}^{\text{NB}} Q_n x = \frac{1}{4\pi} x^T Q x, \tag{22}$$

where $Q = \sum Q_n$ is the total stopband energy matrix and NB is the number of stopbands in the interval $[0, \pi]$. Since each Q_n is a positive definite symmetric matrix, Q is also positive definite and symmetric.

For an odd decimation ratio M,

$$\text{NB} = (M-1)/2. \tag{23}$$

Resampling with an even decimation ratio implies that

$$\text{NB} = M/2, \tag{24}$$

with the exception that

$$E_{\mathrm{NB}} = \frac{1}{2} \cdot \frac{1}{4\pi} x^T Q_{\mathrm{NB}} x, \tag{25}$$

since the last stopband is symmetric about π radians per sample.

Performing the summation in (22), the total stopband energy is given by

$$E_s = \frac{1}{4\pi} x^T Q x, \tag{26}$$

where $Q = [q_{km}]$. For Case 1 filters and an odd decimation ratio M,

$$q_{km} = \begin{cases} 4\,\mathrm{NB}\,\Omega_p, & k = m = 0 \\ 2\,\mathrm{NB}\,\Omega_p + \left[\sum_{n=1}^{\mathrm{NB}} \cos\left(\frac{4\pi}{M} nk\right)\right] \frac{\sin(2\Omega_p m)}{m}, & k = m \neq 0 \\ 2 \sum_{n=1}^{\mathrm{NB}} \cos\left[\frac{2\pi}{M} n(k+m)\right] \frac{\sin[(k+m)\Omega_p)}{(k+m)} & \\ \quad + 2 \sum_{n=1}^{\mathrm{NB}} \cos\left[\frac{2\pi}{M} n(k-m)\right] \frac{\sin[(k-m)\Omega_p]}{(k-m)}, & k \neq m. \end{cases} \tag{27}$$

Case 1 filters with an even decimation ratio M must be slightly modified as noted to produce

$$q_{km} = \begin{cases} 4(\mathrm{NB} - \tfrac{1}{2})\Omega_p, & k = m = 0 \\ 2\left(\mathrm{NB} - \frac{1}{2}\right)\Omega_p + \left[\sum_{n=1}^{\mathrm{NB}} \cos\left(\frac{4\pi}{M} nk\right) - \frac{1}{2}\right] \frac{\sin(2\Omega_p m)}{m}, & k = m \neq 0 \\ 2\left[\sum_{n=1}^{\mathrm{NB}} \cos\left[\frac{2\pi}{M} n(k+m)\right] - \frac{1}{2}\cos(k+m)\pi\right] \frac{\sin[(k+m)\Omega_p]}{(k+m)} & \\ + 2\left[\sum_{n=1}^{\mathrm{NB}} \cos\left[\frac{2\pi}{M} n(k-m)\right] - \frac{1}{2}\cos(k-m)\pi\right] \frac{\sin[(k-m)\Omega_p]}{(k-m)} & k \neq m. \end{cases} \tag{28}$$

Likewise for Case 2 filters, we have for M odd

$$q_{km} = \begin{cases} 2\,\mathrm{NB}\,\Omega_p + 2 \sum_{n=1}^{\mathrm{NB}} \cos\left[(2k-1)\frac{2\pi}{M} n\right] \frac{\sin[(2k-1)\Omega_p]}{(2k-1)}, & k = m \\ 2 \sum_{n=1}^{\mathrm{NB}} \cos\left[\frac{2\pi}{M} n(k+m-1)\right] \frac{\sin[(k+m-1)\Omega_p]}{(k+m-1]} & \\ \quad + 2 \sum_{n=1}^{\mathrm{NB}} \cos\left[\frac{2\pi}{M} n(k-m)\right] \frac{\sin[(k-m)\Omega_p]}{(k-m)}, & k \neq m. \end{cases} \tag{29}$$

For M even,

$$q_{km} = \begin{cases} (2\,\mathrm{NB} - 1)\Omega_p + \left[2 \sum_{n=1}^{\mathrm{NB}} \cos\left[(2k-1)\frac{2\pi}{M} n \right] + 1 \right] \\ \quad \times \frac{\sin[(2k-1)\Omega_p]}{(2k-1)}, \qquad k = m \\ \left[2 \sum_{n=1}^{\mathrm{NB}} \cos\left[\frac{2\pi}{M} n(k+m-1) \right] - \cos[(k+m-1)\pi] \right] \\ \quad \times \frac{\sin[(k+m-1]\Omega_p]}{(k+m-1]} \\ \quad + \left[2 \sum_{n=1}^{\mathrm{NB}} \cos\left[\frac{2\pi}{M} n(k-m) \right] - \cos[(k-m)\pi] \right] \\ \quad \times \frac{\sin[(k-m)\Omega_p]}{(k-m)}, \qquad k \neq m. \end{cases} \tag{30}$$

Thus, we have shown that the total stopband energy can be written as

$$E_s = \frac{1}{4\pi} x^T Q x, \tag{31}$$

where Q is a positive definite, symmetric matrix.

To achieve desirable filter passband performance, magnitude and derivative constraints can be imposed at a discrete set of points $\{\omega_0, \omega_1, \ldots, \omega_{M-1}\}$ in the interval $[0, \Omega_p]$. The constraints

$$\left. \frac{d^n H}{d\omega^n} \right|_{\omega = \omega_m} = k_{nm}, \tag{32}$$

for $n = 0, 1, \ldots, N-1$, become

$$\left(\left. \frac{d^n s}{d\omega^n} \right|_{\omega - \omega_m} \right)^T x = k_{nm} \tag{33}$$

in vector notation. Defining the constraint vector,

$$k = [k_{00}, k_{10}, \ldots, k_{N-1\ 0}, \ldots, k_{0\ M-1}, \ldots, k_{N-1\ M-1}]^T,$$

and the constraint matrix,

$$C = \begin{pmatrix} s^T(\omega_0) \\ \left(\left. \frac{ds}{d\omega} \right|_{\omega = \omega_0} \right)^T \\ \vdots \\ \left(\left. \frac{d^{N-1} s}{d\omega^{N-1}} \right|_{\omega = \omega_{M-1}} \right)^T \end{pmatrix}, \tag{34}$$

the passband constraints can be written as a linear system of equations

$$Cx = k. \tag{35}$$

The optimal stopband energy filtering problem may now be stated as a quadratic programming problem:

$$\text{Minimize} \quad E_s = \frac{1}{4\pi} x^T Q x \qquad \text{subject to} \quad Cx = k. \tag{36}$$

Note that the symmetric Hessian matrix Q is positive definite, which establishes the strict convexity of E_s. The equality constraints define a convex feasible region so that the quadratic programming problem is also convex [14].

VI. SOLUTION BY THE METHOD OF LAGRANGE MULTIPLIERS

The optimal filtering problem stated in Section V can be solved by the association of a Lagrange multiplier λ_{ij} with each passband constraint. The Lagrange multiplier vector is then

$$\lambda = [\lambda_{00} \quad \cdots \quad \lambda_{N-1 \ M-1}]^T. \tag{37}$$

Defining the Lagrangian function Λ, we have

$$\Lambda(x, \lambda) \equiv \frac{1}{4\pi} x^T Q x - \lambda^T (Cx - k). \tag{38}$$

The necessary and sufficient conditions for optimality are

$$\nabla_x \Lambda = 0$$

and

$$\nabla_\lambda \Lambda = 0. \tag{39}$$

The conditions (39) lead to the linear system

$$\begin{bmatrix} \frac{1}{2\pi} Q & -C^T \\ -C & 0 \end{bmatrix} \begin{pmatrix} x \\ \lambda \end{pmatrix} = \begin{pmatrix} 0 \\ -k \end{pmatrix}, \tag{40}$$

which can be solved to yield the optimal filter and the Lagrange multipliers

$$x = (Q^{-1}C^T)(CQ^{-1}C^T)^{-1}k$$

$$\lambda = \frac{1}{2\pi}(CQ^{-1}C^T)^{-}1k. \tag{41}$$

Furthermore, it can be shown that the optimal stopband energy is given by

$$E_s = \frac{1}{4\pi} k^T (CQ^{-1}C^T)^{-1} k. \tag{42}$$

Equations (41)–(42) state that the Lagrange multiplier vector λ is a linear combination of the column vectors in the matrix $1/2\pi(CQ^{-1}C^T)^{-1}$ and the passband constraint values $k_{00}, k_{10}, \ldots, k_{N-1\ M-1}$. They also state that the optimal filter is a linear combination of the column vectors in the matrix $(Q^{-1}C^T)$ and the Lagrange multipliers $\lambda_{00}, \ldots, \lambda_{N-1\ M-1}$. The equations also show that the optimal stopband energy is a quadratic form in the constrained passband values.

The previously stated quadratic programming problem may be infeasible or the solution may be unbounded. In most physically meaningful applications, it is assumed that an optimal solution exists. Since Q is positive definite, its inverse exists. If the linear passband constraints are independent, then the inverse of $CQ^{-1}C^T$ also exists. It is shown in Section VII that for the important case of maximally flat low-pass filters the passband constraints are independent. Under these conditions, the optimal solution exists and is given by (41)–(42). Fletcher discusses in detail the existence issue in [14].

The Lagrange multipliers have an important interpretation. They indicate the rate of change of the optimal energy as the constrained passband magnitude values change [12]; that is,

$$\nabla_k E_s = \lambda. \tag{43}$$

This is important and useful information since it indicates the sensitivity of the optimal energy to changes in the constrained passband values. Some design examples are presented in Section XI that demonstrate the success of the Lagrange multiplier technique.

VII. MAXIMALLY FLAT MULTIRATE FINITE IMPULSE RESPONSE DIGITAL FILTERS

We now focus our attention on FIR low-pass filters that are maximally flat at DC. For maximally flat filters at DC the magnitude and derivative constraints are imposed at only one passband point $\omega_0 = 0$. The magnitude constraint is typically taken equal to unity, that is,

$$k_{00} = 1. \tag{44}$$

The derivative constraints are all taken equal to zero, that is,

$$k_{10} = k_{20} = \cdots = k_{N-1\ 0} = 0. \tag{45}$$

The constraint vector is simply given by

$$k = [1 \quad 0 \quad \cdots \quad 0]^T. \tag{46}$$

For maximally flat passbands at DC, the odd-order derivatives are all identically equal to zero since the Fourier transform kernel vector $s(\omega)$ contains only cosine terms. The rows appearing in the constraint matrix C correspond to the magnitude constraint and even-order derivative constraints at DC. For Case 1 filters, the constraint matrix is of the form

$$C = \begin{pmatrix} 1 & 1 & 1 & \cdots & 1 \\ 0 & -1 & -4 & \cdots & -\left(\frac{L-1}{2}\right)^2 \\ \vdots & \vdots & \vdots & & \vdots \\ 0 & (-1)^n 1^{2n} & (-1)^n 2^{2n} & \cdots & (-1)^n\left(\frac{L-1}{2}\right)^{2n} \end{pmatrix}, \tag{47}$$

where $2n$ is the highest even-order derivative constrained to be zero. For Case 2 filters, the constraint matrix becomes

$$C = \begin{pmatrix} 1 & 1 & \cdots & 1 \\ -\left(\frac{1}{2}\right)^2 & -\left(1\frac{1}{2}\right)^2 & \cdots & -\left(\frac{L-1}{2}\right)^2 \\ \vdots & \vdots & & \vdots \\ (-1)^n\left(\frac{1}{2}\right)^{2n} & (-1)^n\left(1\frac{1}{2}\right)^{2n} & \cdots & (-1)^n\left(\frac{L-1}{2}\right)^{2n} \end{pmatrix}, \tag{48}$$

where $2n$ is the highest even-order derivative constrained to be zero. The rows of these matrices are obtained from signed alternate columns of the following square matrix

$$V = \begin{pmatrix} 1 & t_1 & t_1^2 & \cdots & t_1^{k-1} \\ 1 & t_2 & t_2^2 & \cdots & t_2^{k-1} \\ \vdots & \vdots & \vdots & & \vdots \\ 1 & t_k & t_k^2 & \cdots & t_k^{k-1} \end{pmatrix}. \tag{49}$$

For Case 1 filters $t_1 = 0, t_2 = 1, t_3 = 2, \ldots$, and for Case 2 filters $t_1 = \frac{1}{2}, t_2 = 1\frac{1}{2}$, $t_3 = 2\frac{1}{2}, \ldots$. The square matrix V is known as the Vandermonde matrix and is nonsingular [15]. Hence, the columns in V are linearly independent, which implies that the rows in C are independent and the inverse of $CQ^{-1}C^T$ exists. So, the optimal maximally flat low-pass filter exists and is given by (41).

Writing the matrix product inverse from the previous section by column vectors, we have

$$(C \quad Q^{-1} \quad C^T)^{-1} = [v^0 \quad v^1 \quad \cdots \quad v^{N-1}]. \tag{50}$$

The solution to the system of equations requires only the first column vector v^0 of the inverse. Hence, the maximally flat, optimal filter and associated Lagrange multiplier vector are

$$x = (Q^{-1} \quad C^T)v^0 \qquad \text{and} \qquad \lambda = \frac{1}{2\pi}v^0. \tag{51}$$

The maximally flat optimal stopband energy is given by

$$E_s = \frac{1}{4\pi}v_0^0, \tag{52}$$

where v_0^0 is the flat component in column vector v^0. Note that the first Lagrange multiplier is twice E_s; that is,

$$\lambda_{00} = \frac{1}{2\pi}v_0^0. \tag{53}$$

The number of derivatives taken equal to zero at DC determines the flatness of the filter passband. The passband width, passband rolloff, stopband width, and stopband gain are related to the filter flatness. An approximate relationship between these parameters is

$$N = \frac{\log \delta_p}{\log(\omega_p/\omega_s)}, \tag{54}$$

where $N - 1$ is the number of derivatives equal to zero, ω_p is the passband edge frequency, δ_p is the passband rolloff at ω_p, and ω_s is the stopband edge frequency. The derivation of this expression is presented in the appendix.

VIII. BANDPASS FINITE IMPULSE RESPONSE DIGITAL FILTERS

Optimal linear phase FIR bandpass filters have also been successfully designed with the technique described in Sections I–VII. A typical bandpass filter has a passband $2\Omega_p$ radians per sample wide centered at ω_c radians per sample. The two individual stopbands occur on the intervals $[0, \omega_c - \Omega_s]$ and $[\omega_c + \Omega_s, \pi]$ radians per sample.

The total stopband energy E_s is the sum of the individual stopband energies and can be written as

$$E_s = E(0, \omega_c - \Omega_s) + E(\omega_c + \Omega_s, \pi) = \frac{1}{4\pi} x^T Q x, \tag{55}$$

where Q is a symmetric, positive definite matrix. The energy matrix Q is a sum of two matrices Q_1 and Q_2, that is,

$$Q = Q_1 + Q_2, \tag{56}$$

where

$$\begin{aligned} Q_1 &= 2\int_0^{\omega_c - \Omega_s} s(\omega)s^T(\omega)\, d\omega \\ Q_2 &= 2\int_{\omega_c + \Omega_s}^{\pi} s(\omega)s^T(\omega)\, d\omega. \end{aligned} \tag{57}$$

For Case 1 filters, the Fourier transform kernel is

$$s(\omega) = [1 \cos\omega \quad \cdots \quad \cos[(L-1)/2]\omega]^T. \tag{58}$$

The energy matrix $\mathbf{Q} = [q_{km}]$ where $k, m = 0, 1, \ldots, (L-1)/2$ and

$$q_{\mathrm{km}} = \begin{cases} 2(\pi - 2\Omega_s), & k = m = 0 \\ \pi - 2\Omega_s - \dfrac{\cos 2k\omega_c \sin 2k\Omega_s}{k}, & k = m \neq 0 \\ -\dfrac{2\sin[(k+m)\Omega_s]\cos[(k+m)\omega_c]}{(k+m)} & \\ \qquad -\dfrac{2\sin[(k-m)\Omega_s]\cos[(k-m)\omega_c]}{(k-m)}, & k \neq m. \end{cases} \tag{59}$$

Similarly, for Case 2 filters we have

$$\begin{aligned} s(\omega) &= [\cos\tfrac{1}{2}\omega \quad \cos 1\tfrac{1}{2}\omega \quad \cdots \quad \cos[(L-1)/2]\omega]^T \\ Q &= [q_{km}] \qquad \text{where} \quad k, m = 1, 2, \ldots, L/2 \\ q_{km} &= \begin{cases} \pi - 2\Omega_s - \dfrac{2\cos(2k-1)\omega_c \sin(2k-1)\Omega_s}{k}, & k = m \\ -\dfrac{2\sin[(k+m-1)\Omega_s]\cos[(k+m-1)\omega_c]}{(k+m-1)} & \\ \qquad -\dfrac{2\sin[(k-m)\Omega_s]\cos[(k-m)\omega_c]}{(k-m)}, & k \neq m. \end{cases} \end{aligned} \tag{60}$$

As shown previously in the case of low-pass filters, desired passband performance may be obtained by imposing magnitude and derivative constraints in the passband. The focus in this section will be on maximally flat passbands. At the passband center ω_c, the magnitude and derivative constraints

$$d^n H/d\omega^n|_{\omega=\omega_c} = k_n \tag{61}$$

for $n = 0, 1, \ldots, N-1$ are

$$(d^n s/d\omega^n|_{\omega=\omega_c})^T x = k_n \tag{62}$$

in vector notation. For Case 2 filters, the odd-order derivatives are

$$\left.\frac{d^{2n+1}s}{d\omega^{2n+1}}\right|_{\omega=\omega_c} = (-1)^{n+1}\left[(\tfrac{1}{2})^{2n+1}\sin\tfrac{1}{2}\omega_c \quad (1\tfrac{1}{2})^{2n+1} \quad \sin 1\tfrac{1}{2}\omega_c \quad \cdots \quad \left(\frac{L-1}{2}\right)^{2n+1}\sin\left(\frac{L-1}{2}\right)\omega_c\right]^T \tag{63}$$

for $n = 0, 1, \ldots,$ and the even-order derivatives are

$$\left.\frac{d^{2n}s}{d\omega^{2n}}\right|_{\omega=\omega_c} = (-1)^n\left[(\tfrac{1}{2})^{2n}\cos\tfrac{1}{2}\omega_c \quad (1\tfrac{1}{2})^{2n} \quad \cos 1\tfrac{1}{2}\omega_c \quad \cdots \quad \left(\frac{L-1}{2}\right)^{2n}\cos\left(\frac{L-1}{2}\right)^{2n}\omega_c\right]^T \tag{64}$$

for $n = 1, 2, \ldots$. Similarly, the Case 1 filter odd-order derivatives are

$$\left.\frac{d^{2n+1}s}{d\omega^{2n+1}}\right|_{\omega=\omega_c} = (-1)^{n+1}\left[0 \quad \sin\omega_c \quad 2^{2n+1} \quad \sin 2\omega_c \quad \cdots \quad \left(\frac{L-1}{2}\right)^{2n+1}\sin\left(\frac{L-1}{2}\right)\omega_c\right]^T \tag{65}$$

for $n = 0, 1, \ldots$ and the even-order derivatives are

$$\left.\frac{d^{2n}s}{d\omega^{2n}}\right|_{\omega=\omega_c} = (-1)^n\left[0 \quad \cos\omega_c \quad 2^{2n}\cos 2\omega_c \quad \cdots \quad \left(\frac{L-1}{2}\right)^{2n}\cos\left(\frac{L-1}{2}\right)\omega_c\right]^T \tag{66}$$

for $n = 1, 2, \ldots$. The passband constraints at ω_c can be summarized in vector–matrix notation as

$$Cx = k,$$

where

$$C = \begin{pmatrix} s^T(\omega_c) \\ \left(\left.\dfrac{ds}{d\omega}\right|_{\omega=\omega_c}\right)^T \\ \vdots \\ \left(\left.\dfrac{d^{N-1}s}{d\omega^{N-1}}\right|_{\omega=\omega_c}\right)^T \end{pmatrix} \tag{67}$$

$$k = [k_0, k_1, \ldots, k_{N-1}]^T,$$

and x is the filter coefficient vector. For maximally flat passbands,

$$k_0 = 1 \qquad \text{and} \qquad k_1 = k_2 = \cdots = k_{N-1} = 0. \tag{68}$$

As in the low-pass filter case, the optimal bandpass filtering problem may be stated as a quadratic programming problem:

$$\text{Minimize} \qquad E_s = \frac{1}{4\pi} x^T Q x \qquad \text{subject to} \qquad Cx = k. \tag{69}$$

The Lagrange multiplier solution to this problem is described in Section VI. Several bandpass filter examples have been included in Section XI.

IX. PASSBAND FINE TUNING

Changes in the filter passband can be made by changing the constraint vector k. In the maximally flat passband formulation, k constrains the magnitude to be unity and derivatives up to a specified order to be zero at the passband center. Changing these values does not require complete recomputation of the optimal filter and Lagrange multipliers. From Section VI the optimal filter and Lagrange multipliers for a given constraint vector k are

$$\begin{aligned} x(k) &= (Q^{-1} \quad C^T)(C \quad Q^{-1} \quad C^T)^{-1}k \\ \lambda(k) &= \frac{1}{2\pi}(C \quad Q^{-1} \quad C^T)^{-1}k. \end{aligned} \tag{70}$$

Writing the matrix $(C \quad Q^{-1} \quad C^T)^{-1}$ by column vectors, we have

$$\begin{aligned} (C \quad Q^{-1} \quad C^T)^{-1} &= [v^0 \quad v^1 \quad \cdots \quad v^{N-1}] \\ (Q^{-1} \quad C^T)(C \quad Q^{-1} \quad C^T)^{-1} &= [u^0 \quad u^1 \quad \cdots \quad u^{N-1}], \end{aligned} \tag{71}$$

where the column vector $u^n = (Q^{-1} \quad C^T)v^n$. In this notation the optimal filter and Lagrange multipliers become

$$\begin{aligned} x(k) &= [u^0 \quad u^1 \quad \cdots \quad u^{N-1}]k = \sum_{n=0}^{N-1} k_n u^n \\ \lambda(k) &= \frac{1}{2\pi}[v^0 \quad v^1 \quad \cdots \quad v^{N-1}]k = \frac{1}{2\pi}\sum_{n=0}^{N-1} k_n v_n. \end{aligned} \tag{72}$$

Computing the fine-tuned filter and corresponding Lagrange multipliers requires only changing the scalar multipliers k_n in sums (72).

The optimal zero-phase frequency response can be rewritten as

$$H(e^{j\omega}) = s^T(\omega)x(k) = \sum_{n=0}^{N-1} k_n s^T(\omega)v^n = \sum_{n=0}^{N-1} k_n H_n(e^{j\omega}), \tag{73}$$

where $H_n(e^{j\omega})$ is the optimal response for the passband constraints $k_0 = k_1 = \cdots = k_{n-1} = k_{n+1} = \cdots = k_{N-1} = 0$ and $k_n = 1$ evaluated at the radian passband frequency ω. In this notation the optimal maximally flat response is $H_0(e^{j\omega})$. The optimal response for constrained passband values is a linear combination of the $H_n(e^{j\omega})$ with the constrained values as scalars. Any passband changes should be made with attention to resulting changes in the stopband energy E_s. The Lagrange multipliers are the instantaneous rates of change for the stopband energy,

$$\partial E_s/\partial k_n = \lambda_n. \tag{74}$$

So, for a small change in k_n, the optimal stopband energy change is approximately

$$\Delta E_s = \lambda_n \Delta k_n. \tag{75}$$

The filter designer can therefore choose to change a constrained passband value according to the sign and magnitude of the corresponding Lagrange multiplier and according to the desired passband performance.

For example, in the case of the maximally flat low-pass filter where the magnitude and second derivative are constrained (the first derivative and all other odd derivatives are automatically zero at DC), the optimal response in terms of k is

$$H(e^{j\omega}) = k_0 H_0(e^{j\omega}) + k_2 H_2(e^{j\omega}),$$

where

$$k_0 = H(e^{j0}) \qquad \text{and} \qquad k_2 = \left.\frac{d^2 H}{d\omega^2}\right|_{\omega=0} \tag{76}$$

Requiring $k_0 = 1$ and evaluating $H(e^{j\omega})$ at the passband edge ω_P, we have

$$k_2 = \frac{H(e^{j\omega_P}) - H_0(e^{j\omega_P})}{H_2(e^{j\omega_P})}. \tag{77}$$

Thus, the passband droop can be slightly adjusted by imposing the constrained value k_2 for the second derivative. An example of a fine-tuned passband is shown in the Section XI.

X. FILTER DESIGN ALGORITHM

We are now in a position to state a design algorithm for optimal FIR filters with narrow maximally flat passbands and minimum stopband energy:

1. Determine the filter requirements ω_p, δ_p, ω_s, and δ_s.
2. Using a standard approximation technique [13], determine the corresponding filter length L.
3. Use the equation derived in the appendix to approximate the passband flatness specified by N the number of derivatives constrained to be zero at the passband center.
4. Compute the corresponding energy matrix Q.
5. Compute the constraint matrix C corresponding to N.
6. Determine the optimal filter, the Lagrange vector, and the optimal stopband energy.
7. Fine tune the filter passband if necessary.

The filter passband can be fine tuned by making changes in the passband constraint vector k as discussed in Section IX. Writing the matrix product inverse by column vectors as shown in (71), the optimal filter was shown to be

$$x = (Q^{-1} \quad C^T) \sum_{i=0}^{N-1} k_i v^i. \tag{78}$$

Similarly, the corresponding Lagrange multiplier vector is

$$\lambda = \frac{1}{2\pi} \sum_{i=0}^{N-1} k_i v^i. \tag{79}$$

The signs and relative magnitudes of the Lagrange multipliers indicate how the optimal stopband energy varies with changes in k.

Equations (78)–(79) show that the optimal filter and corresponding Lagrange multiplier vector are simple to recompute for any variation in the constrained passband values. The constrained passband values appear only as scalars multiplying the appropriate column vectors. Any changes made in the passband values should be done with attention to the stopband energy sensitivity. It should be emphasized that large changes in the passband may produce a significant change in the total stopband energy. In this case the total change in stopband energy will be the integrated result of Lagrange multipliers as a function of the constrained passband values.

XI. DESIGN EXAMPLES

Several optimal low-pass prototype filters for a FFT filter band of 64 bandpass channels were designed using the Lagrange multiplier technique. The first filter example has a single stopband. It was designed to handle any

TABLE I. SINGLE-STOPBAND EXAMPLE: FILTER SPECIFICATIONS

Impulse response length	= 128
Passband edge frequency	= 1/128
Stopband edge frequency	= 3/128
Passband rolloff	= 0.7 dB

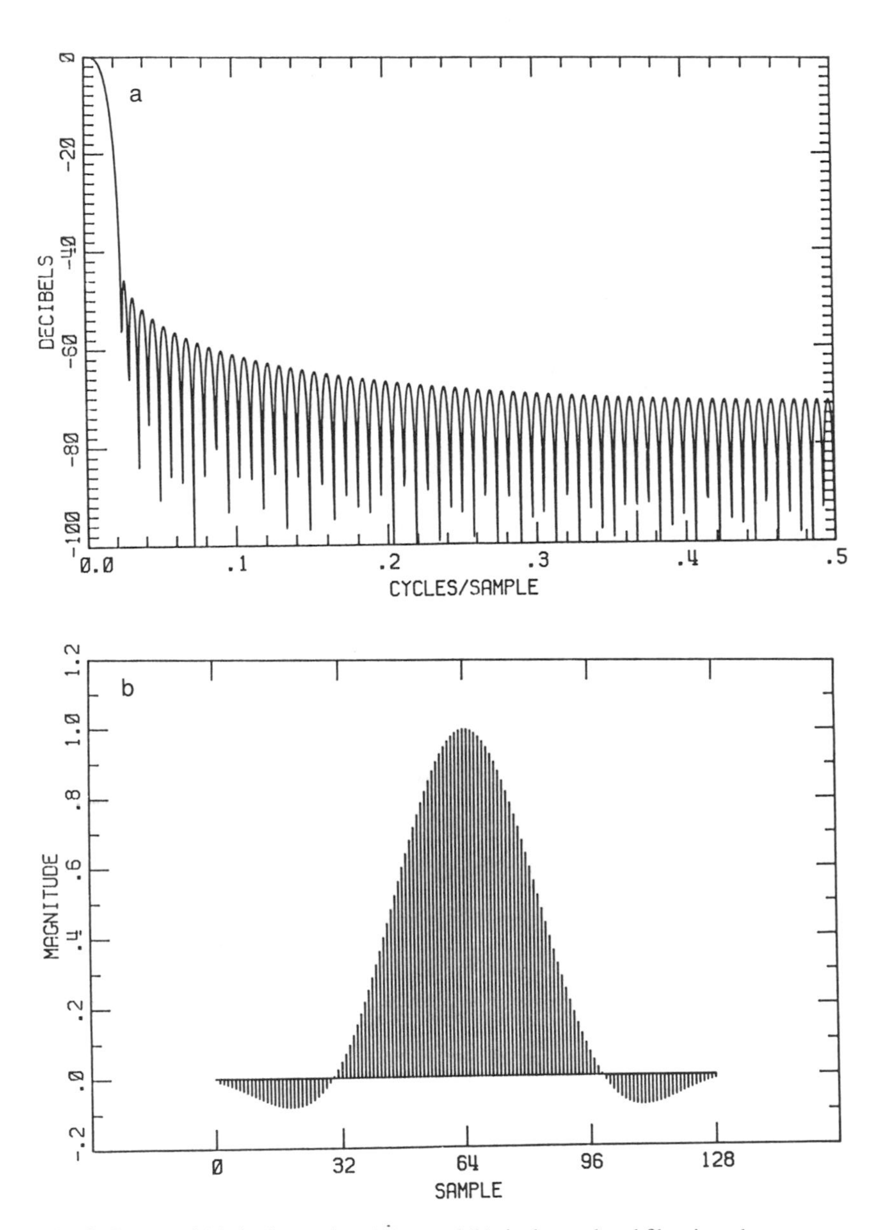

Fig. 2. Proposed (a) single-stopband filter and (b) single-stopband filter impulse response.

decimation ratio of 32:1 or less and to have a passband rolloff of 0.7 dB. In addition, the filter phase response was required to be linear. The design parameters are summarized in Table I and the frequency response is shown in Fig. 2a. The filter peak side-lobe level is approximately −46 dB and the asymptotic side-lobe level is approximately −72 dB. The CPU (central

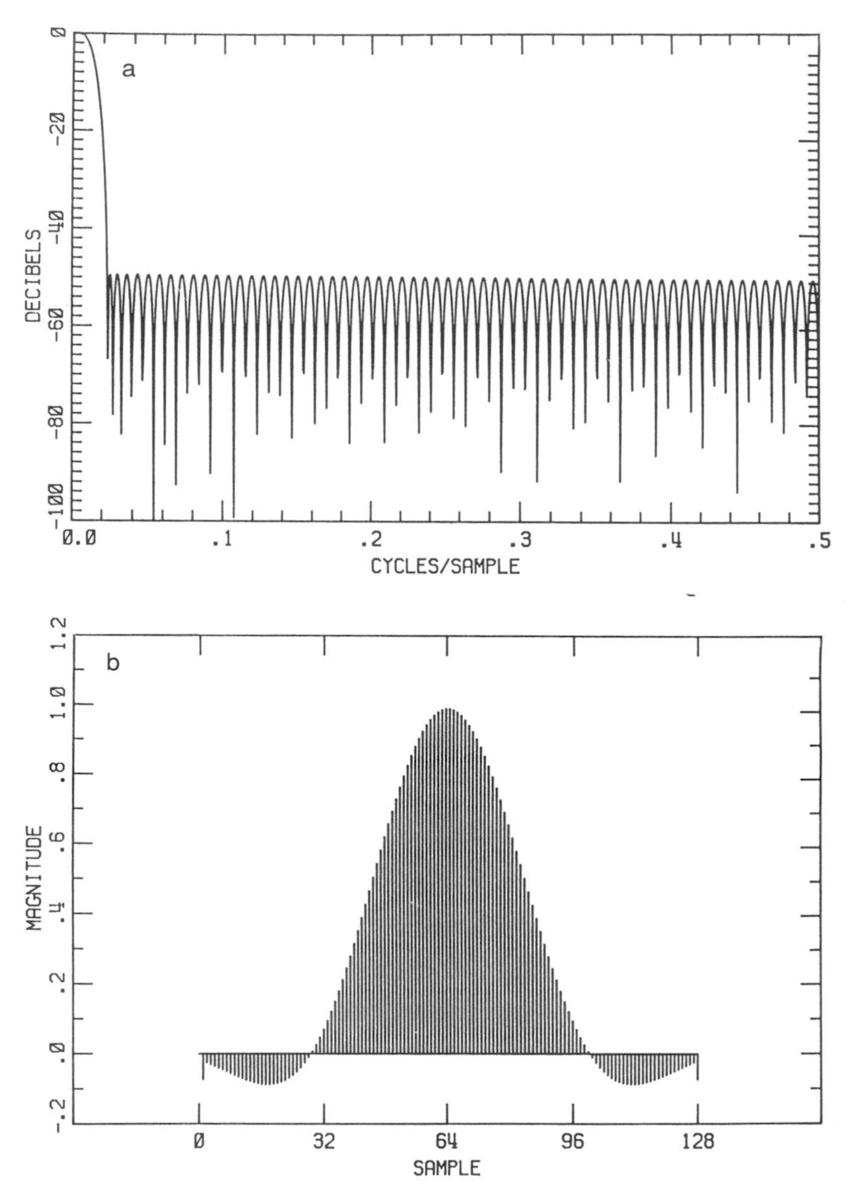

Fig. 3. Equiripple (a) single-stopband filter and (b) single-stopband filter impulse response.

processing unit) time required to calculate the filter coefficients was 7 sec for a VAX 11/780 Computer. The symmetric impulse response for the proposed filter is of length 128 and is shown in Fig. 2b. The peak value of the impulse response has been scaled to one in Fig. 2b.

For comparison, an equiripple filter was also designed according to the requirements in Table I. It was designed with the computer program discussed in [8]. The equiripple side-lobe level is -48.5 dB and the frequency response is plotted in Fig. 3a. The equiripple filter required 20 CPU seconds for computation. The scaled equiripple filter impulse response is shown in Fig. 3b.

Defining the integrated side-lobe ratio (ISLR) as the ratio of the stopband energy to the passband energy, we find that ISLR for the proposed filter $=$ -44.37 dB and ISLR for the equiripple filter $= -34.48$ dB. The stopband energy for the proposed filter is significantly less than the equiripple filter. The improvement is by an order of magnitude (i.e., by a margin of ~ 10 dB in this case). Table II summarizes the results for the two single-stopband filters. Figure 4a is an overlay of the proposed and equiripple frequency responses. Figure 4b shows detailed plots of the proposed filter passband and the equiripple filter passband. The curves show very little difference in the filter passbands.

A multiple-stopband filter was designed by the Lagrange technique. The filter was designed to handle only one decimation ratio of 32:1. Its length and passband specifications were chosen to match the single-stopband case and are shown in Table III. The frequency response in Fig. 5a clearly depicts the 16 individual stopbands separated by "don't care" regions. The scaled impulse response for the multiple-stopband filter is shown in Fig. 5b.

A multiple-stopband equiripple filter with the same specifications in Table III was designed for comparison. The design was accomplished by a slightly modified version of the computer program in [8]. Its frequency response is shown in Fig. 6a and its scaled impulse response is shown in Fig. 6b. Table IV is a summary of results for the multiple-stopband filters. Again, the proposed filter ISLR is ~ 10 dB better than the conventional equiripple case. Figure 7a shows both the proposed and equiripple frequency responses. The filter passbands are shown in Fig. 7b. The plot indicates the passbands are virtually identical.

TABLE II. SINGLE-STOPBAND FILTER SUMMARY

	Peak side-lobe level (dB)	ISLR (dB)	Computation time (sec)
Proposed filter	-46	-44.37	7
Minimax filter	-49	-34.38	20

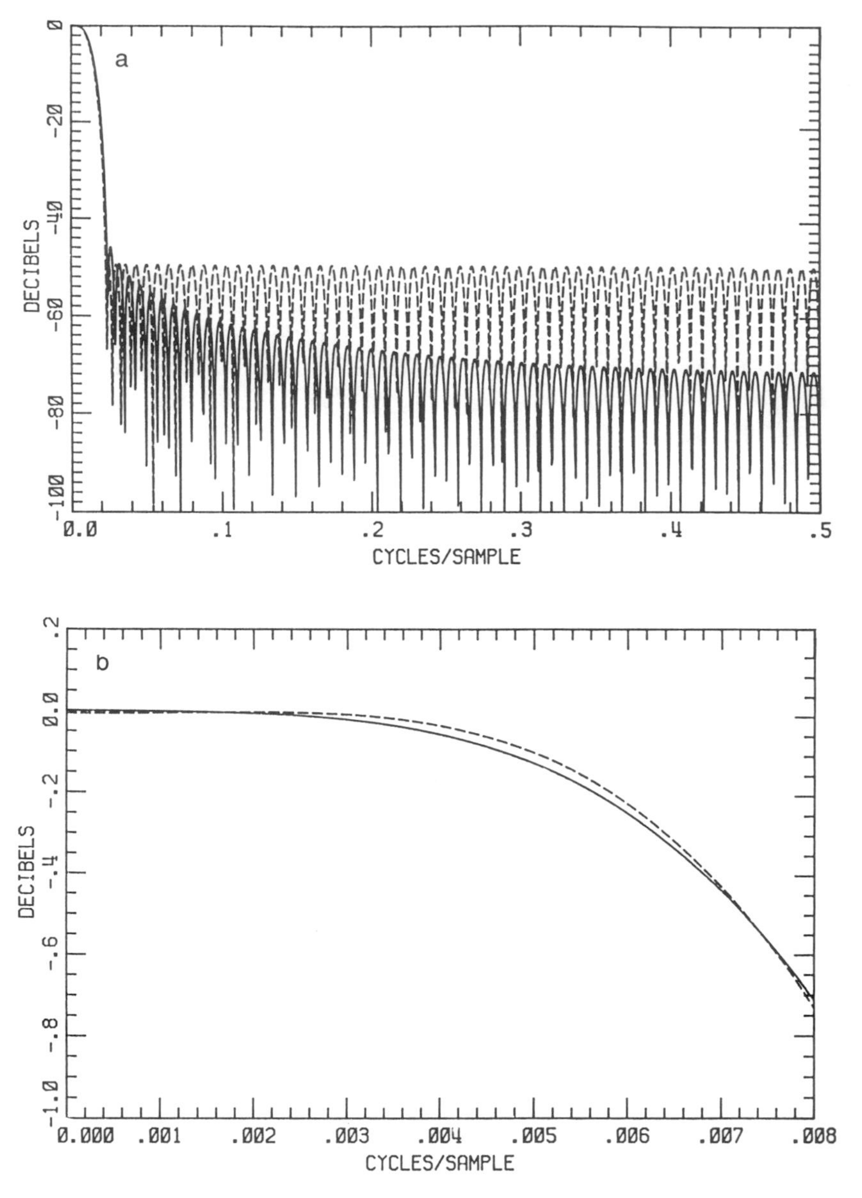

Fig. 4. (a) Single-stopband filters, and (b) passband details for a single-stopband filters: proposed (——) and equiripple (- - - -).

TABLE III. MULTIPLE-STOPBAND EXAMPLE: FILTER SPECIFICATIONS

Impulse response length	= 128
Passband edge frequency	= 1/128
Passband rolloff	= 0.7 dB
Decimation ratio	= 32 : 1

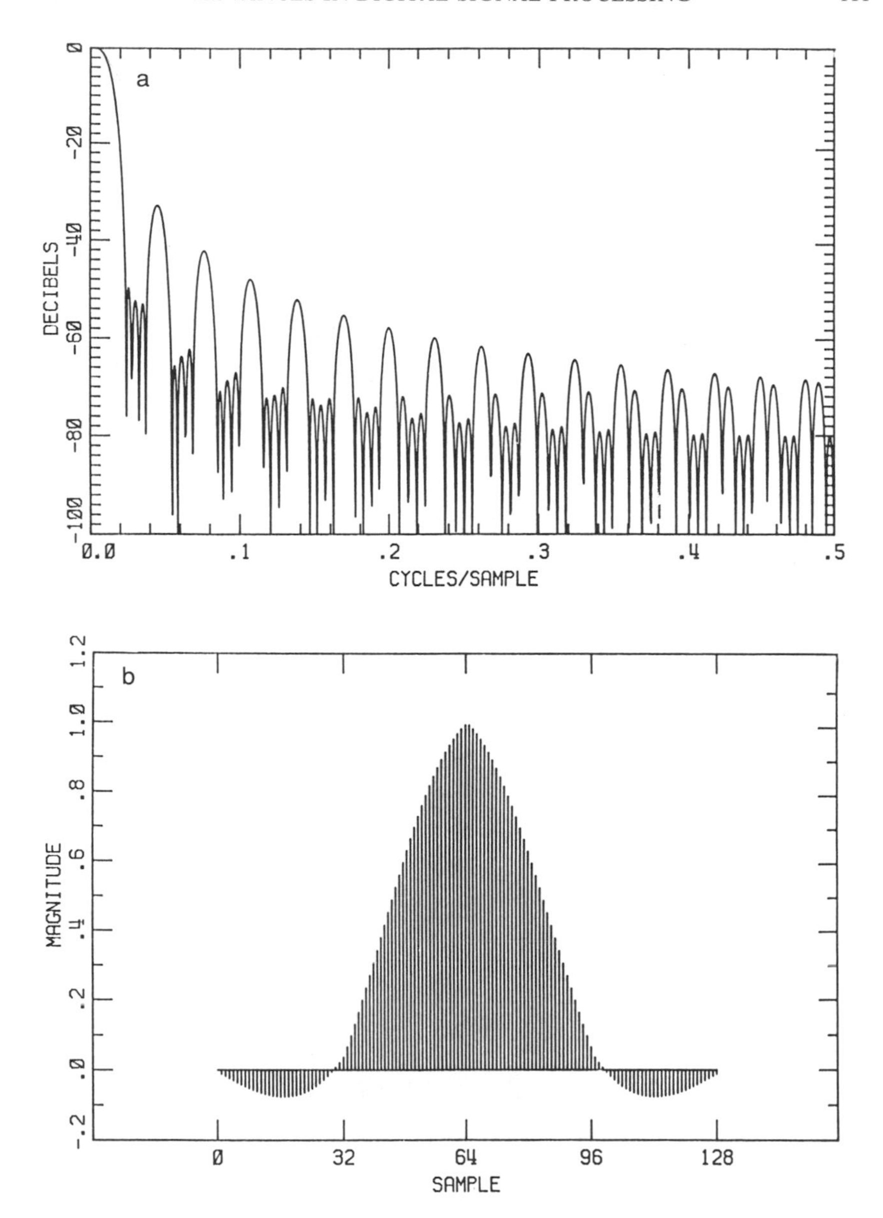

Fig. 5. Proposed (a) multiple-stopband filter and (b) multiple-stopband filter impulse response.

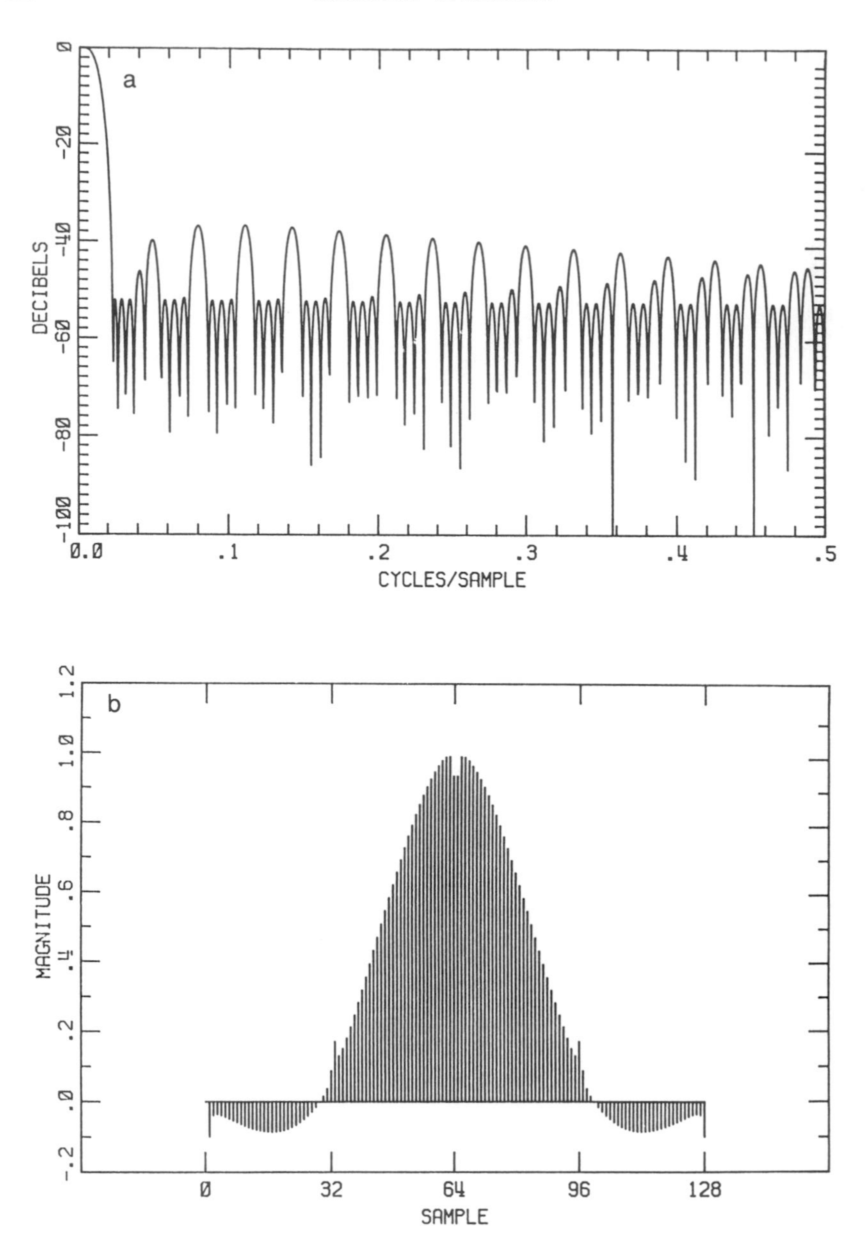

Fig. 6. Equiripple (a) multiple-stopband filter and (b) multiple-stopband filter impulse response.

TABLE IV. MULTIPLE-STOPBAND FILTER SUMMARY

	Peak stopband side-lobe level (dB)	ISLR (dB)	CPU time (sec)
Proposed filter	−50.1	−49.62	9
Minimax filter	−51.6	−39.83	12

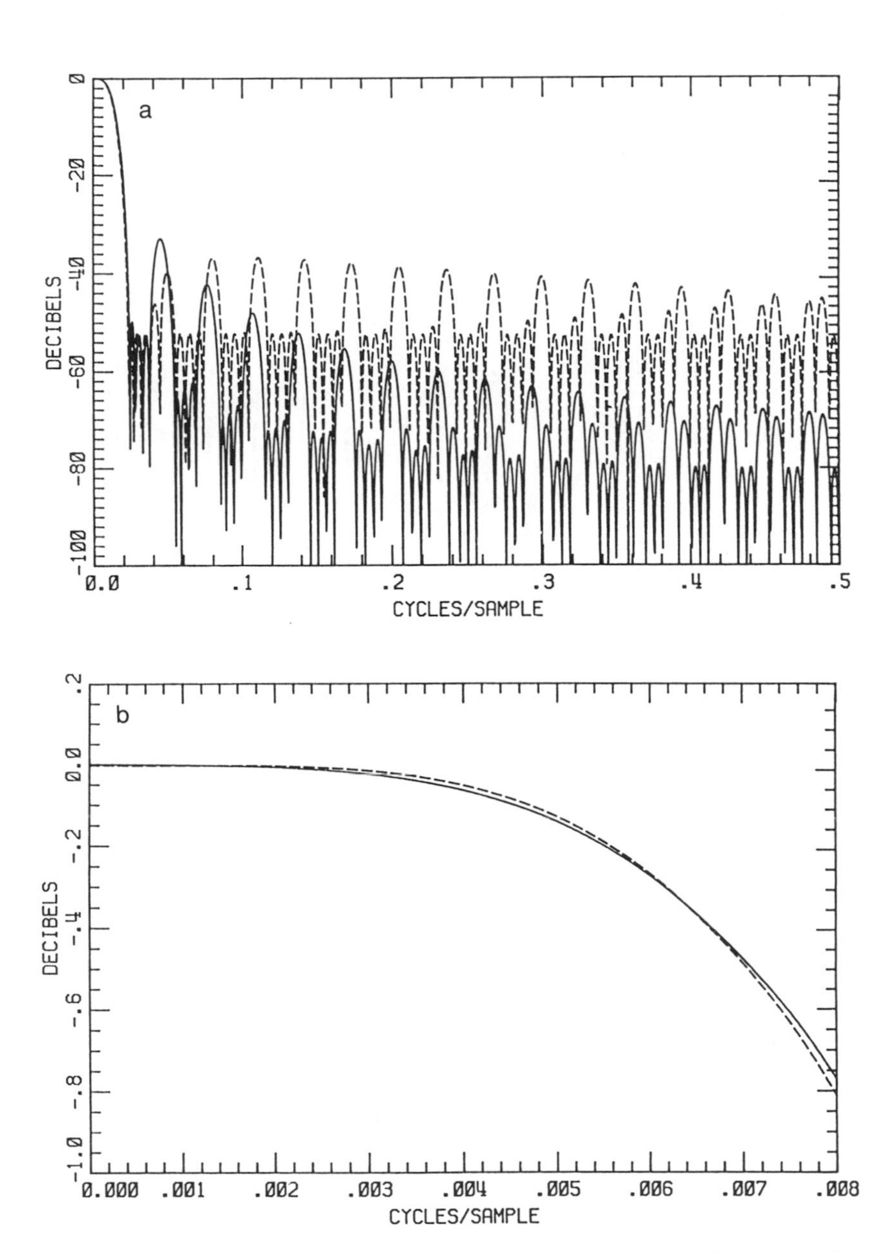

Fig. 7. (a) Multiple-stopband filters and (b) passband details for a multiple-stopband filters: proposed (——) and equiripple (- - - -).

As a last low-pass design example, Figure 8a shows the frequency response of an optimal low-pass filter with a fine-tuned passband. The initial design specifications are the same as the single-stopband filter in Table I. The filter passband has been fine tuned to produce a passband droop of only 0.5 dB. This passband droop was attained by adjusting the constrained value of the

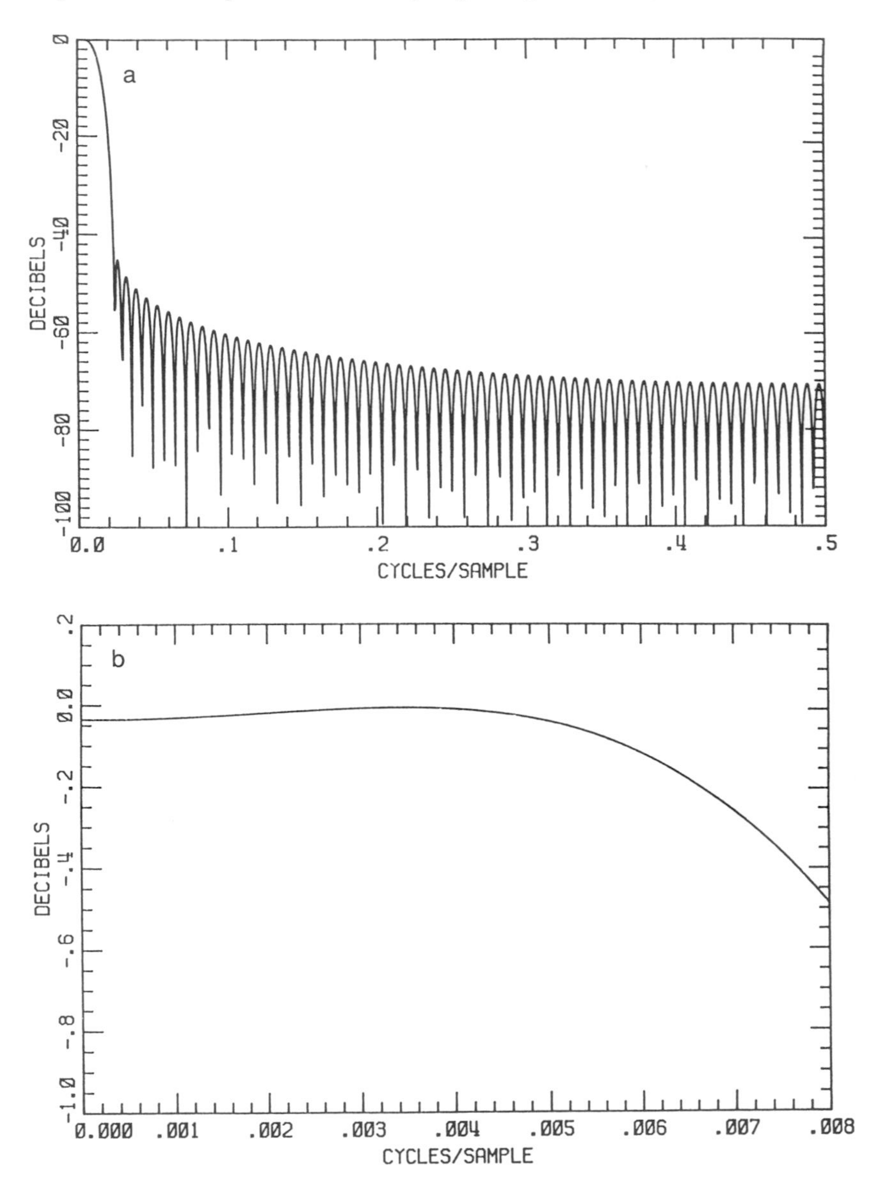

Fig. 8. Fine-tuned passband (a) example and (b) detail.

TABLE V. BANDPASS EXAMPLE: FILTER SPECIFICATIONS

Impulse response length	= 128
Passband center	= 0.3
Passband frequency $\Omega_p/2\pi$	= 1/128
Stopband frequency $\Omega_s/2\pi$	= 3/128
Passband rolloff	= 0.75 DB

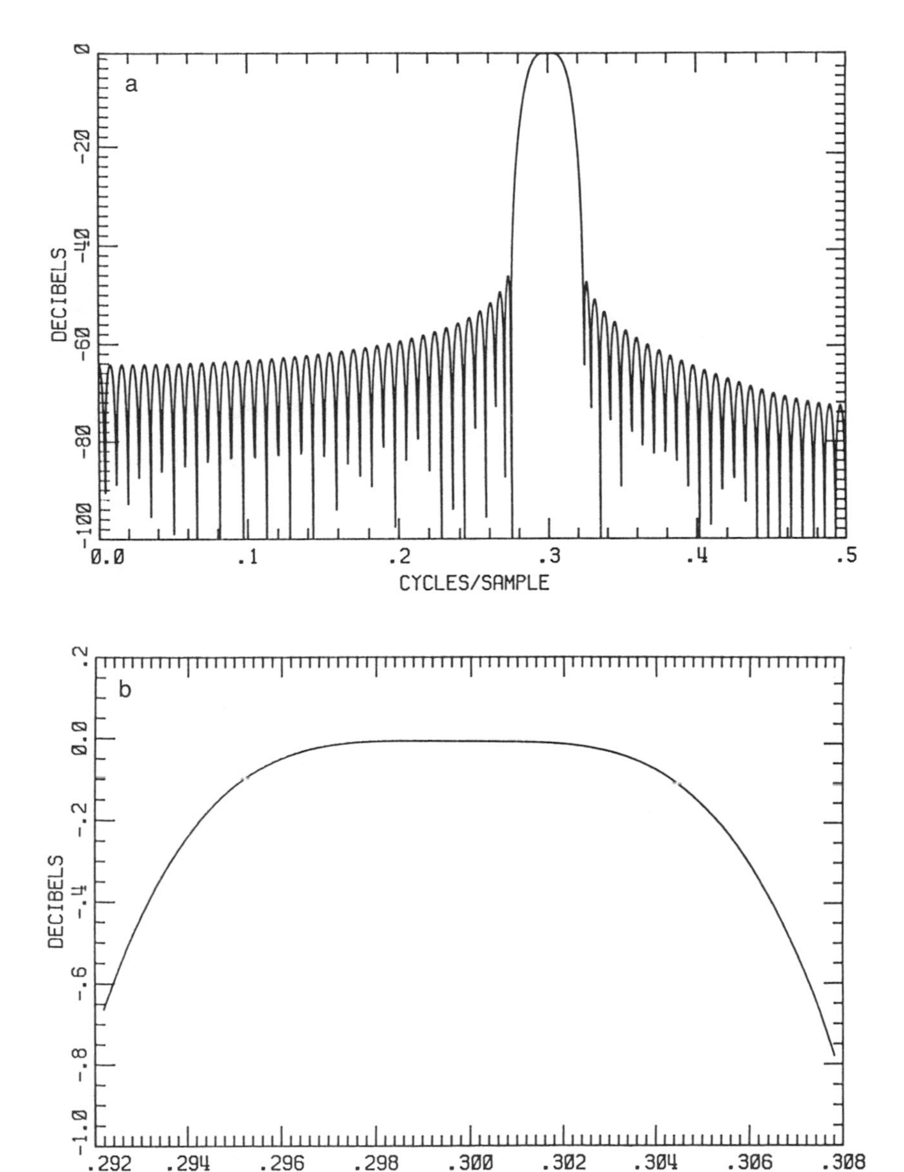

Fig. 9. (a) Two-derivative bandpass filter and (b) passband detail for a two-derivative filter.

second derivative at DC in a fashion similar to the example discussed at the end of Section IX. The passband adjustment was performed only once to obtain the desired rolloff. The filter that was fine tuned is shown in Fig. 2a. Originally, it was constrained to have unity magnitude and a zero second derivative at DC—all odd-order derivatives are automatically zero at DC. Changing the passband rolloff to 0.5 DB, required changing the second derivative to 33 (i.e., the fine tuned $k_2 = 33$). The passband detail is shown in Fig. 8b. As discussed in Section IX, the computation of the fine-tuned filter impulse response required only a linear combination of the impulse responses $H_0(e^{j\omega})$ and $H_2(e^{j\omega})$ with $k_0 = 1$ and $k_2 = 33$. The ISLR for the fine-tuned filter is -43.77 dB and the peak side-lobe level is -45 dB. This represents only a slight deviation from the original filter results shown in Table II.

Three optimal bandpass prototype filters were also designed using the Lagrange multiplier technique. The center frequency, passband width, and stopbands are the same for each filter and those values are shown in Table V.

The first bandpass filter frequency response is shown in Fig. 9a and was designed with the first and second derivatives constrained to be zero. The peak side-lobe level is -45.8 DB and the ISLR is -44.8 DB. Figure 9b is a detailed plot of the filter passband centered at 0.3 cycles per sample. The passband appears flat at the center and droops to ~ 0.75 DB at the edges. Figure 10 is a

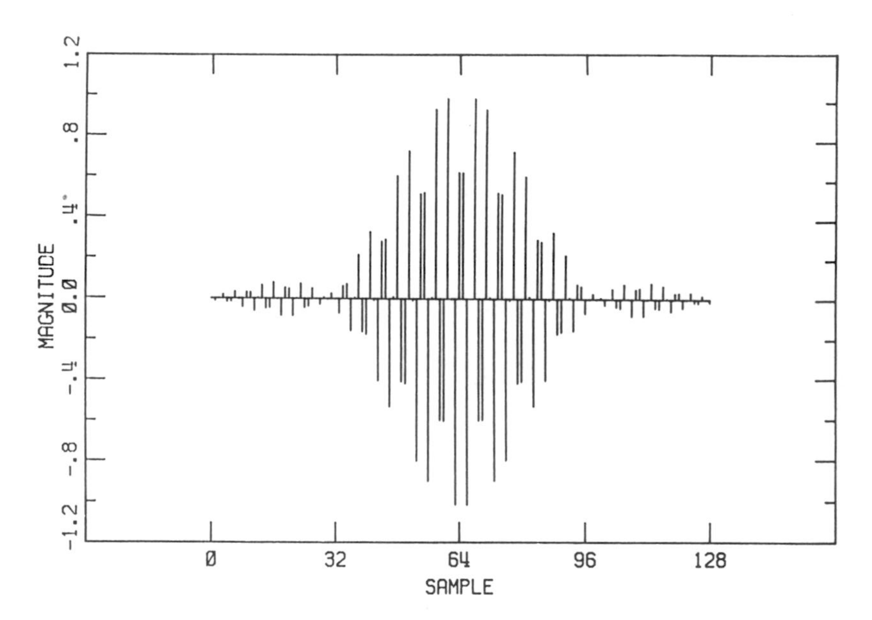

Fig. 10. Impulse response for a two-derivative filter.

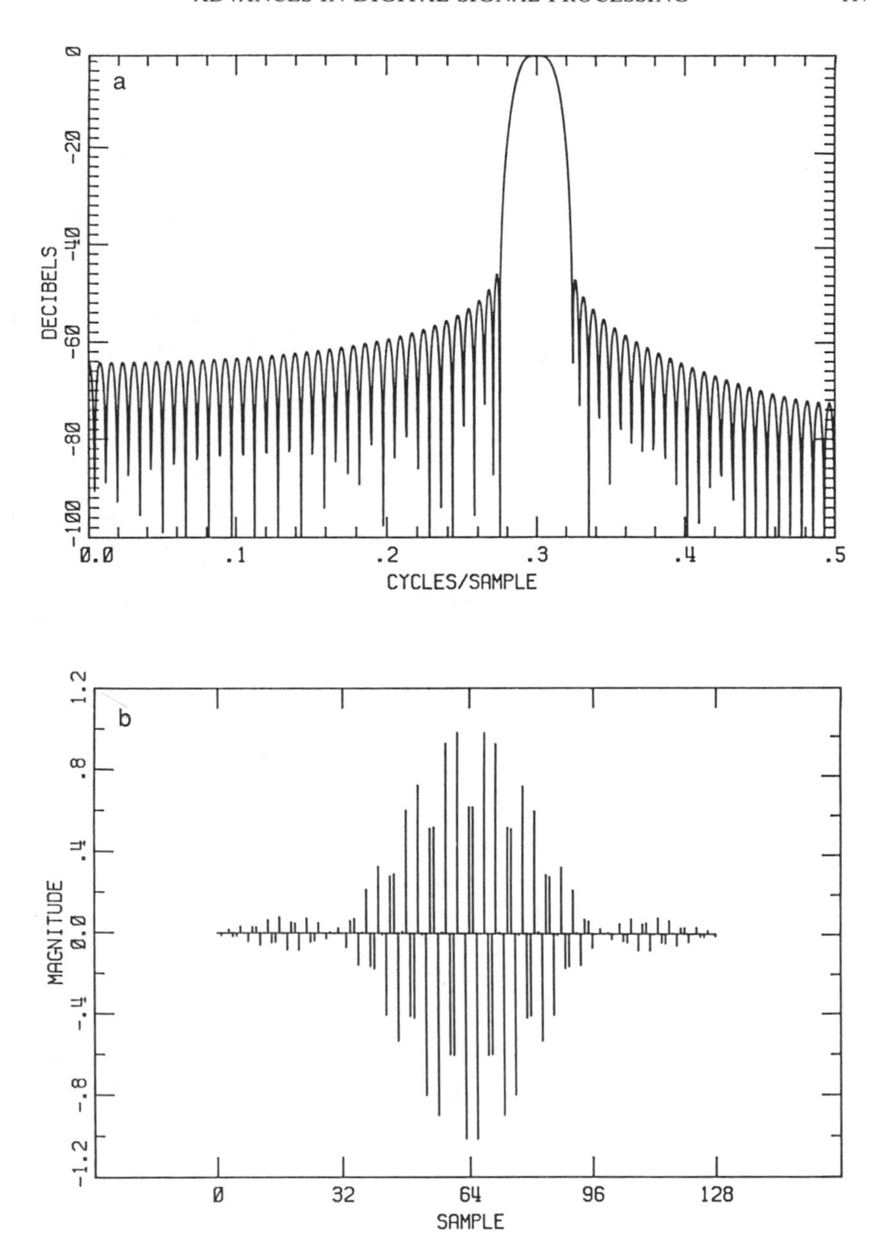

Fig. 11. H_0 (a) frequency response and (b) impulse response.

plot of the impulse response for the bandpass filter that has been scaled so that the peak value is unity.

As an example of the frequency responses denoted by $H_0(e^{j\omega})$, $H_1(e^{j\omega}), \ldots, H_{N-1}(e^{j\omega})$ and discussed in the previous section on fine tuning, the three plots for this bandpass example have been included. Figure 11a is a plot of $H_0(e^{j\omega})$. It is just the maximally flat filter with unity magnitude and the first two derivatives constrained to be zero. Its scaled impulse response is shown in Fig. 11b. Figure 12a is a plot of $H_1(e^{j\omega})$, which is constrained to have zero magnitude, zero second derivative, and unity first derivative at 0.3 cycles per sample. Figure 12b is a plot showing the corresponding impulse response. $H_2(e^{j\omega})$ is shown in Fig. 13a. The magnitude and first derivative are constrained to be zero and the second derivative is constrained to be unity at the passband center. Its impulse response is shown in Fig. 13b. Fine tuning the passband of the filter in Fig. 9a would involve determining a suitable linear combination of the three filters with responses $H_0(e^{j\omega})$, $H_1(e^{j\omega})$, and $H_2(e^{j\omega})$.

For comparison, an optimal equiripple bandpass filter was designed according to the specifications in Table V. The passband droop was designed to match that of the previous filter. Its frequency response, passband detail, and impulse response are shown in Fig. 14a–c, respectively. The equiripple filter has a peak side-lobe level of -49.2 DB, an ISLR of -37.47 DB. Figure 15a and b are overlays of the proposed and equiripple filter frequency responses and passband details, respectively. Figure 15a shows the peak side lobe of the proposed filter extending ~ 3 DB above the equiripple sidelobe level. The passband details in Fig. 15b show the passbands to be nearly identical with the exception that the proposed passband is flat at the passband center.

Figures 16a–c show the frequency response, passband detail, and normalized impulse response, respectively, for an optimal bandpass filter with the first three derivatives constrained to be zero and designed with the specifications in Table V. The peak side-lobe level is -46.03 DB and the ISLR is -44.69 DB. These parameters and the passband detail are very nearly identical to the previous filter with the first two derivatives constrained to be zero. In this case the optimal filter with two derivative constraints happens to have a small third derivative so that there is little change by requiring three zero derivatives at the passband center.

Figures 17a–c are the frequency response, passband detail, and scaled impulse response, respectively, for an optimal bandpass filter with four derivatives at 0.3 cycles per sample constrained to be zero. The filter was designed to the specifications in Table V. The four-derivative case has a peak side-lobe level of -29.3 DB and an ISLR of -27.20 DB. The passband in

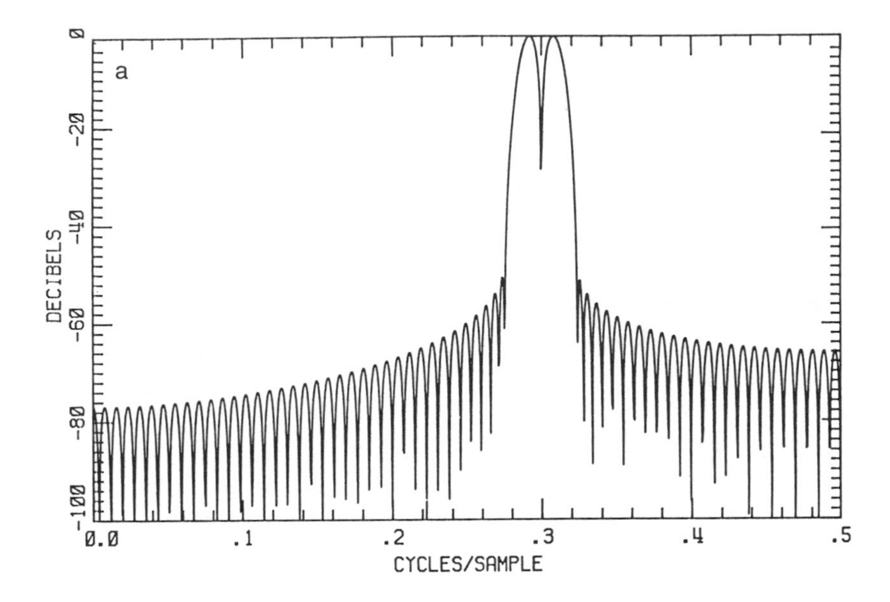

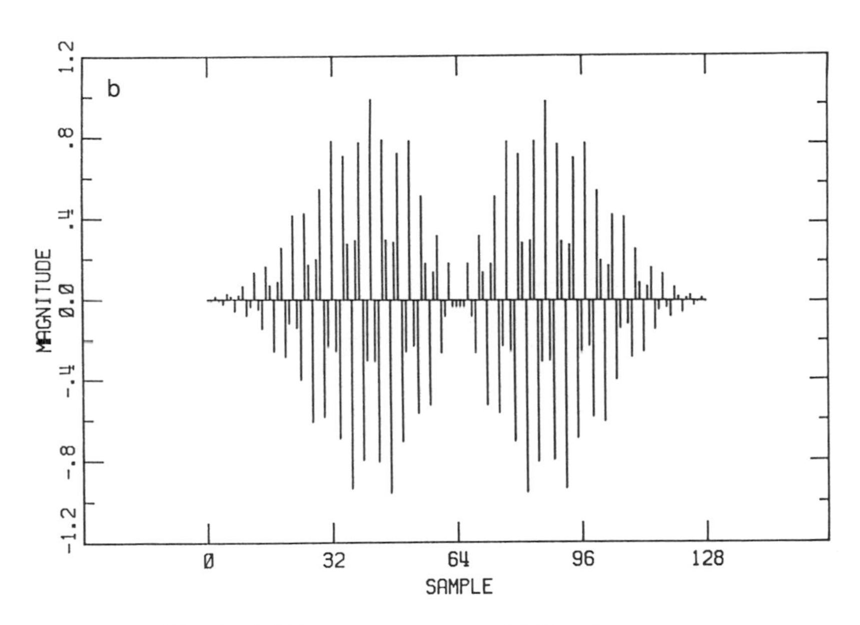

Fig. 12. H_1 (a) frequency response and (b) impulse response.

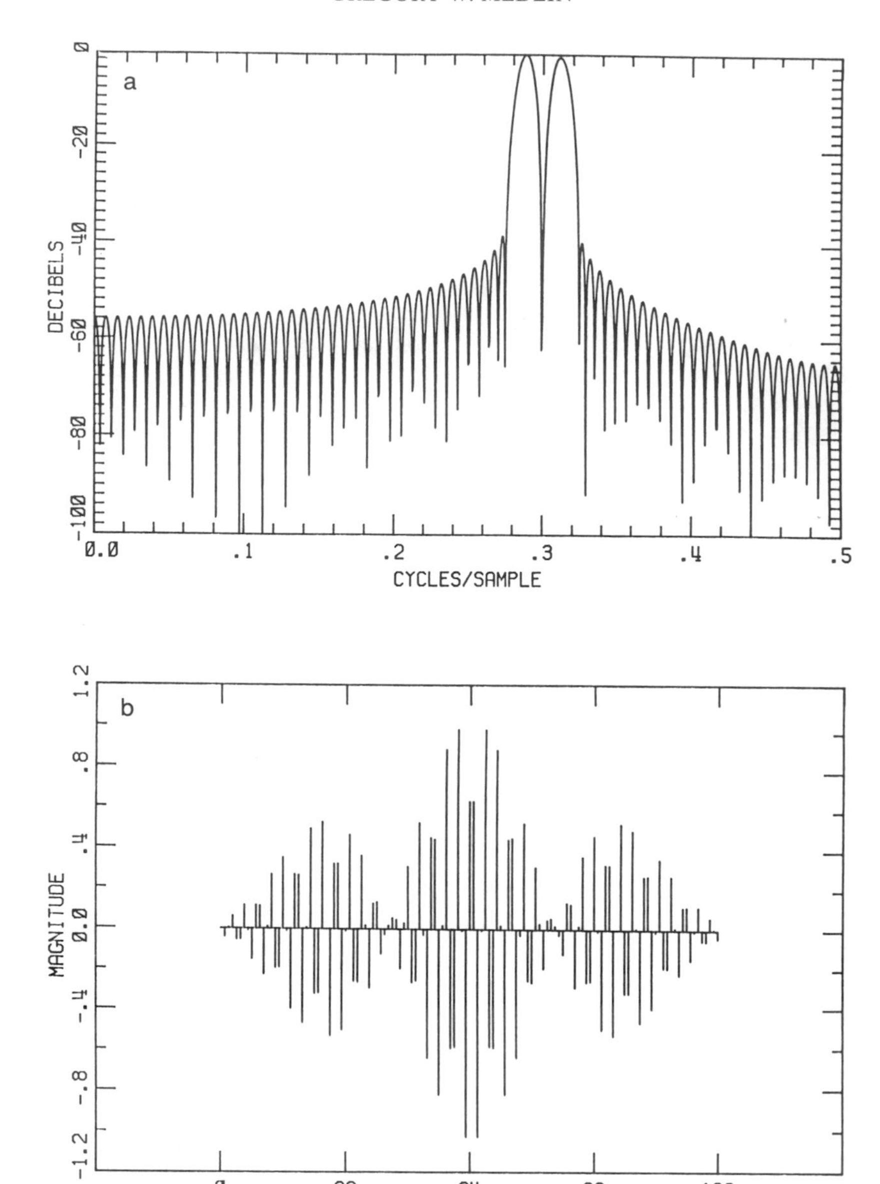

Fig. 13. H_2 (a) frequency response and (b) impulse response.

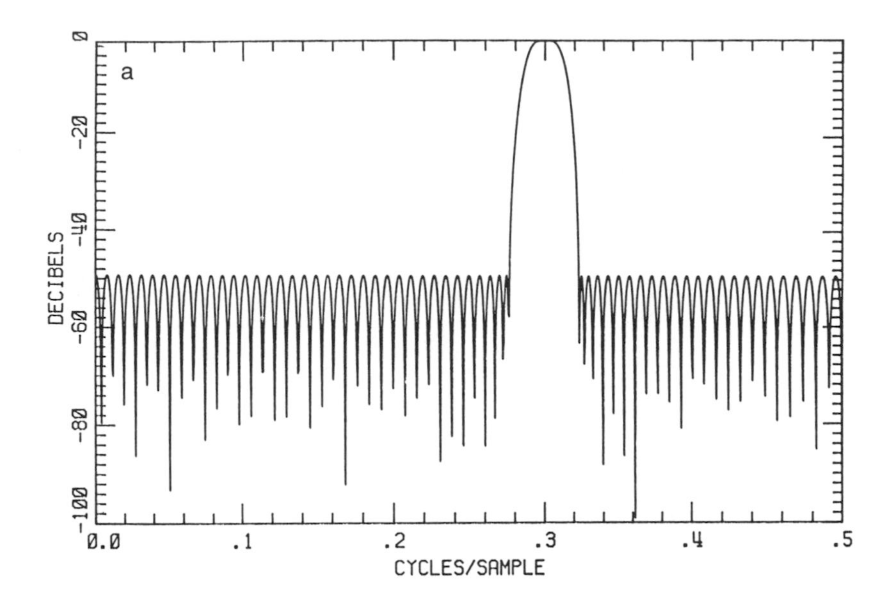

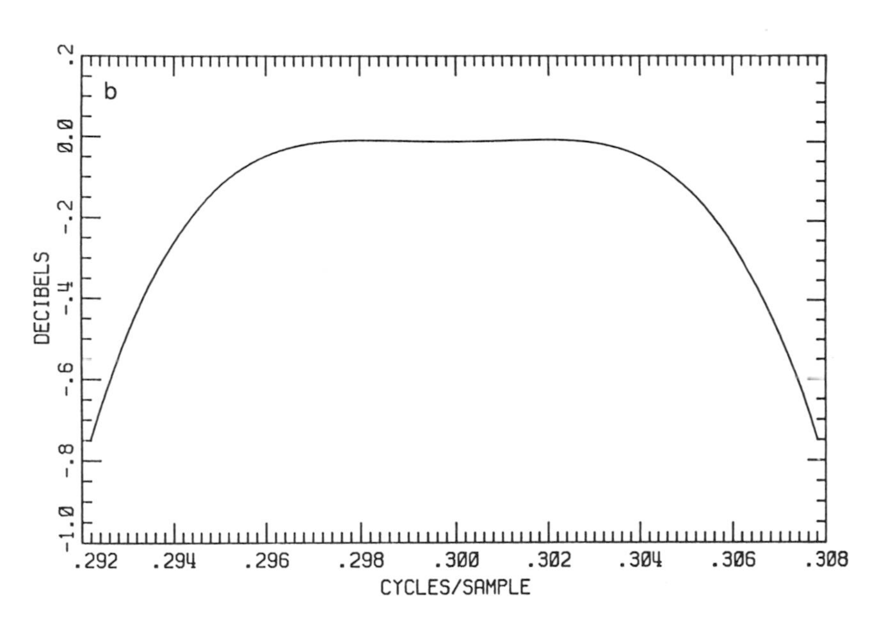

Fig. 14. Equiripple (a) bandpass filter, (b) bandpass detail for a bandpass filter, and (c) bandpass filter impulse response.

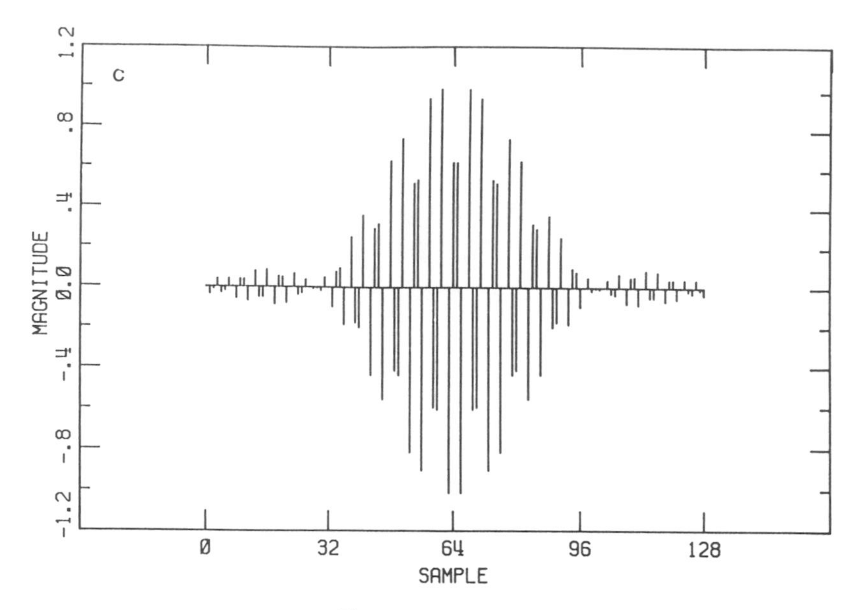

Fig. 14. (*Continued.*)

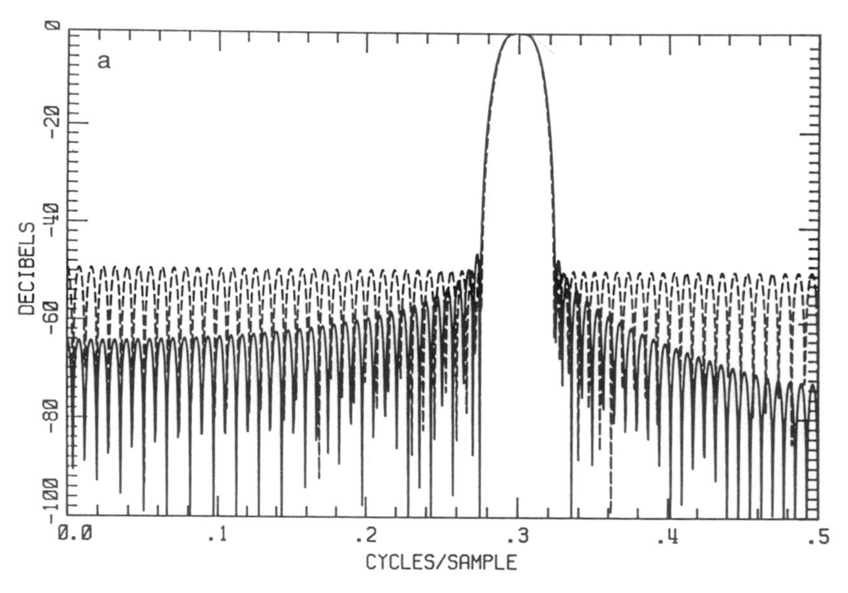

Fig. 15. (a) Bandpass filters and (b) passband details for bandpass filters: proposed (——) and equiripple (- - -).

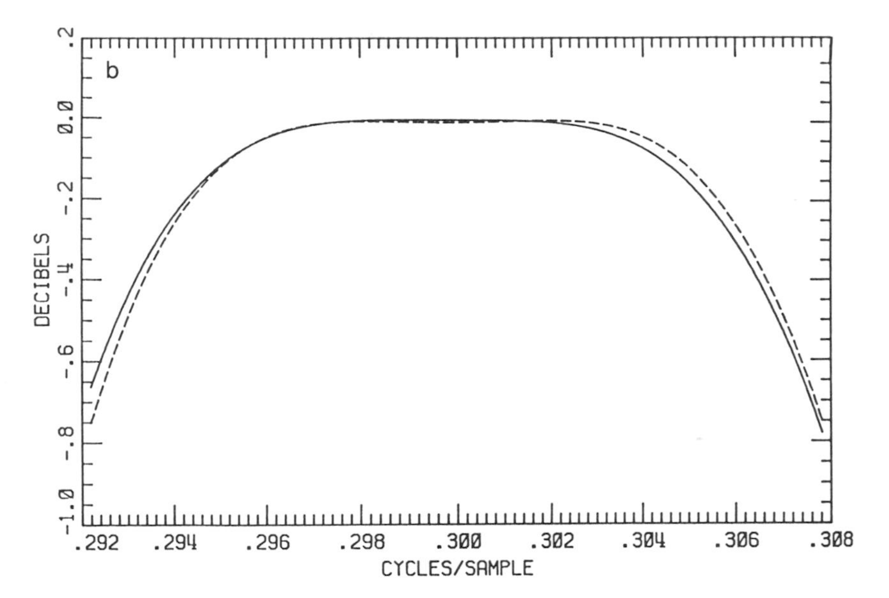

Fig. 15. (*Continued.*)

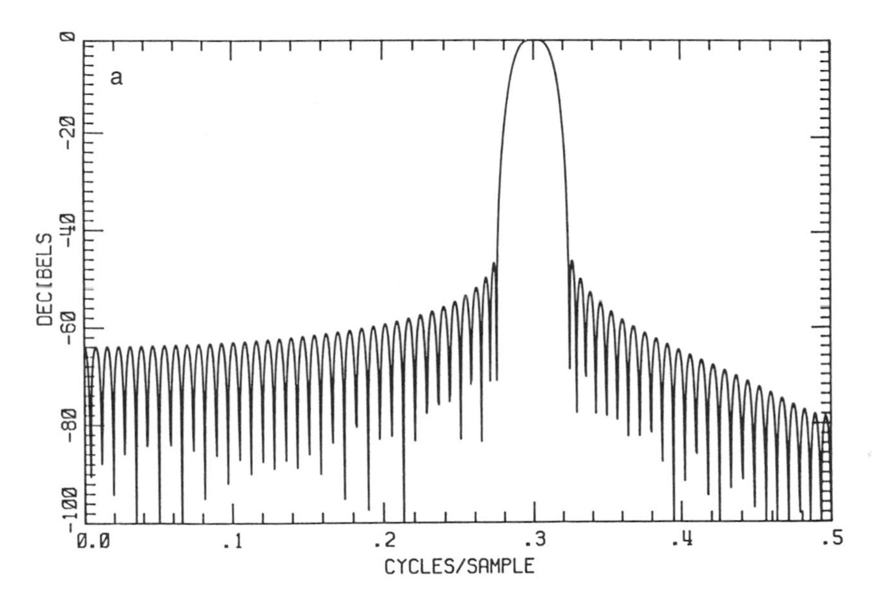

Fig. 16. (a) Three-derivative bandpass filter, (b) passband detail for a three-derivative filter, and (c) impulse response for a three-derivative filter.

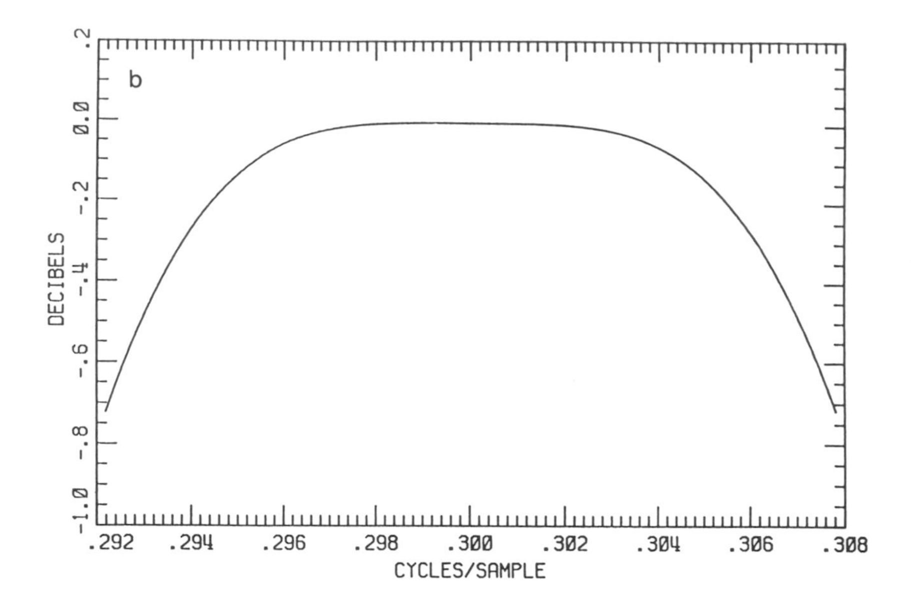

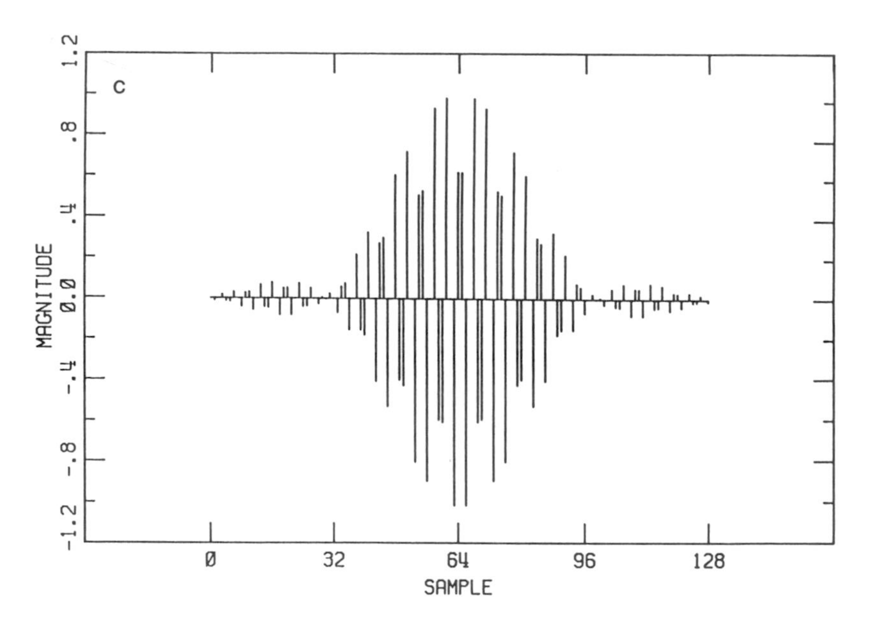

Fig. 16. (*Continued.*)

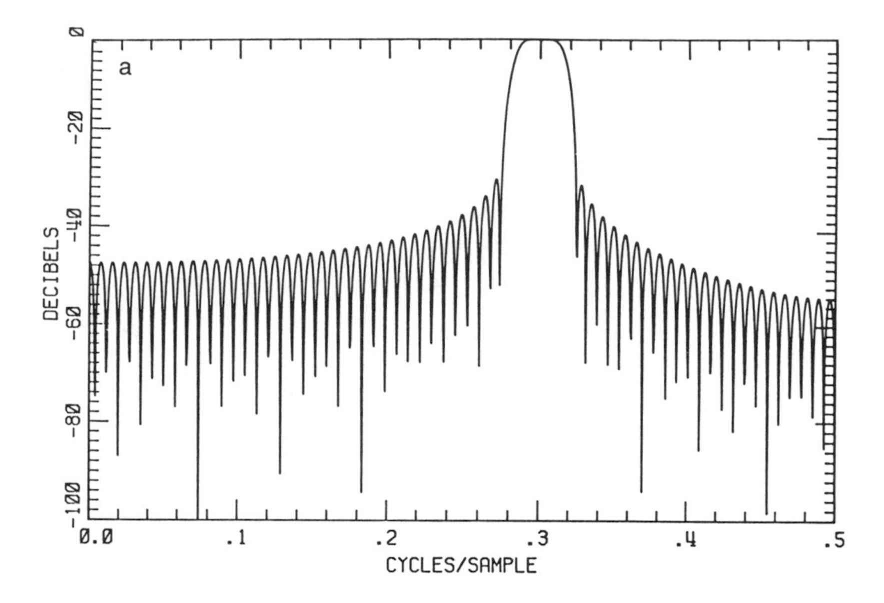

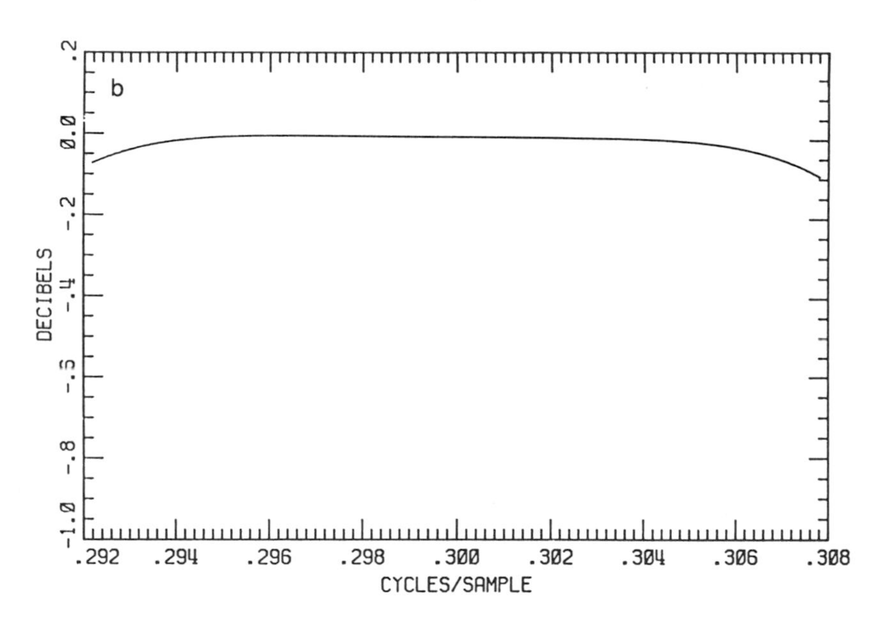

Fig. 17. (a) Four-derivative bandpass filter, (b) passband detail for a four-derivative filter, and (c) impulse response for a four-derivative filter.

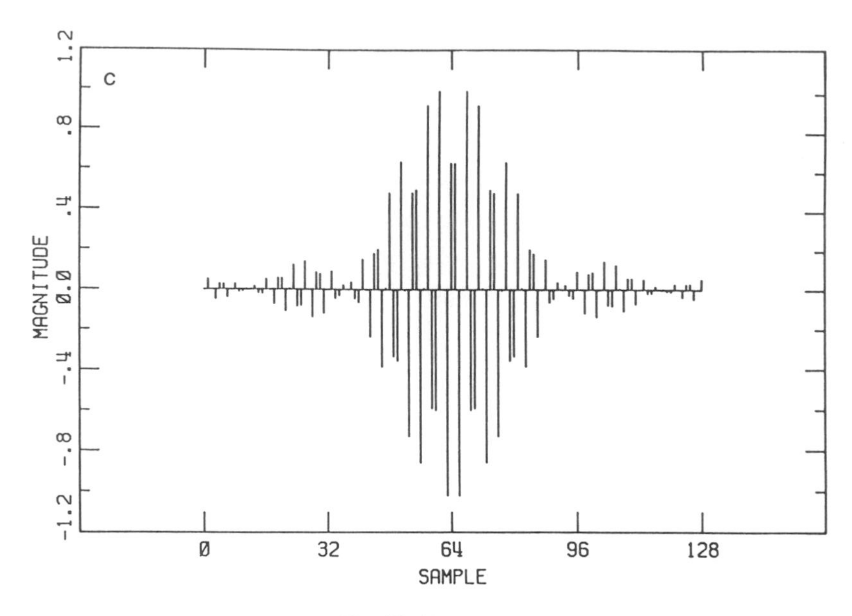

Fig. 17. (*Continued.*)

TABLE VI. BANDPASS FILTER SUMMARY

	Peak stopband side-lobe level (dB)	ISLR (dB)	Computation time (sec)
Proposed two-derivative filter	−45.8	−44.8	9
Proposed three-derivative filter	−46.0	−44.69	10
Proposed four-derivative filter	−29.3	−27.20	10
Equiripple filter	−49.2	−37.47	29

Fig. 17b is very flat with a passband droop of only ~0.1 DB. The results for these three proposed optimal filters and the comparison equiripple filter are summarized in Table VI.

XII. SUMMARY

A new approach to the design of optimal filters for multirate digital signal processing is presented. The optimal filter design problem is formulated as a quadratic programming problem with equality constraints. The closed-form solution is obtained by the Lagrange multiplier method. The Lagrange multipliers are interpreted as the sensitivity of the optimal stopband energy to changes in the constrained passband values.

A design algorithm is presented that provides a practical estimate for the passband flatness necessary to obtain a specified passband rolloff. It is noted that the filter passband can be fine tuned by changing the constrained passband values without complete recomputation of the optimal filter.

Examples are provided that demonstrate that the new filters have significantly reduced stopband energy compared to equiripple filters. This reduction in stopband energy is obtained at the expense of only a minor increase in the peak side-lobe level.

APPENDIX. DERIVATION OF EQUATION (54)

In this section a derivation of (54) in Section VII is presented. Given the zero-phase frequency response

$$H(e^{j\omega}) = \sum_{n=0}^{L-1} h(n)e^{-j\omega n}, \tag{80}$$

where $h(n)$ is the impulse response and ω is the radian frequency, an infinite Taylor series about $\omega = \omega_c$ may be written as

$$H(e^{j\omega}) = \sum_{n=0}^{\infty} C_n(\omega - \omega_c)^n, \tag{81}$$

where

$$C_n = \frac{1}{n!} \frac{d^nH}{d\omega^n}\bigg|_{\omega=\omega_c}. \tag{82}$$

For filters with maximally flat passbands we have

$$H(e^{j\omega}) = 1 + C_N(\omega - \omega_c)^N + C_{N+1}(\omega - \omega_c)^{N+1} + \cdots. \tag{83}$$

Specifying the passband rolloff at $\omega_c + \omega_p$ to be δ_p and the stopband gain at $\omega_c + \omega_s$ to be δ_s, (83) gives the two following results:

$$\begin{aligned} -\delta_p &= C_N\omega_p^N + C_{N+1}\omega_p^{N+1} + \cdots \\ \delta_s &= 1 + C_N\omega_s^N + C_{N+1}\omega_s^{N+1} + \cdots. \end{aligned} \tag{84}$$

Using these two equations, the constant C_N can be eliminated and we find

$$\delta_s\omega_p^N + \delta_p\omega_s^N = \omega_p^N + C_{N+1}(\omega_p^N\omega_s^{N+1} - \omega_s^N\omega_p^{N+1}) + \text{HOT}. \tag{85}$$

Ignoring terms in ω of order $2N$ and higher, (85) becomes

$$\delta_s \omega_p^N + \delta_p \omega_s^N = \omega_p^N.$$

Solving for N we find

$$N = \left\{ \log\left(\frac{\delta_p}{1 - \delta_s}\right) \Big/ \log\left(\frac{\omega_p}{\omega_s}\right) \right\}.$$

Since typically $\delta_s \ll 1$ we may simplify and obtain:

$$N = \log \delta_p \Big/ \log\left(\frac{\omega_p}{\omega_s}\right).$$

REFERENCES

1. R. E. CROCHIERE and L. R. RABINER, "*Multirate Digital Signal Processing*." Prentice-Hall, Englewood Cliffs, New Jersey, 1983.
2. R. E. CROCHIERE and L. R. RABINER, "Interpolation and Decimation of Digital Signals—A Tutorial Review," *Proc. IEEE* **69**, 300–331 (1981).
3. R. W. SCHAFER and L. R. RABINER, "A Digital Signal Processing Approach to Interpolation," *Proc. IEEE* **61**, 692–702 (1973).
4. R. E. CROCHIERE and L. R. RABINER, "Optimum FIR Digital Filter Implementations for Decimation, Interpolation, and Narrow-band Filtering," *IEEE Trans. Acoust., Speech, Signal Process.* **ASSP-23**, 444–456 (1975).
5. R. R. SHIVELY, "On Multistage Finite Impulse Response Filters with Decimation," *IEEE Trans. Acoust., Speech, Signal Process.* **ASSP-23**, 353–357 (1975).
6. A. PELED and S. WINOGRAD, "TDM-FDM Conversion Requiring Reduced Computational Complexity," *IEEE Trans. Commun.* **COM-26**, 707–719 (1978).
7. R. E. CROCHIERE and L. R. RABINER, "Further Considerations in the Design of Decimators and Interpolators," *IEEE Trans. Acoust., Speech, Signal Process.* **ASSP-24**, 296–311 (1976).
8. J. H. McCLELLAN, T. W. PARKS, and L. R. RABINER, "A Computer Program for Designing Optimum FIR Linear Phase Digital Filters," *IEEE Trans. Audio Electroacoust.* **AU-21**, 506–526 (1973).
9. J. F. KAISER and K. STEIGLITZ, "Design of FIR Filters with Flatness Constraints," *IEEE Int. Symp. Acoust., Speech, Signal Process., Boston, Massachusetts*, pp. 197–199 (1983).
10. R. W. HAMMING, "Digital Filters." Prentice-Hall, Englewood Cliffs, New Jersey, 1983.
11. L. R. RABINER and B. GOLD, "Theory and Application of Digital Signal Processing". Prentice-Hall, Englewood Cliffs, New Jersey, 1975.
12. W. I. ZANGWILL, "Nonlinear Programming: A Unified Approach," pp. 66–68. Prentice-Hall, Englewood Cliffs, New Jersey, 1969.
13. L. R. RABINER, "Approximate Design Relationships for Lowpass FIR Digital Filters." *IEEE Trans. Audio Electroacoust.* **AU-21**, 456–460 (1973).
14. R. FLETCHER, "Practical Methods of Optimization," Vol. 2. Wiley (Interscience), New Yorks, 1981.
15. G. STRANG, "Linear Algebra and its Applications." Academic Press, New York, 1976.

METHODOLOGY FOR THE ANALYTICAL ASSESSMENT OF AIRCRAFT HANDLING QUALITIES

RONALD A. HESS

Department of Mechanical Engineering
Division of Aeronautical Science and Engineering
University of California
Davis, California 95616

I. INTRODUCTION AND BACKGROUND

Aircraft handling qualities can be defined as "those qualities or characteristics of an aircraft that govern the ease and precision with which a pilot is able to perform the tasks required in support of an aircraft role" [1]. From the advent of manned flight, pilot evaluation has remained the most reliable means of evaluating aircraft handling qualities [2]. Pilot evaluation refers to the subjective assessment of aircraft handling qualities by pilots and is separated into two areas—pilot comments and pilot ratings. Pilot comments refer to those detailed remarks that pilots make and record regarding vehicle handling qualities in a given experiment. Pilot ratings refer to the numerical ratings that are assigned a vehicle and task in a given experiment and are the end product of the evaluation process.

Since 1969, the standard rating scale, which has been employed in the assessment of aircraft handling qualities, is the Cooper–Harper pilot rating scale, shown in Fig. 1. This scale is actually a decision tree that the pilot employs in evaluating the handling qualities of a vehicle and task. After a series of dichotomous decisions, a numerical rating from 1 to 10 is assigned to the vehicle and task. The adjectives appearing in the scale have been carefully selected and emphasize the two important facets of aircraft handling qualities evaluations—pilot–vehicle performance and required pilot compensation. In

Copyright © 1990 by Academic Press, Inc.
All rights of reproduction in any form reserved.

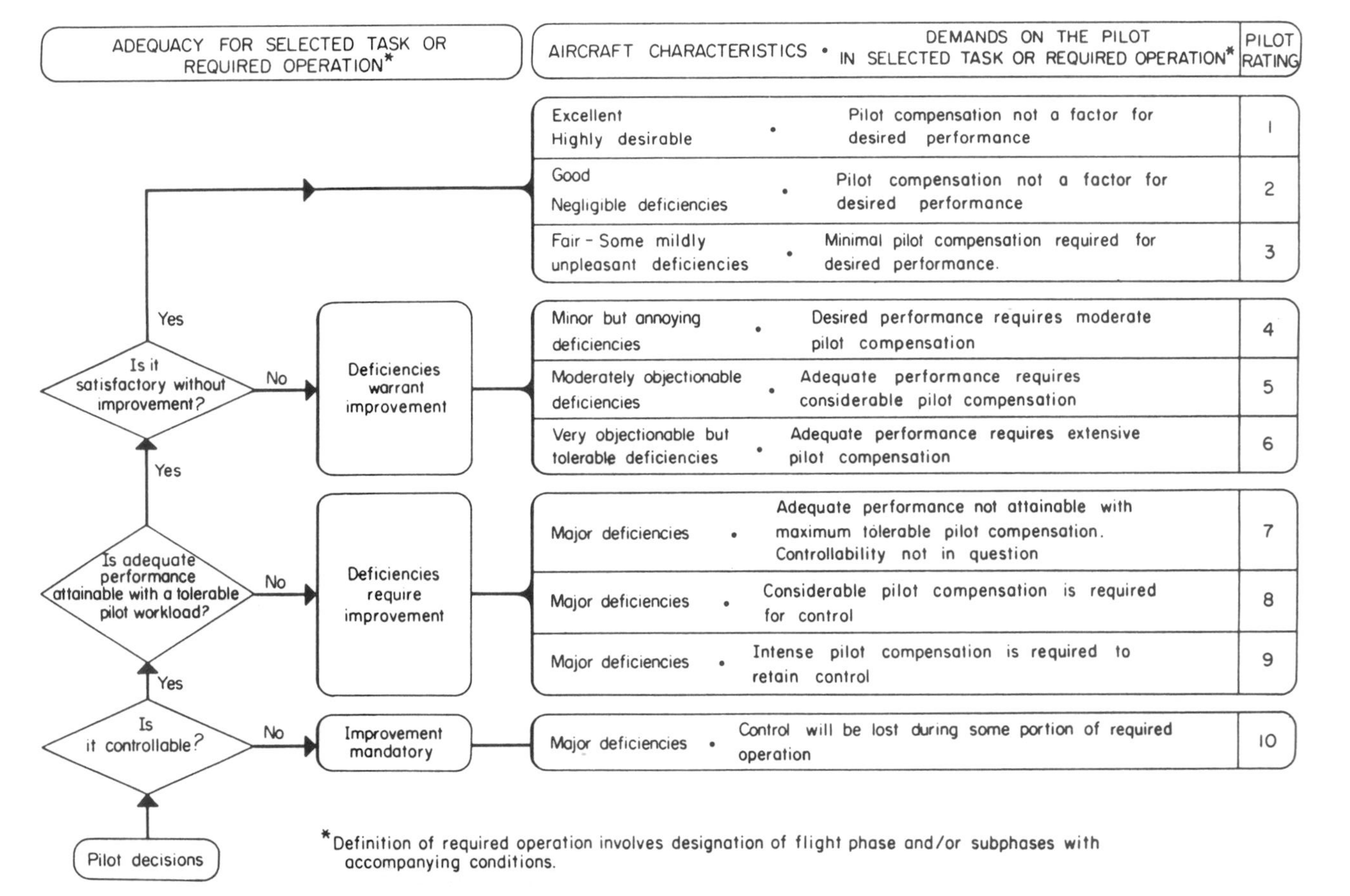

Fig. 1. The Cooper–Harper pilot rating scale.

using the scale, it is essential to communicate to the pilot precisely what constitutes "adequate" performance and "desirable" performance in the task at hand. It falls upon the pilot to determine the amount of compensation required to achieve adequate and/or desired performance.

Compensation refers generally to the mental and physical resources that the pilot must devote to the task. The term *work load* is often used synonymously with compensation in the literature. While objective performance can be readily measured in a well-designed flight or simulation experiment, the objective determination of the amount of pilot compensation or work load that is occurring is, with a few exceptions, quite difficult, although this remains an area of active research (e.g., [3]). The exceptions refer to those instances where measurements of closed-loop pilot–vehicle dynamics allow the determination of pilot compensation in transfer function form (see Hess, "Identification of Pilot/Vehicle Dynamics from Simulation and Flight Test," this volume).

The advances in aircraft performance, which are apparent in reviewing the accomplishments during the past 70 years of powered flight, have been made possible by the introduction and eventual widespread use of theoretical techniques for analysis and design. This is particularly true in the area of aircraft dynamics and flight control systems [4]. However, the primary adaptive element in manned aircraft, the human pilot, has not been as amenable to theoretical treatment as have the inanimate elements of flight control systems.

Beginning with the seminal work of Tustin [5] shortly after World War II, engineers began modeling human manual control activity by comparing it to the action of an inanimate servomechanism. The control theoretic paradigm for the human pilot, which has evolved in the intervening years, has been quite useful and indeed has become the fundamental mode of thinking on the part of most manual control practitioners [6]. One of the first attempts at the development of a theory for aircraft handling qualities relying heavily on a control theoretic model of the human pilot was that of [7]. The goals of this early study are worth repeating since they remain as valid today as they were when first formulated:

1. To focus attention on and explain the connections between subjective pilot opinion and pilot–vehicle system performance.
2. To form a foundation for the insights required to determine airframe configuration variants, which offer possible flight qualities improvements.
3. To provide a basis for deriving handling criteria for configurations with novel dynamic characteristics.
4. To provide a unifying structure for the large amount of dynamic data previously treated as unconnected.

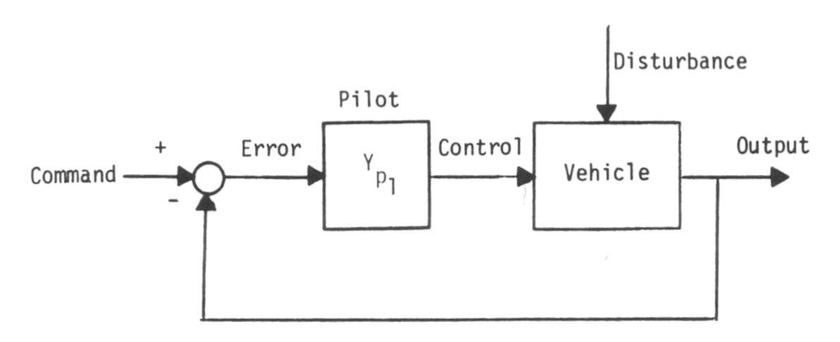

Fig. 2. A single-loop pilot–vehicle system.

The work in [7] and the related studies, which followed, emphasized the fact that flight tasks involving regulation and precision flying (as opposed to unattended operation and large amplitude maneuvers) were often the most critical in revealing handling qualities deficiencies. In addition, it was found that in manual flight control situations, which involved multiple feedbacks with a single control (e.g., altitude and attitude regulation with the elevator), a series rather than a parallel-loop structure was a more realistic representation of the activity of the human pilot [8]. Such a series structure is often referred to as a multiloop as opposed to a multiple-loop manual control problem.

The mid-1960s saw the development of the crossover model of the human pilot to model pilot–vehicle dynamic characteristics in single-loop tasks or in the inner loop of multiloop tasks. Referring to Fig. 2 and summarizing briefly, this model shows that in the frequency region of open-loop frequency crossover the product of the pilot the vehicle transfer functions (Y_{p_1} and Y_c, respectively) can be given by

$$Y_{p_1} Y_c = [\omega_c \exp\{-\tau_e s\}]/s, \tag{1}$$

where ω_c is the crossover frequency and τ_e is an effective time delay, which includes the integrated effect of actual human time delays and also the phase lags attributable to higher frequency dynamics in the human neuromuscular system [9].

The crossover model helped solidify the early handling qualities theory proposed in [7] by allowing the analyst to predict the type of equalization that the pilot was required to provide for a given vehicle and task. This was important, for it was found, for example, that lead generation on the part of the pilot led to handling qualities degradations as compared to identical tasks in which only gain equalization was required.

A decade after the publication of [7], and in anticipation of the impending arrival of a handling qualities specification for vertical/short takeoff and landing (V/STOL) aircraft, Anderson [10] introduced a "minimum pilot rating concept," which postulated that, given a particular vehicle and task, the pilot adopts a control strategy that maximizes his impression of the vehicle's handling qualities, or equivalently minimizes his pilot rating. This idea led to a computer algorithm called paper pilot. Paper pilot represented one of the more successful and utilitarian applications of what might be called classical pilot models to handling qualities prediction. Following the lead of Anderson, Hess turned to modern control theory and utilized the optimal control model (OCM) of the human pilot to predict pilot opinion ratings [11]. Both of these automated handling qualities prediction schemes exhibited limitations that could be traced to the theoretical underpinnings, which supported the pilot rating concepts.

In the mid-1970s, Smith proposed a theory for pilot opinion ratings, which postulated that, in any closed-loop tracking task (such as aircraft pitch-attitude regulation in turbulence), pitch-rate control is of fundamental importance from the standpoint of preceived vehicle handling qualities [12]. Smith held that a physiological measure for pilot opinion ratings is the rate at which nerve impulses (or an equivalent measure) arrive at the point within the central system where all signals due to rate control are summed or operated upon. In Smith's view, the relation between pilot opinion rating and the nerve impulse rate is fixed for each pilot, and is dependent on his piloting experience, training, and personal interpretation of the rating scale. Finally, it was Smith's contention that a model for human pilot dynamics that structurally matches the human physiology in the tracking process will lead to a natural and physical measure for pilot opinion rating.

Smith's theory is intriguing since it hypothesizes a physiological source for the compensation or work load, which is required of the pilot in a flight task and which is a fundamental part of pilot opinion ratings obtained from the Cooper–Harper scale. Given a tractable analytical pilot model, which explicitly accounted for the rate-tracking activity that Smith hypothesized to be a fundamental part of the pilot rating process, it appears that a handling qualities prediction scheme might be created that did not exhibit the limitations of previous methods. This leads naturally to the discussion of a structural model of the human pilot and its use in a handling qualities theory based on Smith's rate-tracking hypothesis.

II. STRUCTURAL MODEL OF THE HUMAN PILOT

Hess has proposed a model of the human pilot, which has its origins in Smith's concept of rate tracking [12]. Consider Fig. 3a, which is a simplified model of the human pilot in a single-loop task or in the inner loop of a

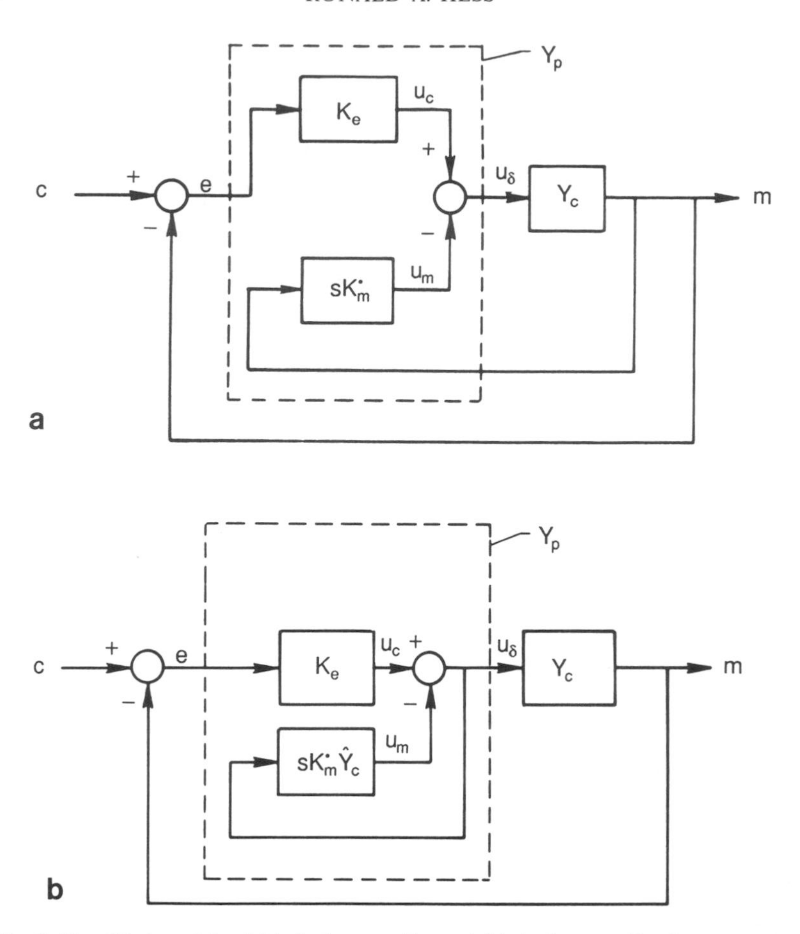

Fig. 3. Simplified models of (a) the human pilot and (b) the human pilot for compensatory tracking (Reprinted from [13].)

multiloop task and is adopted from Smith. The inner feedback loop in this model involves a signal u_m, which is proportional to output rate $\dot{m}$. Figure 3b is a modification of Fig. 3a, in which the accessible control input u_δ is fed back through an element with transfer function $sK_{\dot{m}}\hat{Y}_c$. The $\hat{Y}_c$ represents an internal model of the vehicle dynamics. As opposed to the model of Fig. 3a, the model of Fig. 3b can be considered as a model for the human pilot for compensatory control since error information, alone, serves as an external stimulus to the pilot and since the inner feedback loop utilizes a control input that can be sensed proprioceptively [6]. As Smith pointed out, by allowing $\hat{Y}_c = Y_c$ and by considering three stereotypical controlled elements, $Y_c = K_c$, $Y_c = K_c/s$, and $Y_c = K/s^2$, three open-loop pilot–vehicle transfer functions

$Y_p Y_c$ can be obtained:

$$Y_p Y_c = \frac{K_e K_c}{(K_m K_c s + 1)} = \frac{K_e K_c/(1 + K_m K_c)}{s} = \frac{(K_e K_c)s}{(s + K_m K_c)}. \tag{2}$$

By proper selection of the gains K_e and $K_{\dot{m}}$, the open-loop transfer function $Y_p Y_c$ can be made similar to the crossover model of (1), where the simple model of Fig. 3b does not yet include a pilot time delay or neuromuscular dynamics.

Although the model of Fig. 3b has been obtained by relying on simple feedback principles, the general structure is very similar to that proposed by other researchers to describe human behavior in man–machine systems with the human operating on visually sensed errors [6]. It is possible to generalize Fig. 3b and develop a more complete model of the human pilot; that is, one which includes a time delay and neuromuscular system dynamics. Such a model is shown in Fig. 4 and will be referred to as the structural model of the human pilot.

The model of Fig. 4 has been divided into "central nervous system," "neuromuscular system," and "vestibular system" components—a division intended to emphasize the nature of the signal processing activity involved. System error $e(t)$ is presented to the pilot and multiplied by the gain K_e. If motion cues are available, output rate is assumed to be sensed by the

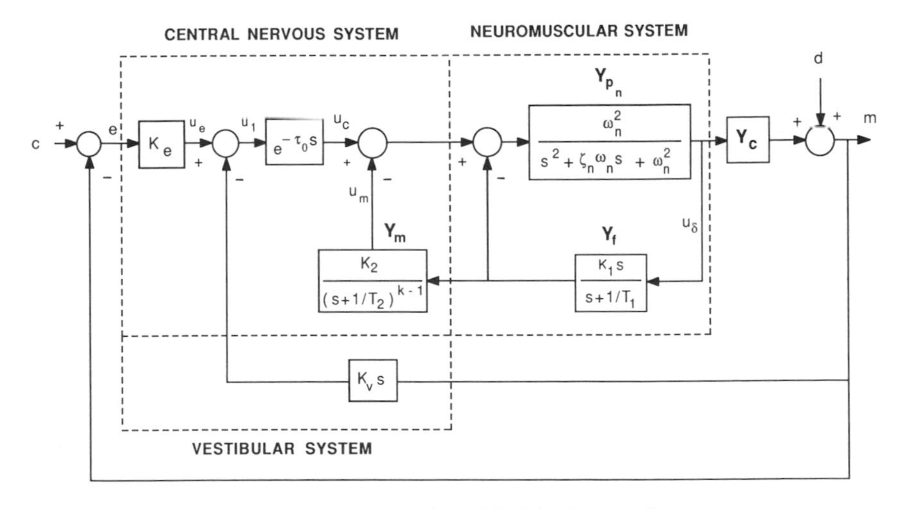

Fig. 4. The structural model of the human pilot.

vestibular system, multiplied by the gain K_v, and subtracted from the signal u_e. The resulting signal u_1 is passed through a central time delay of τ_0 sec, included to account for the effects of latencies in the visual process sensing $e(t)$, motor nerve conduction times, etc.

The signal $u_c(t)$ provides a command to a closed-loop system, which consists of a model of the open-loop neuromuscular dynamics Y_{p_n} of the particular limb driving the manipulator (control stick) and elements Y_f and Y_m, which emulate, in approximate fashion, the combined effects of the muscle spindles, Golgi tendon organs, joint-angle receptors, and the dynamics associated with higher level signal processing [13]. The form of Y_m is determined by the order k of the controlled-element dynamics Y_c in the region of the crossover frequency ω_c. Thus k is the integer that is obtained by representing Y_c as

$$Y_c = K_c/(j\omega)^k. \tag{3}$$

It should be noted that, although the u_m/u_δ transfer function in Fig. 4 is not identical to that in Fig. 3 at all frequencies, the two are very similar in form in the region of frequency crossover for parameter values that yield acceptable matches between model and experiment [13]. For example, consider $Y_c = K_c/s^2$. Here, $k = 2$, and from Fig. 4, with $K_v = 0$ (no motion),

$$\frac{u_m}{u_\delta} = \frac{K_1 K_2 s}{[s + (1/T_1)][s + (1/T_2)]}. \tag{4}$$

If $1/T_1$ and $1/T_2$ are significantly less than the crossover frequency ω_c (which they will be for acceptable matches to measured human pilot transfer functions [13]), then in Fig. 4

$$u_m/u_\delta \cong K_1 K_2/s \tag{5}$$

around ω_c, and in Fig. 3

$$u_m/u_\delta = (1/K_{\dot{m}})/s. \tag{6}$$

This means that the signal u_m in Fig 4 will be proportional to output rate due to control activity and can be considered a candidate for a handling qualities metric in Smith's theory.

III. HANDLING QUALITIES METRIC

Hess and Sunyoto [14] interpreted Smith's handling qualities theory by using the structural model of Fig. 4. They employed values for the structural model parameters that yielded model-generated transfer function Y_p, which closely matched measured human operator transfer functions for six different controlled elements (Y_c); these could then be confidently rank ordered in terms of level of tracking difficulty. Then, selecting the gain K_e so that a common crossover frequency of 2.0 rad/sec was evidence for all the $Y_p Y_c$s, they

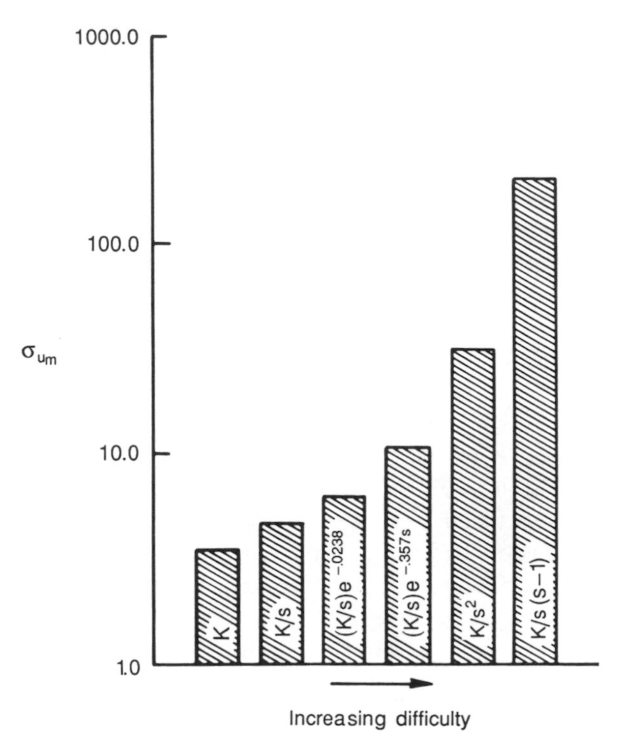

Fig. 5. Root-mean-square value of u_m for six controlled elements of varying tracking difficulty. (Reprinted from [13].)

demonstrated that the root-mean-square value of the signal u_m (σu_m) varied in monotonic fashion with the predetermined level of tracking difficulty when a common command input was used. Figure 5 shows the results of this study. As the controlled elements ranged from an easily controlled pure gain ($Y_c = K$) to an unstable second-order system $\{Y_c = K/[s(s-1)]\}$, the σu_m values increased as shown in Fig. 5.

The approach just summarized can be considered an interpretation of Smith's theory since it paralleled the modeling approach he undertook using the OCM as reported in [12]. Here, σu_m is a representation of the rate tracking activity that occurs when the human operator or pilot completes a given task with a specified level of performance. It is the specification of desired and adequate levels of performance that can be the source of difficulty in analytical handling qualities studies, such as that being discussed here. Such performance criteria have often not been explicitly reported in handling qualities studies in the past. This leaves the analyst with the task of trying to estimate just what the desired and/or adequate performance might have been. The alternative chosen by this author is to use a common performance capability in the systems and tasks being modeled by forcing the open-loop crossover frequency (or its approximate equivalent, the closed-loop bandwidth) to remain invariant across any group of controlled elements whose handling qualities are being evaluated. This was the reason for the selection of the 2.0 rad/sec crossover frequency in the study of [14].

Hess [15] extended the results of [14] to more realistic simulations of pilot–vehicle systems, including multiloop tasks and related $\sigma^2 u_m$, to the numerical pilot ratings received by the configurations and tasks in those simulations. The square of σu_m, rather that σu_m, itself, was chosen merely out of convenience. A total of 35 vehicle configurations, which were evaluated in manned simulations, are analyzed in [15]. The tasks ranged from simple single-axis single-loop pitch attitude tracking to precision hover and to landing approach. In the latter two tasks, control of both vehicle attitude and position was required. As opposed to the experimental data analyzed in [14], those used in [15] did not include measured values of pilot transfer functions. This meant that some method for the *a priori* selection of structural model parameters had to be employed, and in multiloop tasks, appropriate outer-loop pilot compensation had to be selected.

The pilot–vehicle systems for the tasks studied in [15] can be summarized by Figs. 2 and 6. In each of these figures, the inner-most loop is referred to as a primary control loop (i.e., a control loop that involves direct manipulative inputs by the pilot). In Figs. 2 and 6, Y_{p_1} represents pilot compensation modeled by the structural model of the human pilot, whereas in Fig. 6, Y_{p_2} represents pilot compensation modeled by a proportional-integral-derivative (PID) controller, where the form of the PID controller is chosen so that K/s-like open-loop characteristics are obtained (with the primary loop closed) about a crossover frequency that is a factor of 4 below that of the primary-loop closure [8]. Corroborative measurements have been reported that

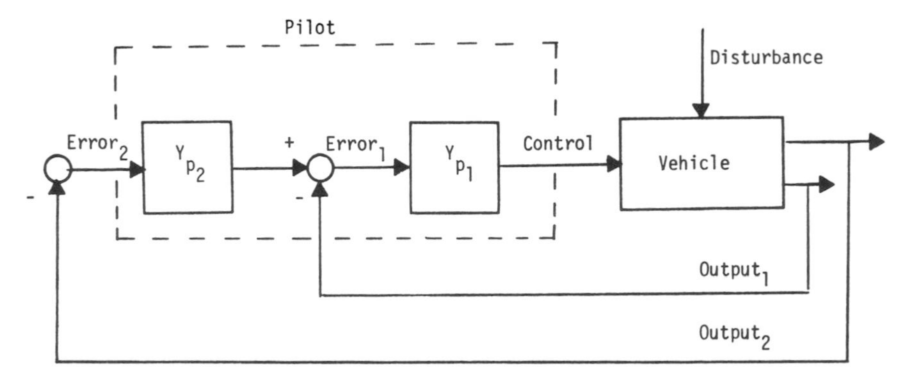

Fig. 6. A multiloop pilot–vehicle system.

TABLE I. NOMINAL PARAMETER VALUES FOR STRUCTURAL MODEL

k	K_e	K_1	K_2	T_1 (sec)	T_2 (sec)	τ_0 (sec)	ζ_n	ω_n (rad/sec)
0	1.0	1.0	2.0	5.0	[a]	0.15	0.707	10.0
1	1.0	1.0	2.0	5.0	[b]	0.15	0.707	10.0
2	1.0	1.0	10.0	2.5	[a]	0.15	0.707	10.0

[a] Selected to achieve K/s-like crossover characteristics.
[b] Parameter not applicable.

support such outer-loop compensation on the part of the pilot [16]. Determination of the values of the ten model parameters can be simplified as follows: First, no motion cues were involved in the manned simulations being analyzed, so $K_v = 0$. Also, in most cases of engineering interest, five of the remaining nine parameters (K_e, K_1, τ_0, ζ_n, ω_n) can be considered fixed at nominal values, three parameters (k, K_2, and T_1) can be determined once the order of the vehicle dynamics (Y_c) in the frequency region of crossover is known, and the final parameter (T_2) can be determined by requiring that the inner-loop pilot vehicle dynamics $Y_{p_1} Y_c$ conform to the crossover model of (1). This information is summarized in Table I.

A. PILOT MODEL FORMULATION

Setting up the structural model for the primary control loop and the PID compensation for any outer loop for the handling qualities investigation to be discussed can be outlined in the following steps.

1. The task is analyzed and appropriate pilot loop closures are selected (e.g., Figs. 2 and 6).
2. The vehicle transfer function Y_{c_1} for the primary control loop is obtained.
3. A primary-loop crossover frequency is selected.
4. Given the primary-loop vehicle transfer function and the specified crossover frequency, the parameter k in the structural model of Fig. 4 is selected. The value of k will depend on whether gain ($k = 1$), lead ($k = 2$), or lag ($k = 0$) compensation is required, as dictated by the crossover model of the pilot and the form of Y_c around the crossover frequency.
5. The remaining structural model parameters, save T_2, are selected from Table I.
6. The parameter T_2 is chosen to ensure K/s-like pilot–vehicle characteristics ($Y_{p_1} Y_c$) around the primary-loop crossover frequency.

7. For reasons to be discussed in the following, it is mandatory to remove control system sensitivity as a variable in the handling qualities investigation. To accomplish this, the analyst adjusts the primary-loop sensitivity so that

$$K_s = \frac{1}{|Y_{p_1}(j\omega)|_{\omega=\omega_c}} \cdot \frac{1}{|Y_c(j\omega)|_{\omega=\omega_c}}. \tag{7}$$

This equation ensures that the desired crossover frequency ω_c will be obtained given the model parameter values of Table I, including $K_e = 1.0$.

8. If the task is multi loop in nature, the pilot compensation in the outer control loop is selected as follows:
   a. The appropriate outer-loop vehicle transfer function is obtained.
   b. The crossover frequency for the outer loop is reduced by a factor of four from that for the inner loop.
   c. Pilot compensation in the outer loop is kept very simple and limited to PID dynamics appropriate for creating K/s-like crossover characteristics for the outer, open-loop transfer function with the inner loop closed.

9. The pilot–vehicle system is simulated using the disturbance and/or command signals employed in the experimental data base for which pilot opinion ratings are available and $\sigma^2 u_m$ is obtained.

It is interesting to compare these steps to those offered in [7] for single-loop tasks:

1. Determination of the effective controlled element dynamics.
2. Estimation of the pilot-adopted describing function form.
3. Adjustment of parameters in the plot-describing-function form consistent with all loop closure rules.
4. Estimation of the closed-loop system dynamics, including the compatibility of $\omega_c > \omega_{c_0}$.
5. Estimation of differential opinion degradation from nonequalized pilot situations.

The general similarity between these steps and those offered previously are noteworthy. The ω_{c_0} mentioned in step (6) is the crossover frequency that the human pilot adopts as the bandwidth of the command input approaches zero.

B. AN EXAMPLE

An example of applying steps (1–9) from the previous section for a particular aircraft configuration and task will now be presented. The manned simulation from which pilot ratings were obtained is that presented in [17]. These nine steps can be summarized for the analysis of configuration PH-39

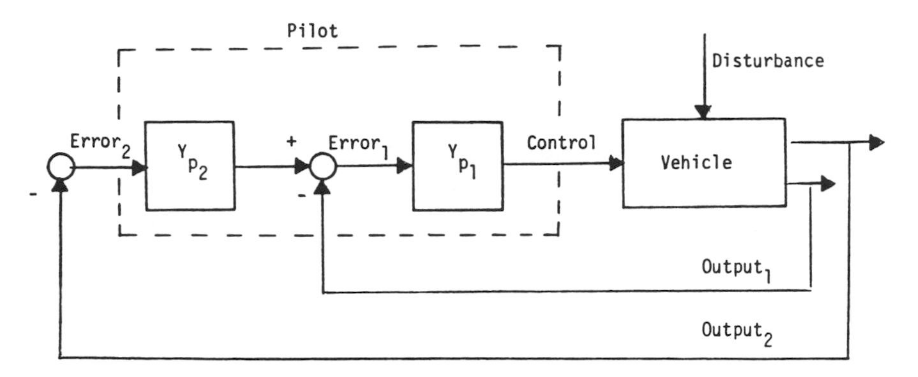

Fig. 7. Hypothesized pilot loop closures for longitudinal hover task. Control is longitudinal cyclic; Y_{p_1} is the structural model of Fig. 1; $Y_{p_2} = K_y(T_L s + 1)$; output$_1$ = pitch attitude θ; output$_2$ = longitudinal vehicle position x; the vehicle = hovering helicopter; and disturbance = longitudinal gust velocity u_g.

from [17] as follows:

1. Figure 7 shows the pilot loop closures for the longitudinal hover task. The primary loop involves pitch attitude command following and regulation, with an outer vehicle position loop closed around the attitude loop.
2. Figure 8a shows the Bode plot of the primary-loop vehicle transfer function for the configuration.
3. A primary-loop crossover frequency of 3.0 rad/sec is selected.
4. Pilot lead compensation is required in order to obtain crossover model characteristics in the region of crossover. The lead time constant is $\sim\frac{1}{3}$ sec and $k = 2$.
5. Table I gives the structural model parameters, save T_2, associated with $k = 2$.
6. T_2 is chosen as 0.33, which places a zero in the structural model at $s = -1/T_2$ and cancels a pole in the aircraft hovering cubic characteristic polynomial. This leads to the desired broad K/s-like region in the open-loop Bode plot of Fig. 8b.
7. The primary-loop control sensitivity is chosen via (7) as $K_s = 123.4$
8. The appropriate outer-loop transfer function for this task can be given as

$$x/\theta = -g/s(s - X_u). \tag{8}$$

The outer-loop crossover frequency is selected as 0.75 rad/sec, which is one-fourth the inner-loop value. The required outer-loop PID

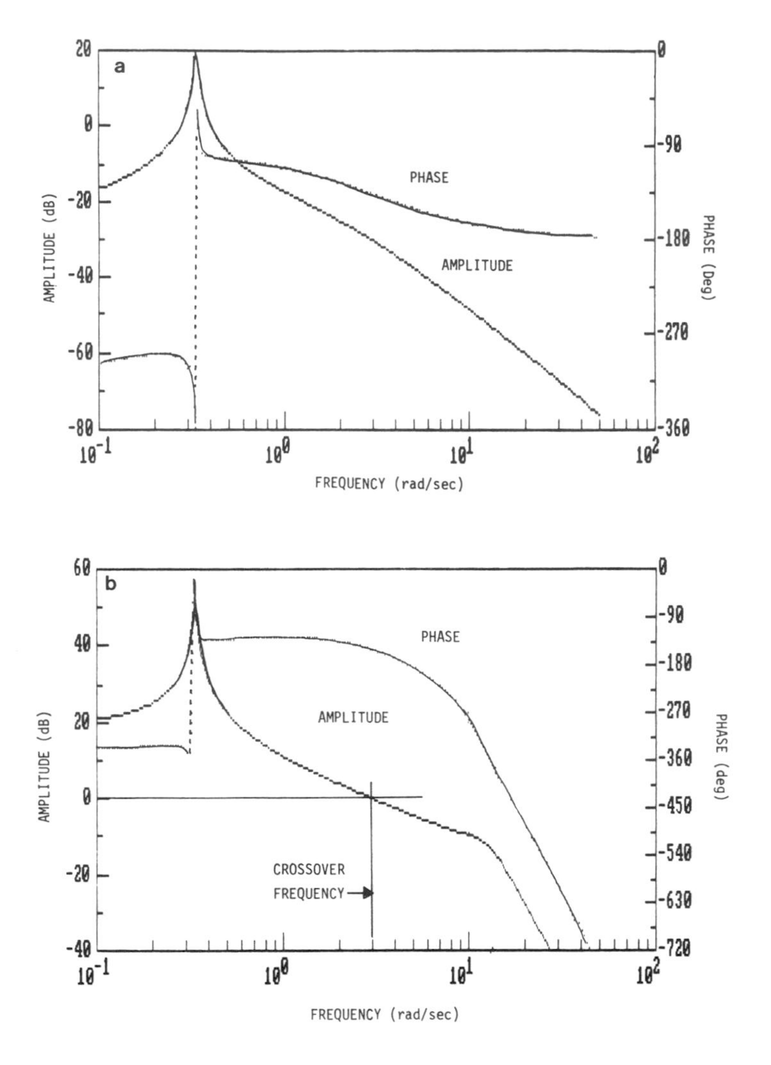

Fig. 8 (a) Y_c for primary loop of longitudinal hover task and (b) $Y_{p_1} Y_c$ for primary loop of longitudinal hover task: PH-29 configuration from [17].

compensation is

$$Y_{p_2} = K_y(T_L s + 1) = -0.0071(3.0s + 1). \tag{9}$$

9. The pilot–vehicle system is then simulated and the value for the handling qualities metric was obtained as $\sigma^2 u_m = 0.17$, with structural model output in the primary loop expressed in degrees of pitch attitude. The reported Cooper rating 4.0 in [17].[1]

C. RESULTS

The procedure outlined in the previous section was applied to 35 configurations whose handling qualities have been evaluated in four manned simulation studies. These included eight configurations from the single-loop laboratory pitch attitude tracking task of McDonnell [18], nine configurations from the single-loop pitch attitude tracking task of Arnold [19], nine configurations from the multiloop longitudinal precision hover task of Miller and Vinje [17], and nine configurations from the multiloop glide-slope tracking task of Franklin [20], as reported by Picha [21]. In each of the simulations, configurations were chosen for analysis with an eye toward predicting as large a range of reported pilot ratings as possible. Table II summarizes the configurations analyzed here. The notation is taken from the references describing the simulations.

TABLE II. VEHICLE CONFIGURATIONS ANALYZED

Miller and Vinje [17]	McDonnell [18][a]	Arnold [19]	Franklin [20] (after Picha [21])
PH-1	$1/s$ (1, 2)	1D	2
PH-13	$1/s(s+1)$ (1, 2)	2D	5
PH-27	$1/s(s+2)$ (1, 2)	3A	11
PH-28	$1/s(s+4)$ (1, 2)	4A	12
PH-29	$1/s^2$ (1, 2)	5A	15
PH-32	$1/s^2$ (2, 2)	10	22
PH-33	$1/s^2$ (2, 3)	11	24
PH-34	$1/(s^2 + 2(0.7)(16)s + 16^2)$ (1, 2)	12	28
PH-37		14	29

[a] Numbers in parentheses refer to pitch command characteristics.

---

[1] The manned simulations of [17] used the older Cooper pilot rating scale as opposed to the Cooper–Harper rating scale that succeeded it. For the purposes of this study, the ratings from this scale were assumed to be identical to those expressed on the Cooper–Harper scale.

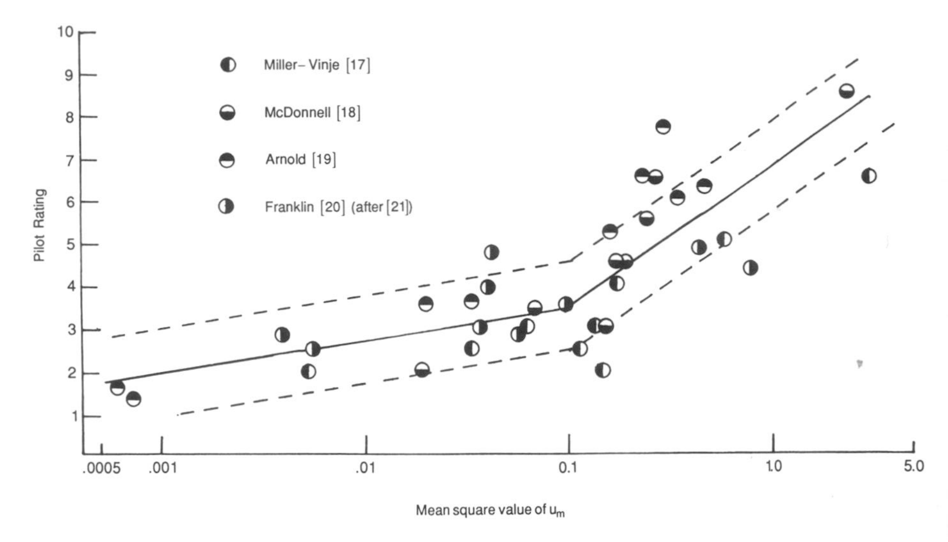

Fig. 9. Pilot ratings versus mean square value of u_m.

Figure 9 shows a plot of $\sigma^2 u_m$ obtained from the modeling procedure versus the reported pilot opinion ratings for the 35 configurations analyzed. The solid line represents hand-faired fit to the results, while the dashed lines represent ± 1 pilot ratings. Approximately 80% of the points lie within the ± 1 rating band. In all cases, the output of the primary control loop is expressed in degrees of pitch attitude. In the case of the McDonnell data the output of the controlled element dynamics in the laboratory tracking task was described in terms of centimeters of display indicator movement on an oscilloscope screen at a specified viewing distance. Output in degrees of vehicle pitch attitude was obtained by considering that the angle in degrees subtended at the pilot's eye by pitch display indicator movement was equivalent to vehicle pitch attitude in degrees. Thus, the display was interpreted as one would a contact analog format. For the pitch attitude command-following and regulation tasks that define the single-loop tasks, the primary-loop crossover frequency was chosen as 4.0 rad/sec, while for the multi loop precision hover and landing approach tasks, the primary control loop crossover frequency was 3.0 rad/sec. In each case, these values were approximately the largest crossover frequencies attainable with closed-loop stability for all the configurations studied.

The results of Fig. 9 indicate that $\sigma^2 u_m$ may indeed be a sensitive measure of the amount of pilot compensation required in the tasks that have been selected for analysis. It can be shown that the mean square value of system error $e(t)$ also correlates well with pilot opinion ratings [15]. The question arises then as to why $\sigma^2 u_m$ should be considered as the handling qualities metric in this methodology. The answer lies in the fact that tracking error is clearly not a measure of pilot compensation. It may reflect the relative efficacy of such compensation, but theoretically it is an effect rather than a cause. This argument is reinforced in the next section. Experimental evidence has shown that the relationship between tracking error and handling qualities is essentially nonexistent when control sensitivity is varied.

IV. CONTROL SENSITIVITY

It is known that, for any particular vehicle dynamics, control manipulator, and task, a control system sensitivity exists that humans find optimum in that larger or smaller values result in degraded pilot opinion ratings. In addition, experimental evidence suggests that the human pilot is quite capable of maintaining nearly constant crossover frequency and tracking performance across a wide range of control sensitivities. In the following the subject of control sensitivity is discussed using the structural model and K/s^2 vehicle dynamics.

Taking into consideration the limitations of the human neuromuscular system in order to model the very precise, low-amplitude, control inputs required for tracking with high control sensitivities, a small residual broad-band motor noise was injected at the summing junction in Fig 4 where u_m was subtracted from u_c. The mean square value of this noise was selected as 1% of the value of the mean square value of u_c that existed when a nominal sensitivity was being used. Here nominal refers to that value selected in (7). With nominal or low control sensitivities, this noise has practically no effect on simulated pilot–vehicle behavior with the structural model. However, at high sensitivities, the residual noise will introduce inaccuracies in the control inputs, which will lead to significant increases in mean square tracking error, barring some sort of pilot adaptation.

The K/s^2-controlled element dynamics used in [18] have been selected for study in this section. In [18], control sensitivities were investigated from 0.1 K_B to 10 K_B, where K_B refers to the best or optimum sensitivity as determined by the subjects involved in the simulation. To begin, assume that the sensitivity in (7) is optimum (i.e, $K_s = K_B$). If one considers decreasing this to 0.1 K_B, the model (and by hypothesis, the human pilot) can compensate by a reciprocal increase in the gain K_e from the nominal value of 1.0 to 10. This results in a maintenance of the 4.0-rad/sec crossover freequency and no change in the tracking performance from that associated with the optimum sensitivity, but a large increase in the value of the handling qualities metric $\sigma^2 u_m$ as compared to that for the optimum sensitivity. According to the results of Fig. 9, this increase would indicate larger pilot opinion ratings.

If one considers increasing the sensitivity to 10 K_B, the model (and by hypothesis, the human pilot) can compensate by a reciprocal decrease in the gain K_e from the nominal 1.0 to 0.1. However, while such a decrease maintains a constant crossover frequency, there is a significant deterioration in tracking performance brought about by the residual motor noise. Since such performance deterioration has not, in general, been noted in experiment, some other mechanism for coping with large control sensitivities must be pursued with the model (and by hypothesis, by the human). One such mechanism is an increase in the gain K_1 in both proprioceptive feedback loops of the structural model of Fig. 4. This increase reduces the amplitude ratio of the transfer function u_δ/u_c, which has the desired result of reducing the effects of the motor noise while not changing the basic compensation provided by the structural model. Although varying the gain K_2, as proposed in [14] can produce similar results, much larger changes occur in the closed-loop neuromuscular system roots. In a physiological sense, increasing K_1 represents an increase in the average tension of the agonist–antagonist muscle groups driving the manipulator.

The large decrease in the amplitude ratio of the u_δ/u_c transfer function, which occurs when K_1 is increased, means that the gain K_e must now be increased to maintain the desired crossover frequency. This overall effect of

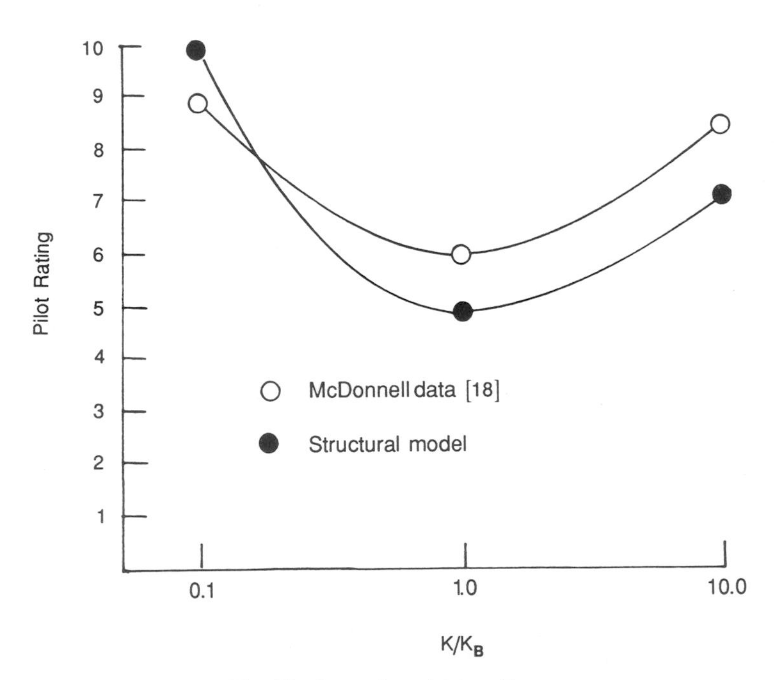

Fig. 10. Control sensitivity effects.

these changes is the maintenance of the 4.0-rad/sec crossover frequency, no deterioration in tracking performance, and, as was the case with the reduced control sensitivity, a large increase in the value of the handling qualities metric $\sigma^2 u_m$ as compared to that for the optimum sensitivity. It should be noted that the choice of the mean square value of the motor noise was arbitrary here, the only stipulation being that it be small relative to the mean square value of the signal u_c when no noise was present. Figure 10 shows pilot opinion ratings from the McDonnell study for the K/s^2-controlled element dynamics for variations in controlled element sensitivity over a range of 100. Also shown are results from the structural model simulations in which $\sigma^2 u_m$ and the solid line in Fig. 10 were used to generate pilot opinion ratings. For the high sensitivity, the model gain K_1 was increased form 1.0 to 25, which significantly reduced the effects of motor noise on tracking performance. The gain K_e was then increased from 1.0 to 2.4 to maintain the desired crossover frequency. It should be emphasized that the model-generated ratings cannot be considered as predictions because of the arbitrary manner in which the mean square value of the motor noise was assigned. It is for this reason that control system sensitivity effects have been eliminated from the modeling procedure via (7). Nonetheless, the structural model has provided a framework within which to interpret Smith's theory in the analysis of the effect of control system sensitivity on pilot opinion ratings.

V. APPLICABILITY OF THE METHODOLOGY

The methodology discussed in this section is currently limited in its applicability because of the appearance of vehicle attitude as the only type of error variable, which appeared in the primary control loop of the structural model that was used in generating Fig. 9. The author has not normalized, or perhaps more appropriately nondimensionalized, the $\sigma^2 u_m$ values in generating Fig. 9. Thus, it may not be possible at present to apply the methodology to a task in which control of a variable, other than vehicle attitude, is the purpose of the primary control loop. Thus, in terms of single-loop tasks, it would be very desirable to be able to normalize each $\sigma^2 u_m$ value by a unique task parameter such as the square of the maximum desirable primary loop error as outlined in the particular task description. Unfortunately, as was pointed out in the Section IV, such performance criteria were not always documented in the data used in this study. Nonetheless, future development of the methodology discussed herein will pursue this approach.

In multiloop tasks, this approach would have to be modified since desired task performance will almost always be specified in terms of outer-loop variables rather than primary-loop variables. For example, in the precision hover task studied here, desired performance would normally be specified in terms of maximum displacement of the vehicle from a desired hover point. This quantity would have to be translated into an attitude performance criterion with which to normalize $\sigma^2 u_m$ for the primary control loop. Means of accomplishing this are currently being investigated.

The necessity for the *a priori* specification of primary- and outer-loop crossover frequencies to interpret desired performance in terms of the structural model is still a necessary part of the methodology. A pair of alternatives can be briefly mentioned. The first is to call upon an optimization procedure to find the lowest primary-loop crossover frequency in some limited frequency range at which the difference between simulated and desired pilot–vehicle performance is minimized. In multiloop tasks, outer-loop crossover frequencies would be maintained a factor of 4 below that of the primary loop in this process. Again, this presupposes that specific and quantitative desired performance levels have been made part of the task description as required by the Cooper–Harper pilot rating scale. The second alternative is to choose primary-loop crossover frequency so that some minimum gain or phase margin is in evidence in each case. The author's arguments against these procedures are as follows: First, from the standpoint of flight safety, disturbance rejection, and rapid response to internally generated commands (often referred to as open-loop control), pilots probably desire high-bandwidth primary-loop closures in addition to the specific performance demands of the task at hand. Second, in handling qualities assessment, selecting primary-loop crossover frequencies solely on the basis of gain and phase margins neglects task performance requirements entirely.

VI. CONCLUSIONS

A theory for aircraft handling qualities has been interpreted in terms of a structural model of the human pilot. In this theory the mean square value of a signal with hypothetical origins in the human proprioceptive system has been shown to be a viable handling qualities metric to quantify the amount of pilot compensation which is required in a given task.

Data from manned simulations involving 35 aircraft configurations in both single and multiloop tasks have been used in the development of the methodology. This methodology provides a framework within which to interpret the pilot's well-documented preference for an optimum control system sensitivity.

Future work will concentrate on extending the applicability of the methodology, particularly in finding a consistent approach to normalizing the handling qualities metric.

REFERENCES

1. G. E. COOPER and R. P. HARPER, "The Use of Pilot Rating in the Evaluation of Aircraft Handling Qualities," NASA TN D-5153 (1969).
2. R. P. HARPER, and G. E. COOPER, "Handling Qualities and Pilot Evaluation," *AIAA J. Guidance, Control Dyn*, **9**, 515–529 (1986).
3. R. C. VAN DE GRAFF, "Investigation of Pilot Workload Measuring Techniques under In-Flight Task Conditions," *Proc. 1986 IEEE Int. Conf. Syst., Man, Cybern*, pp. 897–901 (1986).
4. D. McRUER and D. GRAHAM, "Eighty Years of Flight Control: Triumphs and Pitfalls of the Systems Approach," *AIAA J. Guidance Control* **4**, 353–362 (1981).
5. A. TUSTIN, "The Nature of the Operator's Response in Manual Control and Its Implication for Controller Design," *J. IEE* **94**, Part IIA, No. 2 (1947).
6. D. T. McRUER, "Human Dynamics in Man–Machine Systems," *Automatica* **26**, 237–253 (1980).
7. D. T. McRUER, I. L., ASHKENAS, and C. L. GUERRE, "A Systems Analysis View of Longitudinal Flying Qualities," WADD TR 60-43 (1960).
8. I. L. ASHKENAS, "Twenty-Five Years of Handling Qualities Research," *J. Aircr.* **21**, 289–301 (1984).
9. D. T. McRUER and E. S. KRENDEL, "Mathematical Models of Human Pilot Behavior," AGARD No. 188 (1974).
10. R. O. ANDERSON, "A New Approach to the Specification and Evaluation of Flying Qualities," AFFDL-TR-69-120 (1970).
11. R. A. HESS, "Prediction of Pilot Opinion Ratings Using an Optimal Pilot Model," *Hum. Factors* **19**, 459–475 (1977).
12. R. H. SMITH "A Unifying Theory for Pilot Opinion Rating," *Proc. 12th Annu. Conf. Manual Control*, NASA TM X-73, 170, pp. 542–558 (1976).
13. R. A. HESS, "A Model-Based Theory for Analyzing Human Control Behavior," *in* "Advances in Man–Machine Systems Research" (W. B. Rouse, ed.), Vol. 2, pp. 129–175. JAI Press, Greenwich, Connecticut, 1985.
14. R. A. HESS and I. SUNYOTO, "Toward a Unifying Theory for Aircraft Handling Qualities," *AIAA. J. Guidance, Control Dyn.* **8**, 440–446 (1985).
15. R. A. HESS, "A Theory for Aircraft Handling Qualities Based Upon a Structural Pilot Model," *AIAA J. Guidance, Control Dyn.* **12**, 792–797 (1989).
16. R. A. HESS and A. BECKMAN, "An Engineering Approach to Determining Visual Information Requirements for Flight Control Tasks," *IEEE Trans. Syst., Man, Cybern.* **SMC-14**, 286–298 (1984).
17. D. P. MILLER and E. W. VINJE, "Fixed-Base Flight Simulator Studies of VTOL Aircraft Handling Qualities in Hovering and Low-Speed Flight," AFFDL-TR-67-152 (1966).
18. J. D. McDONNELL, "Pilot Rating Techniques for the Estimation and Evaluation of Handling Qualities," AFFDL-TR-68-76 (1968).
19. J. D. ARNOLD, "An Improved Method of Predicting Aircraft Longitudinal Handling Qualities Based on the Minimum Pilot Rating Concept," M.S. Thesis, Air Force Institute of Technology, GGC/MA/73-1 (1973).
20. J. A. FRANKLIN, "Turbulence and Longitudinal Flying Qualities," NASA CR-1821 (1971).
21. D. G. PICHA, "Predicting Handling Qualities for Glide Slope Tracking Using the Minimum Pilot Rating Concept," M.S. Thesis, Air Force Institute of Technology, GAW/MA/74M-1 (1974).

CONTROL AND DYNAMIC SYSTEMS, VOL. 33

IDENTIFICATION OF PILOT–VEHICLE DYNAMICS FROM SIMULATION AND FLIGHT TEST

RONALD A. HESS

Department of Mechanical Engineering
Division of Aeronautical Science and Engineering
University of California
Davis, California 95616

I. INTRODUCTION

A. BACKGROUND

Manned simulation and flight test are obviously important research, development, and training tools in the aerospace industry. The costs of these endeavors make it mandatory that researchers be able to extract as much information as possible regarding the pilot–vehicle system from each simulation and flight test experiment. It is particularly useful to be able to identify pilot–vehicle dynamics as part of any simulation or flight test experiment. Since the control theoretic model of the human pilot [1] has become the fundamental mode of thinking on the part of researchers involved in analyzing pilot–vehicle systems, it is this model that is the object of most identification efforts.

The identification problem to be discussed in this article is one in which a basic feedback control structure, or *pilot control strategy*, has been hypothesized, and identification algorithms are employed to determine the particular form of pilot equalization in each feedback loop. Consider the simple piloting task of following a commanded pitch attitude. The pilot control strategy is summarized by the single-loop feedback structure of Fig. 1a, where a describing function model of the human pilot is represented by a

Copyright © 1990 by Academic Press, Inc.
All rights of reproduction in any form reserved.

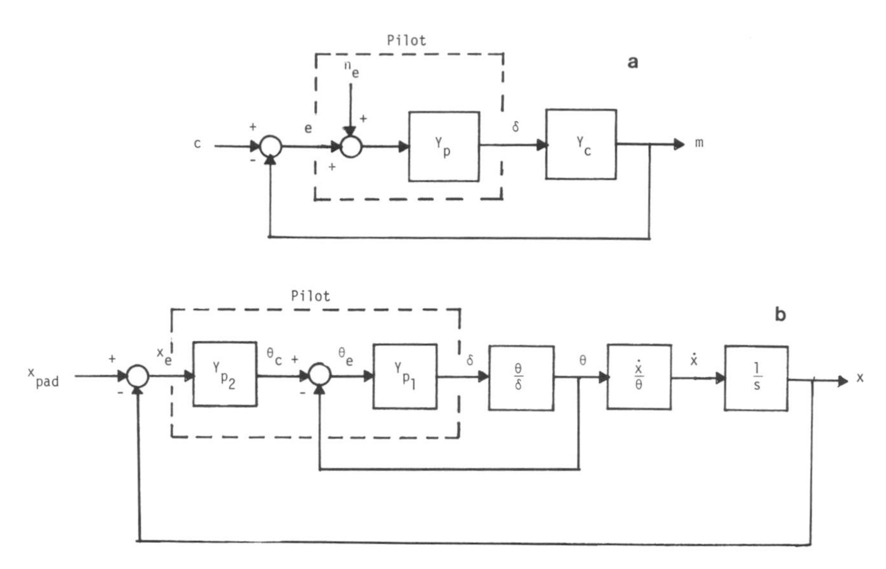

Fig. 1. (a) Single-loop pilot–vehicle system. (b) Multiloop pilot–vehicle system for hover task.

linear transfer function Y_p, and a remnant signal $n_e(t)$. The remnant is that portion of the pilot's output $\delta(t)$ that is not linearly correlated with the input. Although the determination of remnant can be an important part of the pilot modeling procedure, it is the linear transfer function Y_p or the open-loop transfer function $Y_p Y_c$ that will be the focus of attention in the identification techniques to be described in this article. These transfer functions are very useful in assessing the effects of display and flight control system designs on pilot dynamics, and, in turn, on work load and performance.

Although only the simplest of piloting tasks are amenable to the description of Fig. 1a, this single-loop structure can serve as a primary control loop for modeling more complex pilot–vehicle tasks. A primary control loop is defined here as one involving direct manipulative input by the pilot (see Hess, "Methodology for the Analytical Assessment of Aircraft Handling Qualities," this volume). Consider Fig. 1b, which could serve as a representation of a longitudinal helicopter hover task. Pilot dynamics are contained in the transfer functions denoted Y_{p1} and Y_{p2}. Remnant signals have been omitted. Here, the primary control loop is one involving pitch-attitude control using longitudinal cyclic inputs. With this inner loop closed, the effective vehicle for the outer loop is given by x/θ_c. Such a series-loop structure is very useful for analyzing more complex piloting tasks than that shown in Fig. 1a and is typically referred to as multiloop. Examples of pilot–vehicle identification in both single-loop and multiloop tasks will be encountered in the following sections.

B. CROSSOVER MODEL

The well-known crossover model for pilot–vehicle dynamics states that in single-loop tasks involving random or random-appearing inputs or disturbances, such as that shown in Fig. 1a, the combination of the pilot and vehicle transfer function can be approximated by

$$Y_p Y_c = \omega_c[\exp(-\tau_e s)]/s \tag{1}$$

in a fairly broad frequency band about the open-loop crossover frequency ω_c [2]. The time delay τ_e is an effective delay and represents not only the integrated effect of actual pilot time delays, but also the phase lags attributable to higher frequency dynamics in the human neuromuscular system. Because of the wide applicability of the crossover model, and the manner in which a variety of data can be made to coalesce through its application, identification of human pilot dynamics in single-loop and multiloop tasks is often synonymous with measurements of $Y_p Y_c$.

II. IDENTIFICATION TECHNIQUES

A. FREQUENCY-DOMAIN APPROACH BY FAST FOURIER TRANSFORM

Given the model of Fig. 1a, spectral or generalized harmonic analysis techniques indicate that the transfer function $Y_p Y_c$ can be given by

$$Y_p Y_c(j\omega) = \Phi_{em}(j\omega)/\Phi_{ee}(j\omega), \tag{2}$$

where $\Phi_{em}(j\omega)$ is the cross-power spectral density between $e(t)$ and $m(t)$ and $\Phi_{ee}(j\omega)$ is the power spectral density of $e(t)$. There is a considerable simplification to be obtained if the command input is represented as a sum of sinusoids, that is,

$$c(t) = \sum A_i \sin(\omega_i t). \tag{3}$$

In this case, the ratio of spectral densities can be replaced by a ratio of Fourier coefficients as

$$Y_p Y_c(j\omega) = C_m(j\omega)/C_e(j\omega), \tag{4}$$

where C_m and C_e represent Fourier coefficients whose real and imaginary parts are defined as

$$\begin{aligned} \mathrm{Re}[C_m(j\omega_i)] &= \frac{1}{T_i}\int_0^{T_i} m(t)\cos(\omega_i t)\,dt \\ \mathrm{Im}[C_m(j\omega_i)] &= \frac{-1}{T_i}\int_0^{T_i} m(t)\sin(\omega_i t)\,dt \end{aligned} \tag{5}$$

where T_i is the period of the ith sinusoid. In implementing this technique, each constituent sine wave in (3) must have an integral number of cycles over the experimental run length, and none should be an integral multiple of another. Finally, the frequencies should be approximately evenly distributed on a logarithmic scale over the frequency range of interest for manual control (i.e., $0.1 < \omega < 20$ rad/sec). When the first requirement is met, one may replace the T_i appearing in (5) with the run length T. Fulfilling the second requirement ensures a random appearing input with an adequate number of constituent sinusoids (as few as five can suffice). Finally, fulfilling the last requirement allows one to fit a continuous transfer function to the measurements obtained at the discrete input frequencies in (3).

Figure 2a is an example of the results of frequency-domain measurements of $Y_p Y_c$ in a single-loop task like that in Fig. 1a [3]. In Fig. 2a, $Y_c = K[\exp(-0.0238s]/s$. The delay here was one deliberately introduced into the vehicle dynamics and should not be confused with the effective delay appearing in the crossover model of (1). The inclusion of 13 frequencies in the sum of sinusoids provides the command input for the task studied in [3]; and their approximate even distribution over a wide frequency range allows the author to fit a transfer function to $Y_p Y_c$, which is more detailed than the simple crossover model. This is shown in Fig. 2b, where the solid line represents $Y_p Y_c$ obtained when Y_p was generated by a structural model of the human pilot [4].

The fast Fourier transform (FFT) capitalizes on the symmetry of the sine and cosine functions and provides a computationally efficient means of obtaining the Fourier transforms needed in (5) [5]. By using the FFT, an additional requirement is added to that outlined in the preceding, namely, that the number of data points used in calculating the FFT must be expressible as a power of 2.

B. TIME-DOMAIN APPROACH BY LEAST-SQUARES ESTIMATION

A relatively straightforward time-domain approach to the problem of pilot–vehicle identification in tasks such as those just discussed can be realized

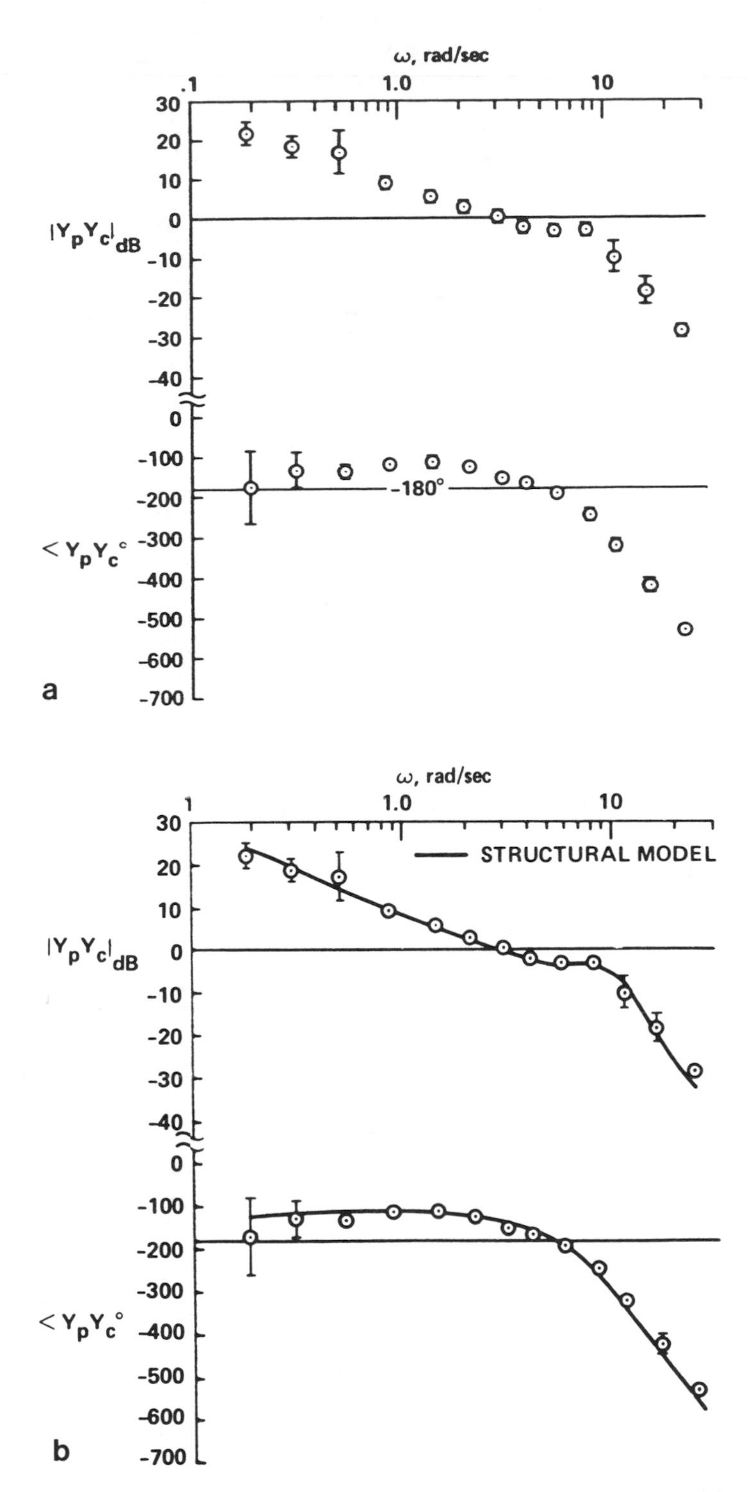

Fig. 2. (a) $Y_p Y_c$ FFT measurements. (b) Data of (a) fitted with structural model Y_p.

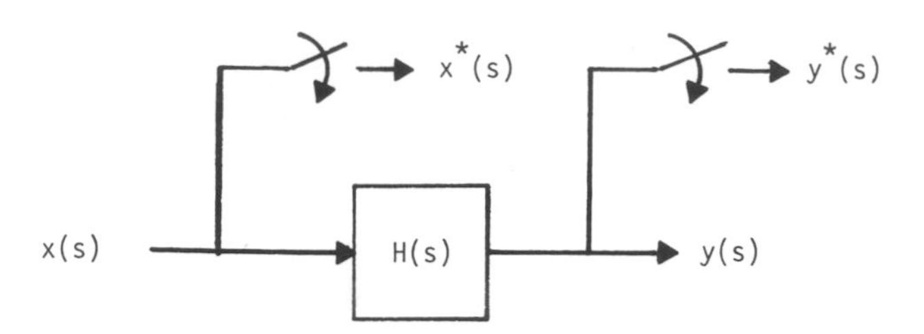

Fig. 3. The sampling process in LSE measurements.

by matching the output of the pilot–vehicle dynamics model to the corresponding output measured in the experiment. The matching process can be accomplished by a least-squares-estimation (LSE) procedure. Unlike the frequency-domain approach just outlined, the LSE approach requires an input–output model of the pilot–vehicle system with undetermined parameters. The least-squares estimates of these parameters are obtained by minimizing the sum of the squares of the errors between model output and measured output over the length of the tracking run. Also, unlike the frequency-domain approach, the LSE technique does not require a sum of sinusoids input or disturbance.

As an example, Fig. 3 shows an unknown pilot–vehicle system to be identified and indicated by $H(s)$, which might represent $Y_p Y_c$ in Fig. 1a. The LSE technique identifies the coefficients of the numerator and denominator of the z transfer function $G(z)$ defined as

$$G(z) = \mathfrak{Z}\left[\frac{1 - e^{-sT}}{s} H(s)\right], \tag{6}$$

where the term $[1 - \exp(-Ts)]/s$ represents the transfer function of the simplest data reconstruction device—the zero-order hold. Table I shows $G(z)$ for a variety of $H(s)$ transfer functions that could serve as models of elements of a pilot–vehicle system. The quality of the LSE identification is ascertained using the same correlation coefficient that defines the output matching process:

$$R^2 = 1.0 - [\sum(y_k - y'_k)^2]/[\sum(y_k)^2], \tag{7}$$

where y_k and y'_k are the measured output and model output, respectively.

Time delays, which are integer multiples of the sampling interval, can be included in the model to be identified. Thus, for example, if one chose a model form as

$$H(s) = Ke^{-ps}/s, \tag{8}$$

TABLE I. Z-PLANE MODEL STRUCTURES

Entry	$G(z)$	$H(s)$
1	$b_1 z^{-1}$	K
2	$\dfrac{b_1 T z^{-1}}{1 - z^{-1}}$ [a]	$\dfrac{K}{s}$
3	$\dfrac{b_1 z^{-1}}{1 - a_1 z^{-1}}$	$\dfrac{Ka}{s + a}$
4	$\dfrac{b_1 z^{-1} + b_2 z^{-2}}{1 - a_1 z^{-1} - a_2 z^{-2}}$	$\dfrac{K\omega^2}{[\zeta; \omega]}$
5	$\dfrac{b_1 z^{-1} + b_2 z^{-2}}{1 - a_1 z^{-1} - a_2 z^{-2}}$	$\dfrac{K(s + a)}{[\zeta; \omega]}$
6	$\dfrac{b_1 z^{-1} + b_2 z^{-2}}{1 - a_1 z^{-1}}$	$\dfrac{K(s + a)}{(s + b)}$
7	$\dfrac{b_1 z^{-1} + b_2 z^{-2}}{1 - a_1 z^{-1} - a_2 z^{-2}}$	$\dfrac{K(s + a)}{s(s + b)}$
8	$\dfrac{b_1 z^{-1} + b_2 z^{-2} + b_3 z^{-3}}{1 - a_1 z^{-1} - a_2 z^{-2} - a_3 z^{-3}}$	$\dfrac{K(s + a)}{(s + b)[\zeta; \omega]}$

[a] T is the sampling period.

where p is an integer multiple of the sampling interval, the corresponding $G(Z)$ would be

$$G(z) = [b_1 T z^{-(1+p)}]/[1 - z^{-1}]. \tag{9}$$

In the LSE procedure, the value of p would change from run to run and the quality of the model fits compared to using (7). The one yielding the value of R^2 closest to unity would represent the best fit.

Since pilot–vehicle dynamics are most easily interpreted in terms of continuous transfer functions in the s domain rather than discrete transfer functions in the z domain, the identified $G(z)$ needs to be related back to an $H(s)$. This is most easily accomplished by taking the w' transform of the identified $G(z)$ [6]. The resulting w' transfer function will closely resemble that of the actual continuous system $H(s)$ as long as the frequency range of interest is well below the sampling frequency. As will be seen, the identified z transfer function coefficients are often of little interest in themselves.

III. EXAMPLES

A. INTRODUCTION

The following examples illustrate the use of the identification techniques just outlined in some selected flight test and simulation studies. In each case the crossover model of the human pilot along with carefully hypothesized single-loop and multiloop control strategies form the basis for the identification work. Thus, the identification techniques serve as tools allowing simulation and flight test results to provide additional insight into the pilot modeling procedure.

B. SINGLE-LOOP TASK FROM FLIGHT TEST

The first example to be considered is a series of flight experiments in which data was generated from the following tasks: a F-14 "pursuer" aircraft was involved in a tail chase of a T-38 "target" aircraft in both level flight and in a 3-g windup turn at a nominal Mach number of 0.55 at an altitude of 10,000 ft with a nominal horizontal separation distance of 800 ft [7]. This flight-path geometry is sketched in Fig. 4a. The F-14 pilot was using a gunsight reticle on a head-up display (HUD), as indicated in simplified form in Fig. 4b. The task of the F-14 pilot was to keep the reticle centered on a point midway between the twin exhausts of the T-38 aircraft throughout the run. In addition to normal disturbances, such as atmospheric turbulence and the wake from the target aircraft, the reticle was driven in the pursuer aircraft $x-z$ plane by a sum of sinusoids. Although the sum of sinusoids was designed to include 11 sine waves in a frequency range from 0.147 to 15.978 rad/sec, an instrumentation problem caused the actual sinusoids to be generated at different frequencies. The frequencies were identified using the FFT technique and were given approximately by 0.18, 0.79, 1.00, 1.49, 2.49, 3.23, 5.09, and 7.0 rad/sec. The fact that the last frequency is an integral multiple of the third had minimal consequences on the results since the amount of power at the last frequency was quite small.

Figure 5 shows the geometry associated with pursuer vehicle attitude and lines of sight. Figure 6 is a simplified block diagram of the F-14 pilot–vehicle system pertinent to the task at hand. Here, β_d represents the sums of sinusoids acting as a reticle disturbance. Both FFT and LSE measurement techniques were called upon in the identification study to determine the pilot–vehicle transfer function $Y_p Y_c$ for these tasks. The sampling interval for the data acquisition was $\frac{1}{12}$ sec (0.08333 sec). The run length for data analysis purposes was 84.485 sec. For the LSE technique, two z transfer functions for identifying

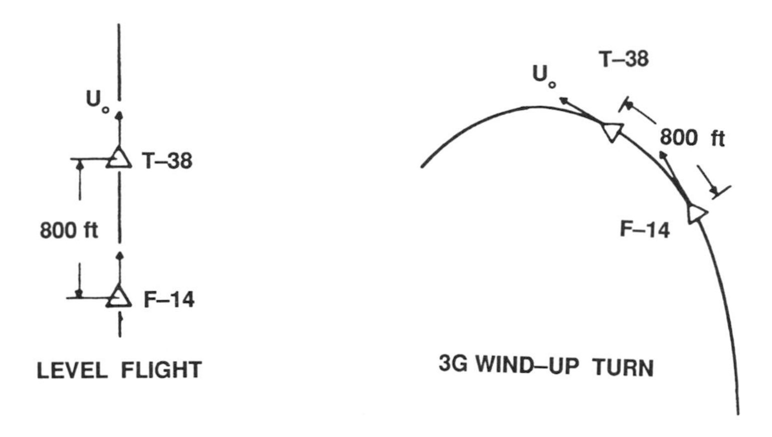

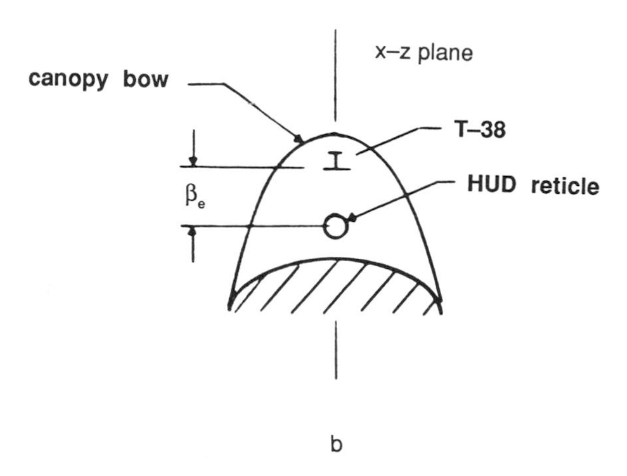

Fig. 4. (a) Tail–chase task geometry. (b) F-14 head-up display.

$Y_p Y_c$ were used: entries 5 and 8 in Table 1. A second-order denominator was included in these models to account for the phase lags associated with time delays, rather that appealing to the procedure outlined in the previous subsection in which time delays expressed as integer multiples of the sampling period were employed in an iterative identification procedure.

Figures 7 and 8 show the identification results for the level flight and 3-g turns for entry 8 in Table I. Table II summarizes the identified LSE

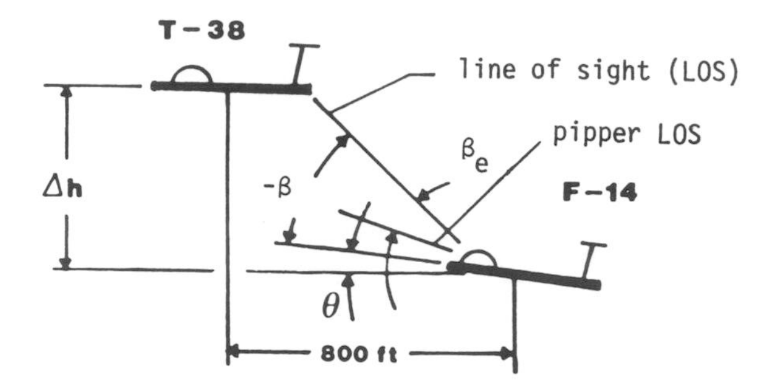

Fig. 5. Tail-chase task attitude and line of sight definitions.

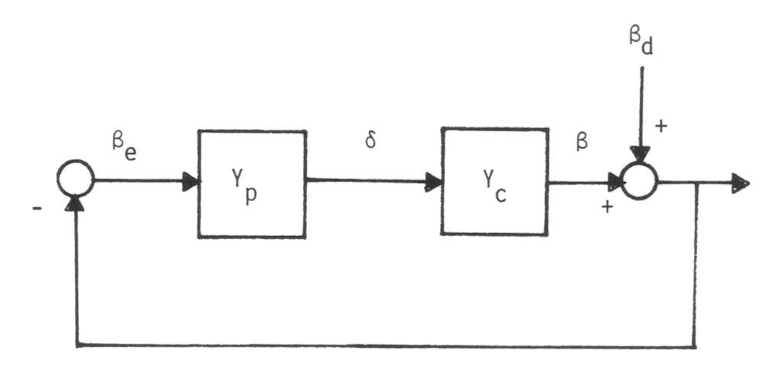

Fig. 6. Single-loop pilot–vehicle system for tail-chase task.

parameters. As mentioned in the previous subsection, these coefficient values are of little interest in themselves since it is the resulting w' transfer functions that provide the pertinent information regarding the form of $Y_p Y_c$ in the tasks. Note the very high correlation coefficient values indicated in the figures. The identification results represent the averages of 43 runs for each task. The vertical bars on the FFT results and the dashed lines on the LSE results both indicate plus or minus one standard deviation. Also included in the LSE identification is a bias term B, which is the entry in the last row of Table II. This term can account for any biases in the recorded data, nonzero "trim" conditions, or unaccounted command inputs to the system. The LSE results were obtained using a program referred to as NIPIP (nonintrusive parameter identification procedure) [8].

A cursory examination of the identification results in Figs. 7 and 8 indicates that they could be fitted with the crossover model of (1). Here crossover frequencies of 1–2 rad/sec are in evidence with effective time delays of ~ 0.2 sec. An interesting comparison between the level-flight and 3-g turn results can be made: The 3-g turn results exhibit lower average crossover fre-

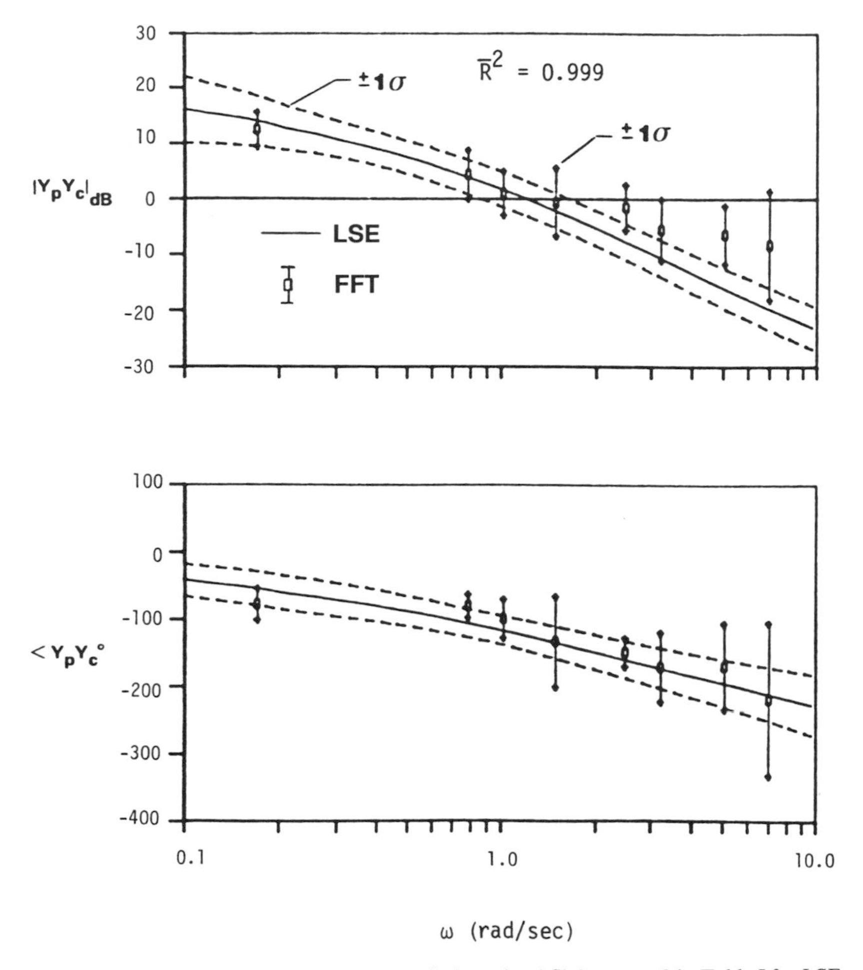

Fig. 7. Y_pY_c FFT and LSE measurements; tail chase, level flight; entry 8 in Table I for LSE.

quencies than those for level flight. Concentrating on the LSE results with the entry 8 model, the average crossover frequency for the 3-g task is 1.71 rad/sec, while that for the level flight is 1.18 rad/sec. These results were found to be statistically significant at the $p = 0.0075$ level using a Wilcoxon's sum of ranks test. This difference should not be overlooked since crossover frequency is perhaps the single most important pilot–vehicle parameter one can measure.

In the flight task studied here, the cause of the crossover regression was traced to the fact that the chase F-14 aircraft was being strongly affected by the wake from the target T-38 in the level flight condition when the HUD reticle motion caused the F-14 flight path to drop considerably below the target

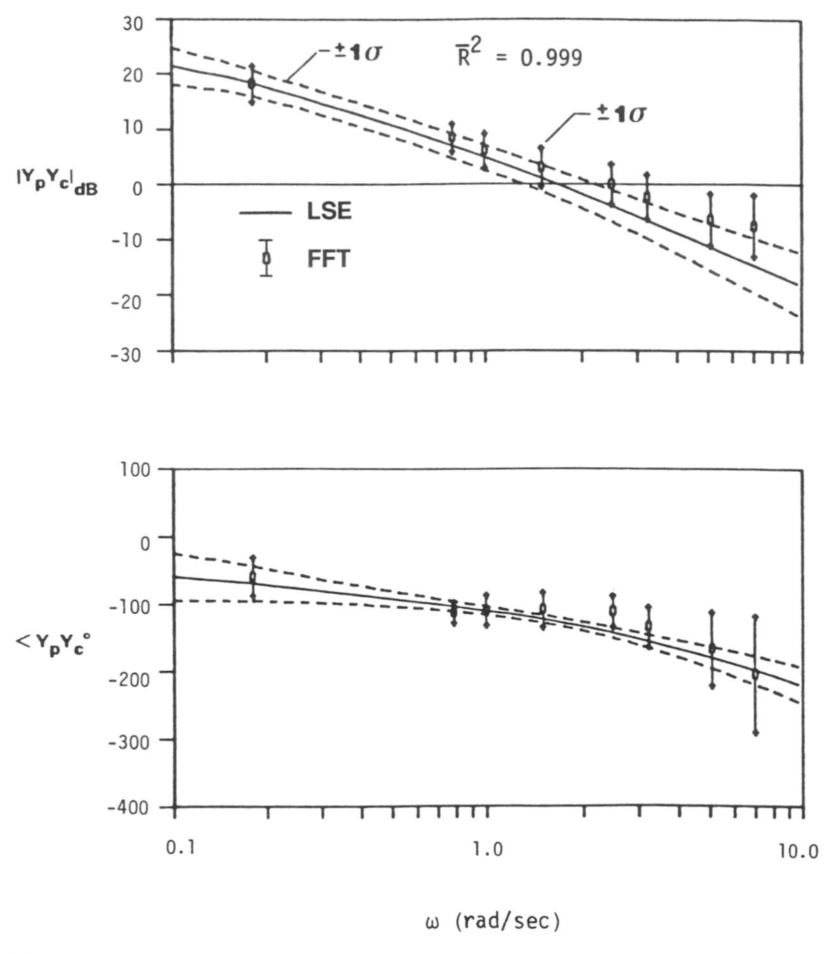

Fig. 8. $Y_p Y_c$ FFT and LSE measurements; tail chase, 3-g turn; entry 8 in Table I for LSE.

TABLE II. LEAST-SQUARES IDENTIFICATION MEANS AND STANDARD DEVIATIONS FOR ENTRY 8

	3 g		Level	
a_1	1.28	(0.211)	1.30	(0.119)
a_2	0.00666	(0.218)	0.119	(0.183)
a_3	−0.288	(0.138)	0.422	(0.0794)
b_1	−0.0858	(0.160)	−0.0321	(0.0293)
b_2	0.119	(0.131)	0.0459	(0.0312)
b_3	0.0317	(0.107)	0.0181	(0.0374)
B	0.253	(0.361)	0.204	(0.145)

vehicle. The F-14 pilot found precision tracking extremely difficult in the presence of this disturbance, which effectively reduced his gain. Had this disturbance existed throughout the task, much lower crossover frequencies would have probably been recorded. No wake encounters occurred in the 3-g turn tasks.

This example demonstrates how pilot–vehicle identification techniques can be incorporated into routine flight tests involving well-defined piloting tasks. The example of the next sub-section demonstrates how *a priori* hypotheses about pilot control strategies and transfer functions in conjunction with a measured pilot–vehicle transfer function will allow the identification of pilot dynamics in a multiloop task.

C. MULTILOOP TASKS

1. Simulated Localizer Capture

This task involves the lateral-directional control of a simulated rotorcraft flying at an airspeed of 60 kn (101 ft/sec) in which the pilot is attempting to follow the commanded ground track y_c shown in Fig. 9a [9]. As simulated, the ground-track task could be thought of as simulating a series of localizer captures in instrument landing system approaches. The rotorcraft was disturbed in roll attitude by a random-appearing sum of 12 sinusoids simulating the effects of atmospheric turbulence. These sinusoids are shown in Table III. The simulation run length was 114.8 sec and only frequency-domain pilot–vehicle identification was undertaken. Figure 9b shows the format of the cockpit display that was used in this fixed-base simulation. It consists of a combined display of roll attitude and ground-track information and the ground-track error was presented in a manner very similar to that used on modern horizontal situation indicators. The simulator in this case was a simple fixed-base laboratory facility consisting of a chair-mounted isometric control stick and a color computer graphics display.

Figure 10a shows the hypothesized pilot-control strategy in this task. This strategy utilizes the multiloop structure outlined in Section I. The forms of the pilot models in the inner and outer loops are given by

$$
\begin{aligned}
Y_{p_\phi} &= \frac{K_\phi e^{-(\tau s - \alpha/s)}}{(s/\omega_n)^2 + (2\xi_n/\omega_n)s + 1} \\
Y_{p_y} &= K_y(T_L s + 1).
\end{aligned}
\tag{10}
$$

The pilot equilization in (10) has been derived from multiple applications of the crossover model of (1), given the vehicle dynamics shown in Fig. 10a. The form of Y_{p_ϕ} was derived from an extended crossover model, which contains a

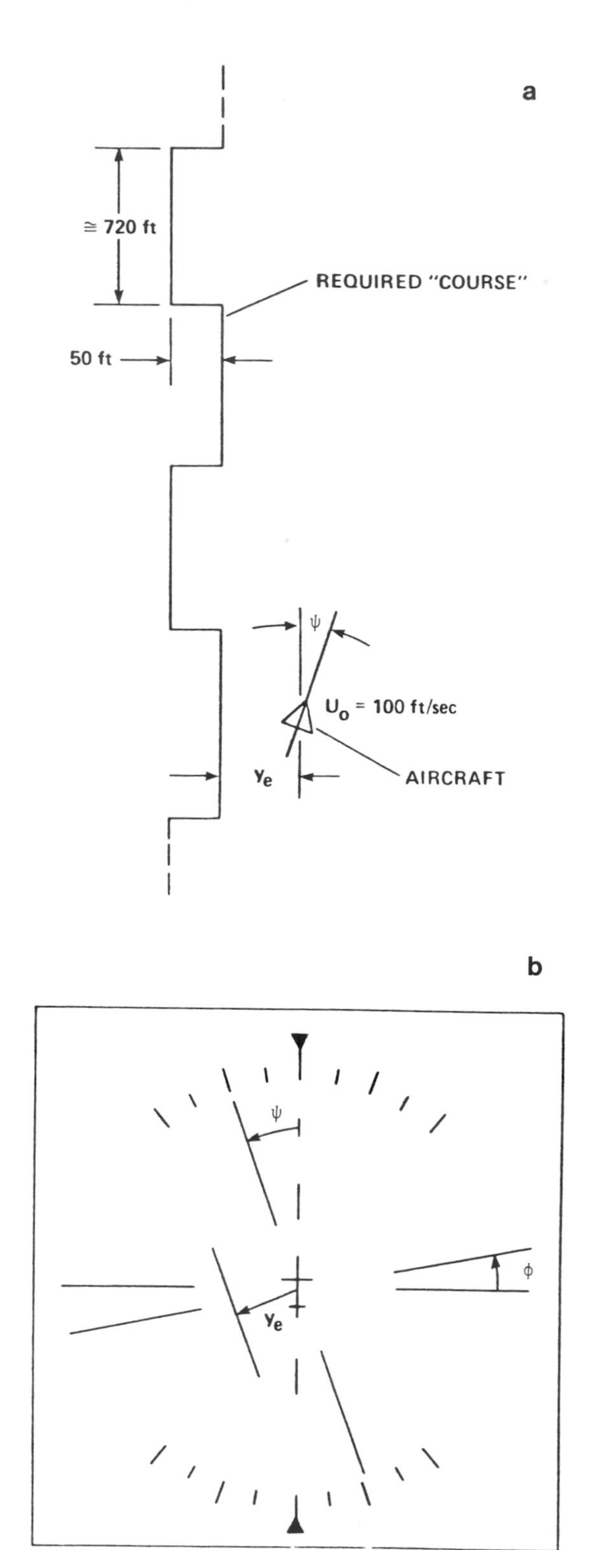

Fig. 9. (a) Command ground track and (b) display symbology for lateral-directional task.

TABLE III. SUM OF SINUSOIDS ROLL DISTURBANCE[a]

ω_i (rad/sec)	A_i/A_1	Number of cycles in run
0.16419	1.0	3
0.27366	1.0	5
0.76624	1.0	14
1.25883	1.0	23
1.86087	1.0	34
2.68185	0.1	49
3.66702	0.1	67
5.03531	0.1	92
7.16984	0.1	131
9.79695	0.1	179
13.73763	0.1	251
20.85273	0.1	381

[a] $\phi_g = \Sigma_{i=1}^{12} A_i \sin(\omega_i t + \theta_i)$, where θ_i are constant phase lags randomly selected between 0 and 2π.

second-order denominator to capture some of the effects of the human neuromuscular system in a more precise manner than a simple effective time delay [2]. The [$\exp(\alpha/s)$] term appearing in Y_{p_ϕ} is intended to capture the effects of low-frequency phase lags (so-called phase droop) often seen in measured human transfer functions [2].

The number of independent pilot transfer functions that can be uniquely identified in the frequency domain using spectral analysis techniques is equal to the number of uncorrelated system inputs (disturbances) multiplied by the number of pilot controls [10]. What is meant by "uniquely identified," is that the amplitude and phase of the pilot transfer functions at each of the input or disturbance frequencies have no dependence on *a priori* assumptions about the form of the transfer functions. Thus, in this simulation, only one pilot transfer function can be uniquely identified. However, the *a priori* assumption of specific pilot transfer functions with undetermined parameters in (10) change this situation. From Fig. 10a it can be easily shown that the ratio of the Fourier coefficients of the control input δ to roll attitude ϕ at each of the 12 disturbance frequences ω_1 yields the following

$$(\delta/\phi)(j\omega_i) = -Y_{p_\phi}\{Y_{p_y}[32.2/(j\omega_i)^2 + 1\}. \tag{11}$$

Thus the measurement of δ/ϕ is equivalent to determining a composite transfer function in which the dynamics of the human pilot (10) are

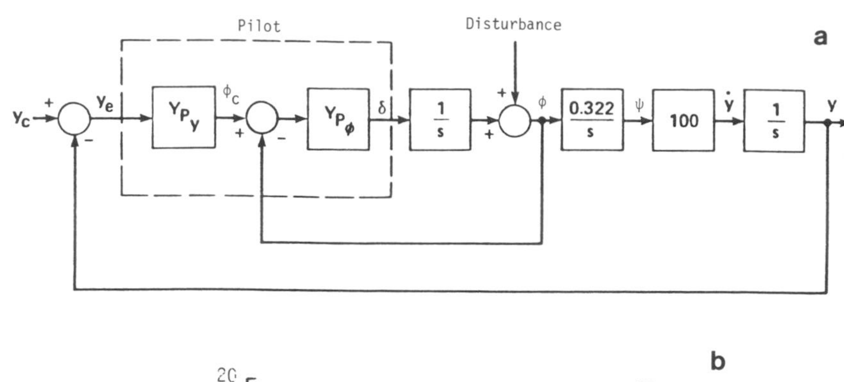

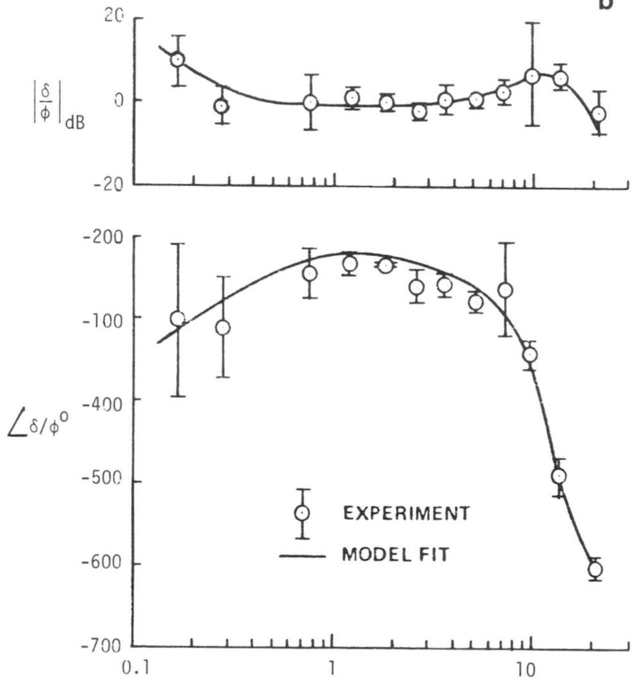

Fig. 10. (a) Multiloop pilot–vehicle system and (b) δ/ϕ FFT measurements for lateral-directional task.

embedded. Equation (11) can be rewritten

$$\frac{\delta}{\phi}(j\omega_i) = -\omega_n^2 K_\phi e^{-[\tau(j\omega_i)-\alpha/j\omega_i]} \frac{(j\omega_i)^2 + 32.2K_y T_L(j\omega_i) + 32.2K_y}{(j\omega_i)^2[(j\omega_i)^2 + 2\xi_n\omega_n(j\omega_i) + \omega_n^2]}. \tag{12}$$

Note that each of the coefficients in the numerator and denominator polynomials in (12) are explicit functions of the pilot-model parameters. A transfer-function fitting routine such as offered in [11] can be employed to find the values of these parameters, which yield acceptable matches to the frequency-domain identification results.

Figure 10b shows the results of this procedure. The solid line represents the transfer function representation of (12), which provided the best fit to the amplitude and phase identification results at the 12 discrete input frequencies. Figure 10b represents the results of five runs from a single, well-trained subject. The resulting pilot transfer functions are

$$Y_{p_\phi} = 2.8e^{-0.15s}/\{(s/12)^2 + [2(0.13)s/12] + 1\} \quad (13)$$

$$Y_{p_\phi} = 0.003(4.5s + 1). \quad (14)$$

This example has demonstrated that, when used in consort with well-established control theoretic models of the human pilot, simple frequency-domain measurements can yield useful quantitative information about pilot dynamics in realistic multiloop tasks. The final example, in the next subsection, demonstrates how pilot–vehicle analyses and identified pilot–vehicle dynamics can support a display evaluation effort involving a multi-loop rotorcraft hovering task.

2. Simulated Hover Task

This task involves the longitudinal control of a hovering rotorcraft in which the pilot is attempting to maintain the vehicle over a translating hover pad [12]. The pilot is using a head-up display with symbology shown in Fig. 11. The hover pad symbol is being driven in the longitudinal direction by a sum of nine sinusoids. These sinusoids are shown in Table IV. The moving pad can represent a task of station keeping over a moving landing pad on a ship or hovering over a fixed pad in the presence of strong atmospheric disturbances. The simulator in this study is unique in that it is the NASA Ames CH-47B variable stability rotorcraft used in a fixed-base simulation mode. The simulation run length was $\sim$2 min in length with 102.4 sec of data used in the pilot–vehicle identifications. As in the single-loop tasks of Section III,A, both FFT and LSE identification techniques were used in the study. The model for the LSE identifications was

$$G(z) = \{[b_1z^{-1} + b_2z^{-2} + b_3z^{-3}]/[1 - a_1z^{-1} - a_2z^{-2} - a_3z^{-3}]\}z^{-p} \quad (15)$$

This model is equivalent to entry 8 in Table I with the addition of the delay term, z^{-p} with p being an integer multiple of the 0.05-sec sampling interval. As discussed in the preceding, the LSE measurements were interpreted in the w' plane, which was considered identical to the s plane.

The purpose of this study was to determine the effect of display changes on pilot work load and pilot–vehicle performance. The display variations

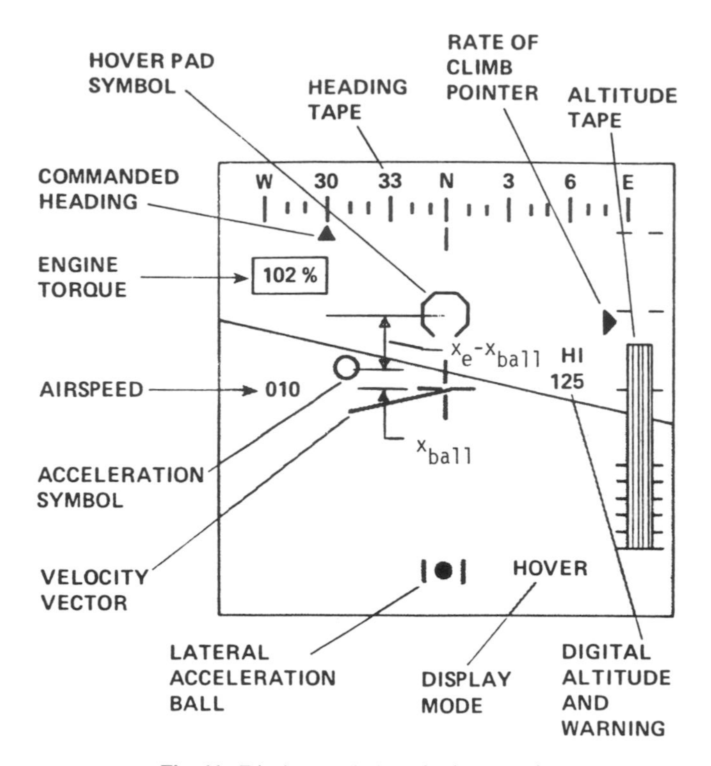

Fig. 11. Display symbology for hover task.

TABLE IV. SUM OF SINUSOIDS FOR MOVING-PAD INPUT[a]

ω_i (rad/sec)	A_i (ft)	Number of cycles in run
.1841	4.4	3
.3068	4.4	5
.4909	4.4	8
.7977	4.4	13
1.166	0.1	19
1.779	0.1	29
2.823	0.1	46
4.663	0.05	76
6.943	0.05	113

[a] $x_{\text{pad}} = \sum_{i=1}^{9} A_i \sin(\omega_i t)$.

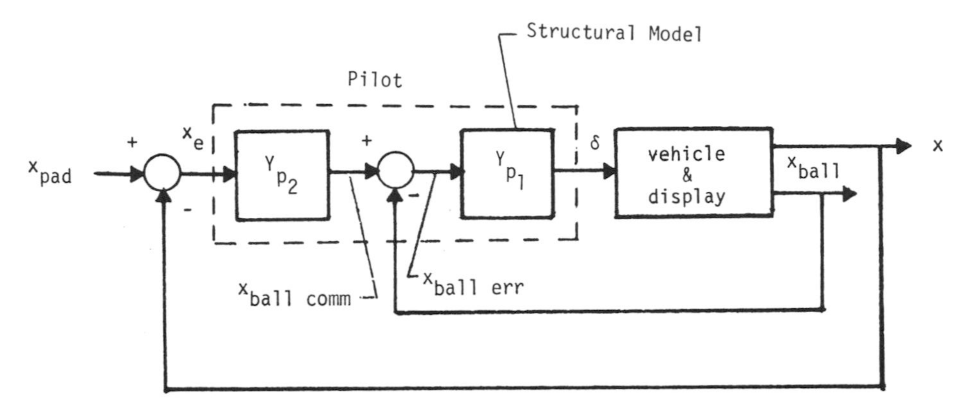

Fig. 12. Multiloop pilot–vehicle system for hover task with acceleration symbol.

centered on the acceleration symbol shown in Fig. 11. This symbol was actually a flight director, which the pilot was to keep centered on the moving-pad symbol by moving the cyclic control. Doing so would cause the rotorcraft to follow the moving pad. Thus, the acceleration symbol had the potential to reduce pilot work load and improve pilot–vehicle performance as it integrated the information needed in the pad-tracking task and presented it in the motion of a single display element.

Figure 12 shows the pilot control strategies hypothesized to be employed with the acceleration symbol. In Fig. 12, x_{pad} refers to the pad position relative to a reference point on the earth in a direction parallel to the instantaneous x body axis of the vehicle; x refers to the vehicle position relative to the same point in the same direction; x_e refers to the relative longitudinal positions of the pad and the vehicle; and x_{ball} is defined in Fig. 11. As in the previous example, only one pilot control and one command input were in evidence here, so only one independent pilot–vehicle transfer function could be uniquely identified. Also, as in the previous example, the *a priori* assumption of specific pilot–vehicle transfer functions will alter this situation. Three transfer functions were measured in the experiments. These were $x_{ball}/(x_e - x_{ball})$, x_{ball}/x_{pad}, and x/x_{pad}. Only the task with the acceleration symbol will be treated here.

In Figure 12, the pilot transfer function Y_{p_1} in the primary control loop was modeled using the structural model of the human pilot, which was also employed in Fig. 2b [4]. This model is shown in Fig. 13. A detailed discussion of this model is beyond the scope of this article. For the purposes of identification for this task, all the model parameters save K_e in Fig. 13 were set to nominal values, which are determined once the dynamics of the primary-loop-controlled element are known in the region of probable crossover [12].

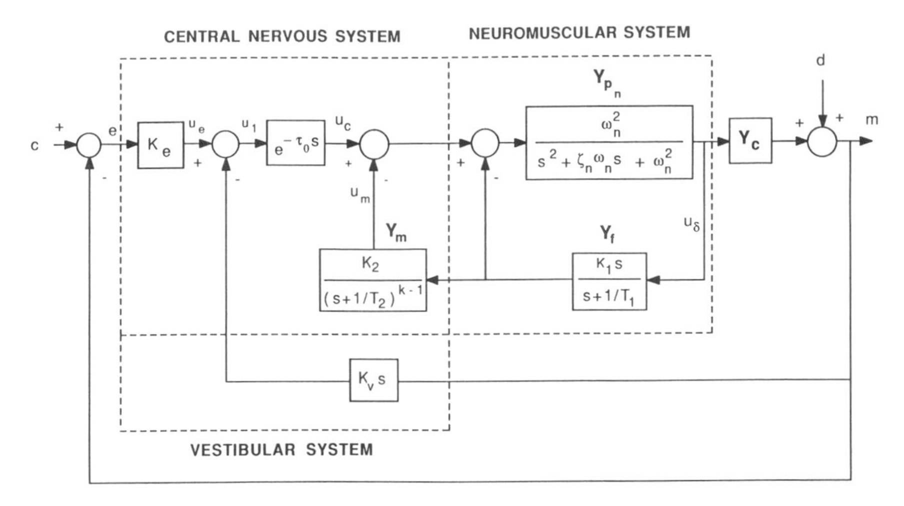

Fig. 13. The structural model of the human pilot.

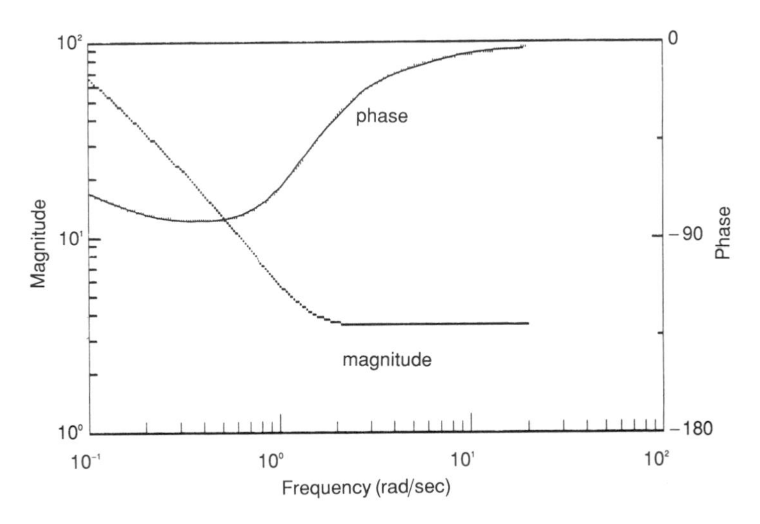

Fig. 14. x_{ball}/δ transfer function for hover task.

Figure 14 shows the transfer function x_{ball}/δ, which defines the primary-loop-controlled element. As can be seen, this transfer function begins to resemble a pure gain for frequencies above 1 rad/sec. The pilot would have to develop lag compensation (transfer function Y_{p_1} in Fig. 14) in the primary control loop over this frequency range in order for the product of pilot and controlled element transfer functions to resemble the crossover model of (1) for frequencies above 1 rad/sec. Table V shows the nominal structural model

TABLE V. STRUCTURAL MODEL PARAMETER VALUES

k	K_e	K_1	K_2	T_1 (sec)	T_2 (sec)	τ_0 (sec)	ω_n (rad/sec)
0	1.57	1.0	2.0	5.0	1.0	0.15	10.0

parameters for a pure-gain controlled element. The K_e value shown merely represents an initial estimate and yields a crossover frequency of 2 rad/sec. An improved estimate of this crossover frequency will be one of the pilot–vehicle characteristics determined from the identification procedure.

As was the case in the identification work of Section III,B,1, the form of the pilot transfer function for the outer loop (Y_{p_2} in Fig. 12) was obtained by an application of the crossover model given the effective controlled element transfer functions for the outer loop, shown in Fig. 15. Separating the initial estimates for the primary- and outer-loop crossover frequencies by a factor of at least four (See Hess, "Methodology for the Analytical Assessment of Aircraft Handling Qualities", this volume) indicates that $Y_{p_2} = K_2$ would be a useful initial model for the outer-loop pilot dynamics.

Figure 16 shows the results of the FFT and LSE identification runs. These results represent five runs made by one of the pilots participating in the simulation. The solid lines in the figures represent each of the LSE identifications, while the FFT results are expressed as means and standard

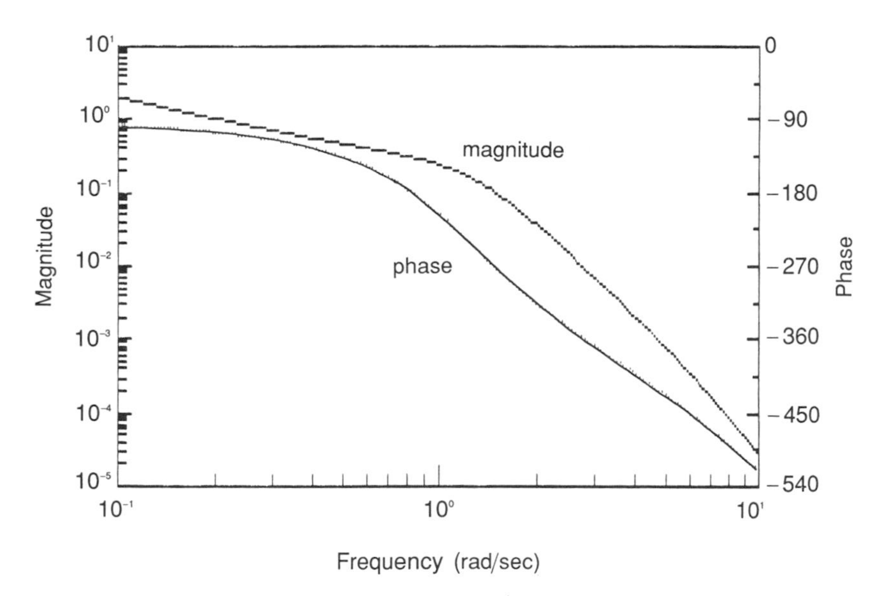

Fig. 15. $x/x_{\text{ball comm}}$ transfer function; primary loop closed with structural model.

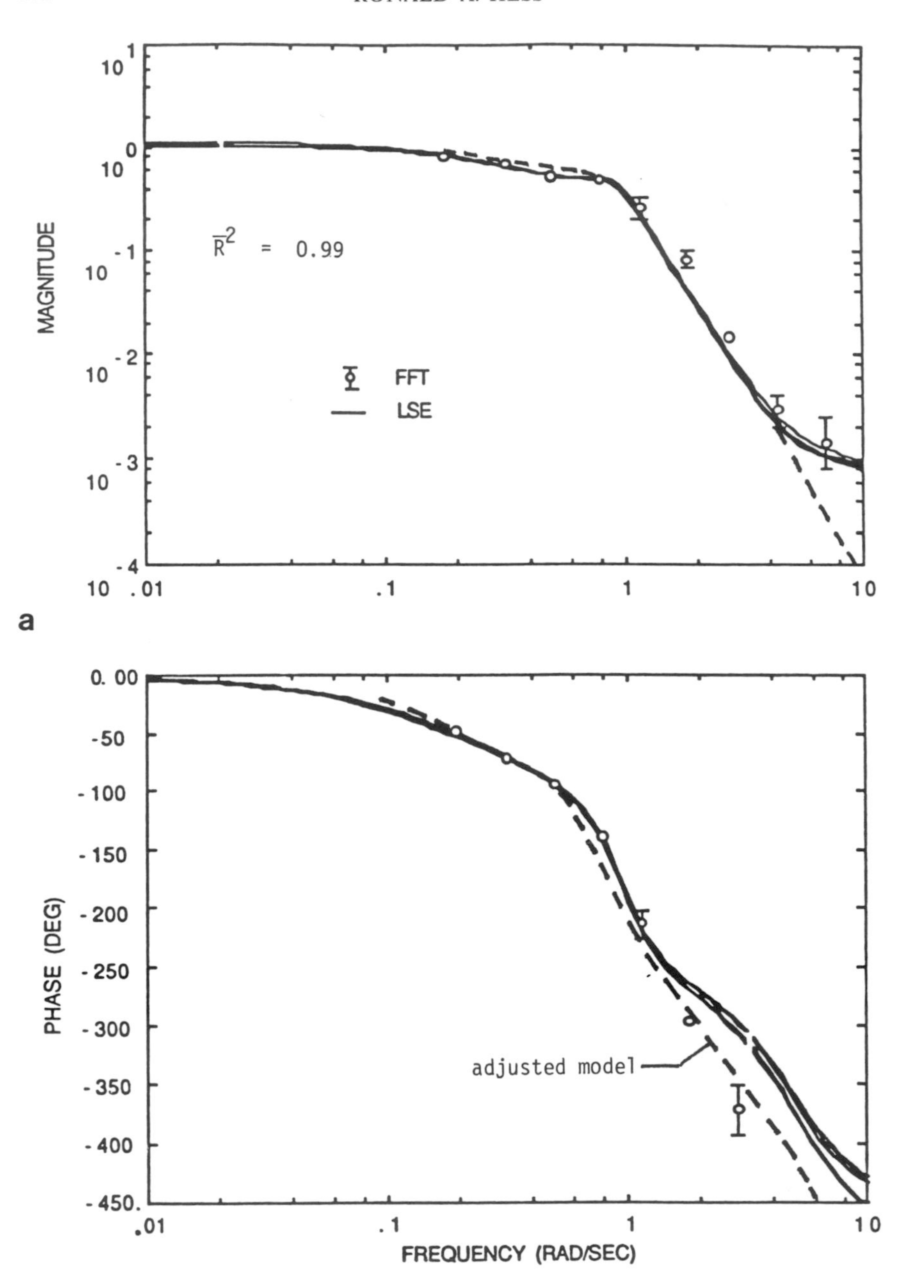

Fig. 16. Measurements for hover task. (a) x/x_{pad} FFT and LSE, (b) $x_{\text{ball}}/x_{\text{pad}}$ FFT and LSE, and (c) $x_{\text{ball}}/(x_{\text{e}} - x_{\text{ball}})$.

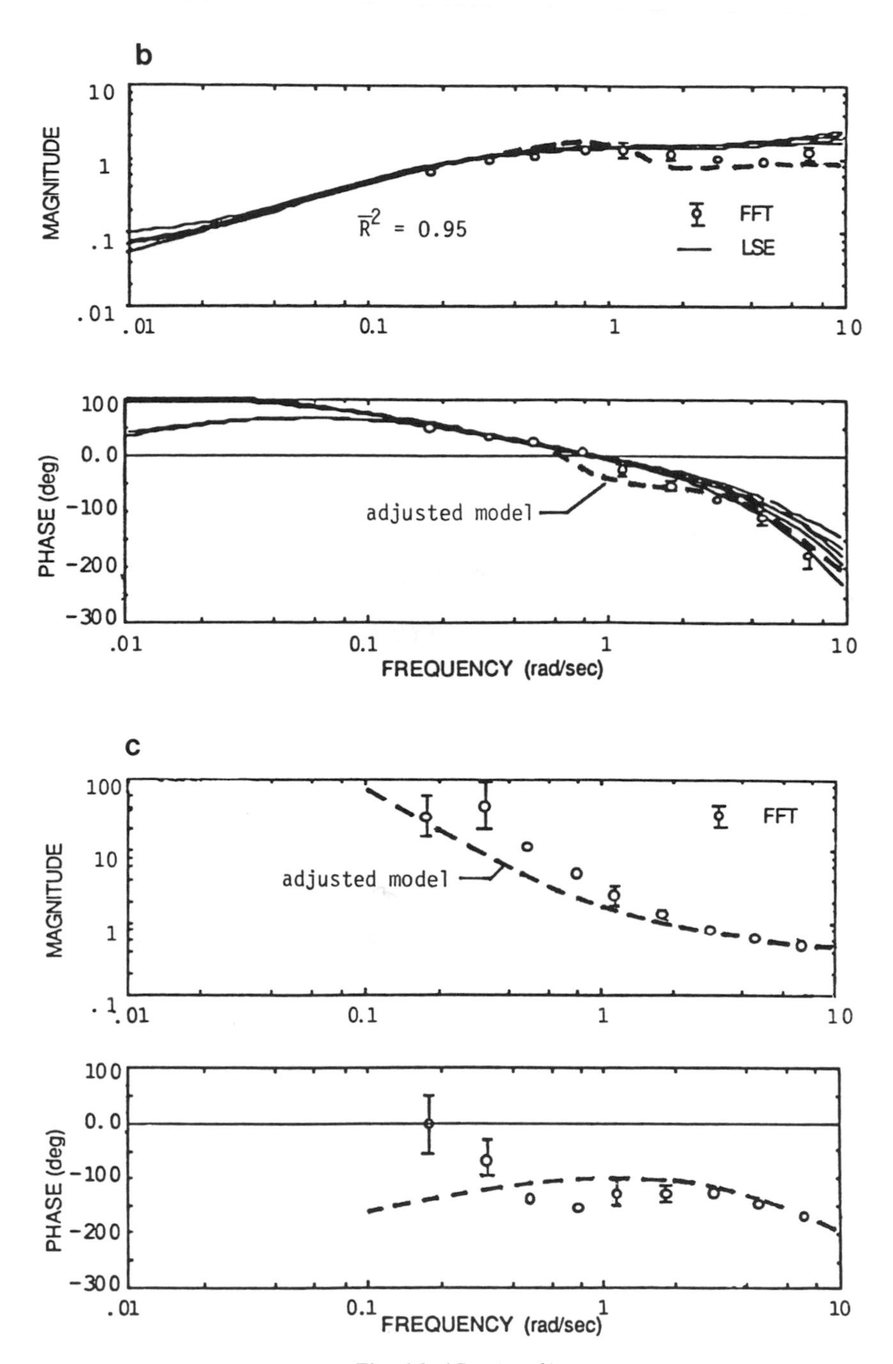

Fig. 16. (*Continued.*)

deviations as in previous figures. The dashed lines represent the results of the model fits. As in the case of Section III,B,1, the identified transfer function $x_{\text{ball}}/(x_e - x_{\text{ball}})$ represents a composite of the pilot transfer functions Y_{p_1} and Y_{p_2}:

$$\frac{x_{\text{ball}}}{(x_e - x_{\text{ball}})} = \frac{Y_{p_1}(x_{\text{ball}}/\delta)\,Y_{p_2}}{1 + Y_{p_1}(x_{\text{ball}}/\delta)(1 - Y_{p_2})}. \tag{16}$$

The LSE identification of the transfer function (16) produced either unstable functions or ones with unacceptably small values of R^2 and for this reason they are not shown in Fig. 16c. This phenomena is probably due to the $(1 - Y_{p_2})$ factor appearing in the denominator of (16). This term can lead to poles of $x_{\text{ball}}/(x_e - x_{\text{ball}})$ in the right-half plane. Of course, the FFT identification approach is immune to these problems since it does not attempt to generate a transfer function, per se, but only transfer-function amplitude and phase information at the discrete input frequencies.

The structural model gain K_e and the form of Y_{p_2} were varied until acceptable matches could be obtained between the model-generated and measured transfer functions of Fig. 16. The quality of these matches were judged by eye rather that by a formal fitting procedure such as that used in the study of Section III,B,1. The resulting model-generated transfer functions are indicated by the dashed lines in these figures. The gain K_e in the structural model was changed from 1.57 to 0.758, which reduced the primary-loop crossover frequency from 2.0 to 1 rad/sec. The pilot transfer function Y_{p_2} was changed from

$$Y_{p_2} = K_2 = 1.0 \tag{17}$$

to

$$Y_{p_2} = (0.5s + 1)e^{-0.2s}. \tag{18}$$

This Y_{p_2} yielded an outer-loop crossover frequency of 0.2 rad/sec, which is a factor of five below that for the primary control loop. It is interesting to note that the lead term in Y_{p_2} is well beyond the crossover frequency in the outer loop and may represent the pilot's attempt to improve outer-loop damping by reducing high-frequency phase lags. The time delay of 0.2 sec may be attributable to the necessity of the pilot obtaining the rate information from the relative displacement of the central stationary cross symbol and the moving-pad symbol on the display of Fig. 11. Such delays have been proposed in rate-sensing manual control models in the past (e.g., [13]).

IV. CONCLUSIONS

The identification of pertinent pilot–vehicle transfer functions can be made of manned simulations and flight tests involving well-defined piloting tasks in a nonintrusive manner. Both frequency- and time-domain identification techniques provide useful information, and the results of the studies summarized here suggest use of both techniques is advisable. The successful identification of the pilot–vehicle dynamics in the examples presented were dependent on the judicious *a priori* selection of pilot control strategies and parameterized pilot models. In the latter case, the utilization of the crossover model of the human pilot in both single loop and multiloop tasks proved to be quite successful.

ACKNOWLEDGMENTS

Many of the results reported in this article were obtained as the result of research grants from NASA Ames Research Center, Dryden Flight Research Facility. Donald T. Berry was the contract technical manager.

REFERENCES

1. D. T. McRUER, "Human Dynamics in Man-Machine Systems," *Automatica* **26**, 237–253 (1980).
2. D. T. McRUER and E. S. KRENDEL, "Mathematical Models of Human Pilot Behavior," AGARD No. 188 (1974).
3. R. A. HESS, "Effects of Time Delays on Systems Subject to Manual Control," *AIAA J. Guidance, Control Dy.* **7**, 416–421 (1984).
4. R. A. HESS, "A Model-Based Theory for Analyzing Human Control Behavior," *in* "Advances in Man–Machine Systems Research" (W. B. Rouse, ed.), pp. 129–175. JAI Press, Greenwich, Connecticut, 1985.
5. J. S. BENDAT and A. G. PIERSOL, "Random Data: Analysis and Measurement Procedures." Wiley, New York, 1971.
6. R. F. WHITBECK and L. G. HOFFMAN, "Digital Control Law Synthesis in the w′ Domain," *AIAA J. Guidance Control* **1**, 319–326 (1978).
7. R. A. HESS and M. A. MNICH, "Identification of Pilot–Vehicle Dynamics from In-Flight Tracking Data," *AIAA J. Guidance, Control Dyn.* **9**, 433–440 (1986).
8. G. D. HANSON and W. F. JEWELL, "Non-Intrusive Parameter Identification Procedure User's Guide," NASA CR-170398 (1983).
9. R. A. HESS and A. A. BECKMAN, "An Engineering Approach to Determining Visual Information Requirements for Flight Control Tasks," *IEEE Trans. Syst., Man, Cybern.* **SMC-14**, 286–298 (1984).
10. R. L. STAPLEFORD, S. J. CRAIG, and J. A. TENNANT, "Measurement of Pilot Describing Functions in Single-Controller Multiloop Tasks," NASA CR-1238 (1969).
11. R. C. SEIDEL, "Transfer-Function-Parameter Estimation from Frequency Response Data—A Fortran Program," NASA TM X-3286 (1975).
12. R. A. HESS and J. T. REEDY, "Determining Pilot Control Strategies in a Simulated Instrument Hover Task," *IEEE Trans. Syst., Man, Cybern.*, to appear.
13. D. T. McRUER, L. G. HOFMANN, H. R. JEX, G. P. MOORE, A. V. PHATAK, D. H. WEIR, and J. WOLKOVITCH, "New Approaches to Human Pilot/Vehicle Dynamic Analysis," AFFDL-TR-67-150 (1968).

STRAPDOWN INERTIAL NAVIGATION SYSTEM REQUIREMENTS IMPOSED BY SYNTHETIC APERTURE RADAR

JAMES L. FARRELL

Development and Engineering Divisions
Westinghouse Defense and Electronic Systems Center
Baltimore, Maryland 21203

I. INTRODUCTION

Knowledge of gimbaled platform requirements for synthetic aperture radar (SAR) mapping, while not universal, is becoming fairly widespread. When there is relative motion and/or large distance between inertial navigation system (INS) and antenna, however, plans to use a strapdown mechanization must account for several error sources (particularly those involving rectification of motion-sensitive degradations) that would normally be ignored in a gimbaled implementation. This article assesses the impact for an extensive set of gyro and accelerometer errors applied to SAR.

The following nomenclature defines the symbols used in this article.

$\tilde{A}$	Acceleration error (ft/sec^2)
$\mathbf{d}$; d	IMU-to-radar displacement in airframe coordinates; magnitude of $\mathbf{d}$ (ft)
g	Acceleration due to gravity (ft/sec^2)
$\tilde{J}$	Jerk error (ft/sec^3)
j/H	Ratio of gyro inertia to angular momentum (sec)
j/P	Ratio of accelerometer inertia to pendulosity (ft)
K^2	Number of error contributions to be root sum squared
L	Static lift (g)
l, m, n	Conventional airframe axes

Copyright © 1990 by Academic Press, Inc.
All rights of reproduction in any form reserved.

N	Number of coherent integration intervals per SAR frame
n_ω	Total angular drift rate (rad/sec)
$\mathbf{R}_R, \mathbf{R}_S$	Position vector of the radar antenna phase center and strapdown IMU, respectively (ft)
$\tilde{t}$	Timing uncertainty (sec)
t_c	Coherent integration interval (sec)
T_M	Mapping segment duration (sec)
T_W	Effective averaging interval of navigation update Kalman filter (sec)
$[T]$	Orthogonal transformation from airframe to reference axes
$\tilde{\mathbf{v}}$	Velocity vector error (ft/sec)
X, Y, Z	Inertial instrument axis directions
$\mathbf{l}_R$	Unit vector along direction of $\underset{\sim}{\mathbf{R}}_R$
Δv	Quantization of integrating accelerometer (ft/sec)
$\delta\{\ \}$	Variation in $\{\ \}$
λ	Radar wavelength (ft)
$\underset{\sim}{\rho}$	Navigation reference frame profile rate (rad/sec)
$\tilde{\phi}$	Motion compensation phase error (rad)
ψ	INS tilt (rad)
$(\dot{\ })$	Time derivative of ( )
$(\ )^T$	Transpose of ( )
$(\tilde{\ })$	Error in ( )
$\lvert\cdot\rvert$	Absolute magnitude of $(\cdot)$

It is noted at the outset that this entire analysis is based on a scheme whereby SAR imaging is allowed to continue undisturbed, without any updating or transfer alignment operations, for durations on the order of a minute or a few minutes. An alternative, permitting adjustments between (and even within) SAR frames, is briefly mentioned after the analysis is presented.

Discussion of these topics would be incomplete without noting several shortcomings of presently accepted procedures. The following items are believed to be critical:

1. Strapdown inertial measuring unit (IMU) specifications quite often stress the usual long-term behavior (e.g., nautical miles per hour) while giving little attention to crucial short-term performance criteria. Explicit requirements for short-term stability of strapdown velocity and attitude errors should be included in applicable specifications. Dependable means for testing short-term performance in flight must be established.

2. Available vibration information is typically sketchy and limited to translational motion. Confidence in design would call for improvements in all of the following aspects of vibration data collection: specific attention to the

sites (airframe stations) where sensors and instruments are located; addition of gyros (as well as accelerometers) near those sites for flight test; and generalization of data processing to include coherence functions (in-phase and quadrature cross-axis correlation for all combinations of waveform histories [1]) rather than restriction to individual waveforms.

3. Attempts to adapt low-cost inertial instruments often lead to compromises in accelerometer quality as well as in gyro performance. This is pennywise; accelerometers generally do not dominate costs. It also defeats the purpose of the design since, in applications where short-term accuracy is paramount, effects of accelerometer degradations are far more serious and more immediate than gyro-drift effects.

4. Inertial instrument biases are often treated as an overall resultant that is carried forward between flight phases having different dynamics. In marked contrast to that approach, recognition of separate contributors and the ramifications of their dependence on the dynamic environment provided much of the motivation for this article.

5. Accelerometer-bias states are often included, somewhat indiscriminately, in transfer alignment formulations. Except during in-flight calibration, with maneuvers taken to isolate these biases, they cannot be separated from tilt effects; augmenting states should never be included in an estimation algorithm when they are unobservable.

6. Except for RLG (ring laser gyro) implementations, typical gyro scale factor accuracies are marginal for dynamics with large rotational excursions. Gyros also need to be better aligned to the corresponding accelerometer sensitive axes. Although some of the degradation can be undone by rotational observations (comparison of attitude changes seen by master and slave), that benefit can be compromised by mission constraints against large rotational excursions, lag, and errors in the master itself (its own scale factor and misalignment if it is strapdown, or pickoff error if it is gimbaled).

7. On-line corrections of inertial instrument error sources, using laboratory-derived coefficients, is acknowledged and enthusiastically endorsed. However, there is no reason to believe that all coefficients age at the same rate. Since imperfections in these corrections of course constitute uncompensated degradations, more detailed information is needed about stability of lab calibration.

With the seven items just listed in mind, attention can be turned to conditions adopted for the analysis to be presented:

1. No autofocus operation is performed on the SAR data; all gate-placement and phase-adjustment commands must depend on on-board inertial sensors.

2. The strapdown IMU, consisting of three gyros and three accelerometers, is mounted on an antenna (gimbaled or electronically steerable) structure.

3. A gimbaled master INS is available for repeated in-flight leveling and azimuth alignment of the strapdown reference. The combination can also be externally updated from radar or other [e.g., global positioning system (GPS)] observations, over intervals denoted T_{W}. Between these update–alignment intervals, there are SAR mapping mission segments of duration T_{M} (in general containing several SAR frames) within which the strapdown IMU acts as an autonomous source of short-term navigation data for gate placement and phase correction.

4. The dynamic conditions experienced during update–alignment segments can, in general, differ from those (both static and vibratory) present during mapping segments. Thus the attempt to learn accelerometer-bias effects in transfer alignment (for subsequent usage as in-flight calibration data during SAR operation) should be exercised only during very limited periods (i.e., during the in-flight calibration mentioned in item 5 previously listed). Throughout this article it is taken for granted that this operation has been completed early in the flight, for the master INS as well as the slave, so that only a smaller uncorrected accelerometer bias remains thereafter. Further elaboration of this point appears at appropriate points in the subsequent text.

5. Each individual strapdown error source is allowed to reach $(1/K)$ of the total error allowed (so that as many as K^2 separate sources of degradation could be root sum squared without exceeding that total).

6. It is recognized that several inertial instrument errors are compensated in the front-end processing; only the uncompensated residual error is being addressed here.

7. With standard (l, m, n) aircraft axis conventions, the Z axes for No. 1, No. 2, and No. 3, inertial instruments under nominal reference conditions will point along (m, l, m), respectively. Thus the lift force will not appear along any inertial instrument Z axis for electronically steerable arrays or, under conditions of zero sightline depression, for gimbaled antennas.

8. SAR motion compensation accuracy requirements are first derived for straight-and-level flight conditions during the mapping segment, and generalization to maneuvering flight is covered in Section IV.

9. Strapdown requirements are based on SAR mapping alone, without regard to other radar modes or operations that may accompany it. In combination with condition 8, it follows that azimuth alignment is not critical in the immediate development. Again, the subsequent consideration of SAR mapping during maneuvers can address this issue separately, and the perspective here will be clearer as a result.

Since the system to be analyzed is hypothetical, nominal parameter values (e.g., 1 ft, 1 g, 1 Hz, 1°) can be assumed:

1. There is one coherent integration time interval per SAR frame.
2. Lever arm separation between IMU and radar antenna is 1 ft.
3. Mean product of specific force components is characterized by superposition of the following two effects:

   $1\ g(\text{rms}) \times 1\ g(\text{rms}) = 1\ g^2$ tightly coupled translational vibration

   and

   (deterministic 1 g lift) × (0.1 g horizontal component) $= 0.1\ g^2$ for cruise.

4. Mean product of angular rates is dominated by narrow-band random components, each in conformance to a 1°-amplitude oscillation at 1-Hz center frequency. This produces rms angular rates of $\sim$0.08 rad/sec per axis and, with tight coupling, essentially 0.006 $(\text{rad/sec})^2$ for angular rate-squared sensitive errors.
5. Wavelength $\lambda = 0.1$ ft.
6. Coherent integration time $t_c = 2$ sec.
7. The maximum duration of the mapping segment T_M is 100 sec.
8. The valuc of K (from condition 5) is 4, so that each separate contributor to the error budget can be one-fourth of the allowable total.

II. BACKGROUND

Instantaneous position vectors of the radar-antenna phase center and the strapdown IMU will be denoted here as $\mathbf{R}_R$ and $\mathbf{R}_S$, respectively, expressed in a reference (e.g., locally level) coordinate frame with the origin at a designated point in the mapped area. If $\mathbf{d}$ and $[T]$ denote, respectively, the IMU-to-radar displacement vector in airframe coordinates, and the orthogonal transformation from airframe to reference axes, then

$$\mathbf{R}_R = \mathbf{R}_S + [T]\mathbf{d}, \tag{1}$$

so that the motion compensation phase error is

$$\tilde{\phi} = \frac{4\pi\delta\{|\mathbf{R}_R|\}}{\lambda} = \frac{4\pi}{\lambda}\delta\{(\mathbf{R}_R^T\mathbf{R}_R)^{1/2}\} = \frac{4\pi}{\lambda}\frac{1}{2}\frac{1}{|\mathbf{R}_R|}\delta\{\mathbf{R}_R^T\mathbf{R}_R\}. \tag{2}$$

To first order, $\tilde{\phi}$ is simply

$$\tilde{\phi} = (4\pi/\lambda)\{\mathbf{I}_R^T\tilde{\mathbf{R}}_S + \mathbf{I}_R^T[\tilde{T}]\mathbf{d} + \mathbf{I}_R^T[T]\tilde{\mathbf{d}}\}, \tag{3}$$

where the variational operator $\delta\{\ \}$ is now replaced by the error notation ($\tilde{\ }$) and $\mathbf{I}_R$ denotes the unit range vector $\mathbf{R}_R/|\mathbf{R}_R|$. The bracketed factor on the right-hand side of (3) contains three terms which, in sequence, represent along-range components of the following effects:

1. *Incremental IMU position error appearing during a SAR frame.*[1] This includes effects of velocity quantization, acceleration error (tilt plus accelerometer errors plus vertical deflections), and contribution of gyro drift to cumulative position uncertainty.
2. *Instantaneous attitude error.* This includes the interaction between rapid rotations and attitude data timing offsets as well as quantization.
3. *IMU/radar displacement error.* This includes unknown vibrations and phase-center wander, plus interactions between rotational dynamics and static displacement uncertainty.

As an illustrative example consider a constant along-range acceleration bias $\tilde{A}$, producing a contribution to the first term in (3),

$$\tilde{\phi} = (4\pi/\lambda)(1/2)\tilde{A}(t_c/2)^2 \tag{4}$$

when the bias is integrated over $(t_c/2)$. The rationale for this is that, with the phase reference set in correspondence to the center of the coherent integration interval, the phase error has a duration of half this interval to accumulate. A common limit of acceptability for this effect is one-fourth cycle of cumulative phase error; thus

$$\frac{(4\pi/\lambda)\tilde{A}(t_c/2)^2}{2} < \frac{\pi}{2} \tag{5}$$

when the number N of coherent integration intervals per SAR frame is unity; otherwise this limit must be divided by $\sqrt{N}$.

A requirement of this type actually dictates the need for both (1) an initial tilt ceiling not to exceed $(1/K)(\tilde{A}/g)$ rad, which really places demands on a master source for transfer alignment and/or aiding signals, and (2) the ability of the IMU to hold an initial alignment within tighter tolerances, not just over the

[1] Only the dynamic variations in radio-frequency phasing are critical for SAR motion compensation. Thus, while static position uncertainties can far exceed the wavelength, changes in those errors must be held to a small fraction of a wavelength during a SAR frame (typically a few seconds). Reference [2] contains a brief analysis on pp. 260–266.

long-term average, but through any interval of mode-duration length (i.e., larger error sequences of alternating sign cannot be averaged). It is the latter that imposes frequently overlooked requirements on the strapdown IMU and, excluding quantization effects (covered in Section III,C and III,E), short-term stability requirements for output errors should become standard specification items for IMUs to be used for such applications.

III. ANALYSIS FOR STRAIGHT AND LEVEL MAPPING SEGMENT

In cruising flight, the contribution of a leveling error (ψ rad) to total uncorrected horizontal acceleration error is simply the product $g\psi$. Thus since, on the basis of (5), maximum acceptable total horizontal error is

$$\tilde{A} = \lambda/t_c^2, \tag{6}$$

and since the tilt effect is only one of K^2 contributors (to be root-sum squared with all accelerometer biases and vertical deflection effects), the allowable tilt is

$$\psi = (1/K)(\tilde{A}/g) = \lambda/(Kgt_c^2). \tag{7}$$

It is reiterated here that, due to condition 4 of Section I, any correlation that may have existed initially between ψ and uncorrected accelerometer bias cannot be assumed to stay maintained. Since the initial correlation is negative (i.e., tilt and accelerometer-bias effects counteract during initial alignment), the root-sum-squaring operation used here is conservative when much of the initial correlation remains. When this is not the case, however, a potential pitfall is present in this approach. Since the resulting problem is most likely to arise under maneuvering conditions, and since the full scope of that problem includes material yet to be covered, the subject is saved for Section IV.

A. INFLUENCE OF NAVIGATION UPDATE–ALIGNMENT PHASE

At this point a value must be established for the duration T_W of the data window for measurements used to achieve the necessary strapdown IMU leveling accuracy. This quantity, controlled by the spectral density of process noise in a Kalman filter (Chapter 5 in [2]), represents a nominal interval over which these measurements are effectively averaged. A cardinal rule for realization of expected performance in a Kalman filter is

$$\text{duration of model fidelity} > T_W > \text{time required for observability.} \tag{8}$$

The first decision made along these lines was to abandon all hope of identifying each separate contributor to gyro and accelerometer error budgets (Sections III,B–III,D) as augmenting states for in-flight calibration. To do that would have required an impractical state dimensionality for operational software and, of greater importance, an impracticable sequence of aerodynamic maneuvers needed to isolate each augmenting state unambiguously. Thus the model used to represent flight dynamics for master INS initialization contains only time-invariant misorientation angles, position, and velocity in three dimensions. For the goal of achieving strapdown leveling accuracy there are three different conditions to be addressed: (1) master INS update (Section III,A,1), (2) transfer of master INS leveling accuracy to the strapdown slave [Section III,A,2), and (3) full transfer alignment, including azimuth axis (Section IV).

1. Initialization

On occasions when master INS leveling accuracy is poorer than that needed for the slave, navigation updata from external aids (e.g., radar and GPS) will clearly be the requisite operation. The position–velocity–angular misorientation state selection just described then calls for the Kalman filter formulation defined in Section 6.5 of [2]. An upper limit for model fidelity duration in (8) is therefore a tenth of the Schuler period, or ~ 500 sec. The other side of this inequality is influenced by the navaid accuracy in the expression

$$\text{navaid error tolerance}/(0.5\, gT_{\mathrm{W}}^2) < \psi; \tag{9}$$

that is, while considerable flexibility exists for update scheduling, there must be enough aiding information collected in each interval of duration T_{W} to deduce tilt to within ψ rad. Typically a satisfactory result can be obtained from differential GPS measurements, with accuracies described in [3], averaged over a data window of a few hundred seconds. At the same time, gyros must not produce enough integrated drift effect to add another ψ rad of tilt within the same period. Allowable total drift (n_ω) is therefore subject to the restriction

$$n_\omega < \psi \quad \text{rad}/T_W \quad \text{sec}. \tag{10}$$

2. Leveling Alignment Transfer

When the master INS is level to within ψ rad but the strapdown is not, the value of T_{W} can be reduced to substantially less than a minute [4]. A value of $T_{\mathrm{M}} = 100$ sec would then allow reasonably efficient utilization ($T_{\mathrm{W}} \ll T_{\mathrm{M}}$ for low fractional dead time between mapping segments). Inequality (10) would then be easily satisfied if

$$n_\omega = \psi / T_M, \tag{11}$$

and, since each individual gyro drift source is again one of K^2 separate contributors, the allowable limit for each [in view of (7)] is

$$\psi/(K T_M) = \lambda/(K^2 g t_c^2 T_M), \tag{12}$$

or ~ 0.5 μrad/sec $\doteq 0.1°$/hr. For thoroughness it is noted that total effective drift about a level axis in straight and level flight [Eqs. (3-39) and (3-46) of [2]] includes the product (azimuth misalignment expressed in rads) $\times$ (angular rate ρ of the navigation reference coordinate frame). Thus (12) also imposes the following upper limit on allowable azimuth misalignment:

$$(1/\rho)\psi/(K T_M) = \lambda/(K^2 g \rho t_c^2 T_M). \tag{13}$$

In practice this restriction may be superseded by requirements of a separate operation or mode other than SAR.

B. ACCELEROMETER REQUIREMENTS DERIVED FROM QUADRATIC PHASE SHIFT

Quadratic and cubic phase terms arise from quasi-static components of acceleration and gyro errors, respectively. Taken directly, these terms produce a power series in time; since this is not an expansion in orthogonal functions, the overall error history is sometimes reexpressed in terms of a Legendre series. That procedure will not be followed here, however, since retention of the original power series produces a slightly conservative requirements spec.

With parameter values from Section I substituted into (6), allowable total horizontal acceleration bias is

$$\tilde{A} = (0.1/2)^2 = 0.025 \quad \text{ft/sec}^2 \doteq 0.0008\ g, \tag{14}$$

which, with $K = 4$ from Section I, imposes a limit of 200 μg for each separate source of acceleration bias (including effects of verticality error as in the present development). Following is a derivation of specifications for each contributor to the integrating accelerometer error budget. The approach uses methods that are fairly well documented (e.g., [5] and Chapter 4 in [2]).

1. *Null bias.* Immediately from above, allowable null bias remaining after in-flight calibration (Section I) is 0.0002 g.

2. *Scale factor.* In straight and level flight, accelerometer scale factor error does not produce any static horizontal acceleration bias.

3. *Misalignment*. It is fairly well recognized (see, e.g., pp. 72 and 111 in [2]) that accelerometer alignment is not at all critical for credibility of the inertial navigation data per se. Departure from an intended aerodynamic reference is not critical, and even input axis (IA) nonorthogonality can be dumped into the altitude channel (which causes it to be largely ineffective). On this basis the allowable misalignment of an accelerometer about the intended direction of each orthogonal axis [output axis (OA) and pendulous axis (PA)] should be dictated by requirements of whatever other system operations accompany SAR mapping.

4. *OA Sensitivity*. The coefficient, denoted j/P, for sensitivity to angular acceleration about the OA, multiplied by the 0.006 $(\text{rad/sec})^2$ established in Section I, must not exceed 0.0002 g. Thus (j/P) can be 0.033 $g/(\text{rad/sec})^2$ at most. It is worth noting that, although the instantaneous degradation from this phenomenon is proportional to angular acceleration, the rectified bias is proportional to mean-squared angular rate. Chapter 4 in [2] describes this and other rectification effects.

5. *Anisoinertia*. The aforementioned 0.006 $(\text{rad/sec})^2$ figure, multiplied by the anisoinertia coefficient, cannot exceed 0.0002 g. Maximum allowed anisoinertia coefficient is then 0.033 $g/(\text{rad/sec})^2$.

6. *Rebalancing loop response delay*. Bandwidth of presently available strapdown accelerometers easily satisfies any conceivable motion spectrum of interest. For completeness another requirement must be imposed on accelerometer output processing, due to unequal bandwidths of gyro and accelerometer loops. With no compensation, velocity increments would be transformed through previous rather than current, gyro-fed attitude matrices. Velocity increments must therefore be delayed to counteract this timing offset and, furthermore, each accelerometer output should be time trimmed to equalize these delays as closely as possible. Any imperfection in this timing adjustment is charged against the overall timing mismatch budget as discussed in the following.

7. *Accelerometer output delay mismatch*. The aforementioned timing mismatch cannot exceed the ratio of (0.0002 g)/(mean product of vibratory angular rate) $\times$ (the translational acceleration along an orthogonal axis). Motion specifications for correlations between translational and rotational vibration components are quite scarce but, even with tight coupling assumed between the 1 g and the 0.08 rad/sec (from Section I), allowable mismatch is essentially 3 msec. With currently available accelerometer loop bandwidths, this requirement is so lenient that it hardly needs to be stated.

8. *Vibropendulosity*. This coefficient multiplied by 1.1 g^2 (parameter value 3 in Section I) cannot exceed 0.0002 g; thus, vibropendulosity < 0.00019 g/g^2.

9. *Size effect*. Distance between accelerometers must not exceed the ratio $0.0002\ g/0.006\ (\text{rad/sec})^2$, or $0.033 \times 32.2 = \sim 1$ ft; this is quite lenient, since strapdown packages have separations measured in inches, not feet.

10. *Residual sculling*. With finite inertial instrument resolution and data rates, some sculling effects remain at the output; in the presence of motion levels already discussed, the residual bias from that source cannot exceed $0.0002\ g$.

The foregoing analysis addresses an extensive, but not exhaustive, list of accelerometer parameters. One additional effect, quantization, allows portions of the motion experienced to go temporarily unnoticed (i.e., until the next quantum threshold is crossed). The next section addresses this phenomenon.

C. VELOCITY QUANTIZATION

One straightforward way to characterize this degradation is to visualize a steady specific force A producing a velocity ramp with a slope of A ft/sec/sec, while the apparent velocity history would be a staircase function just below, and tangent to, the ramp. Velocity error would then be a sawtooth function with frequency f and amplitude equal to the integrating accelerometer resolution (Δv); furthermore,

$$f = A/\Delta v \tag{15}$$

and the amplitude of this spectral component in the sawtooth is $(\Delta v/\pi)$. In accordance with the velocity error expressions on p. 264 in [2], this produces a normalized error of order

$$(\Delta v/\pi)/(f\lambda) = (\Delta v)^2/(\pi \lambda A). \tag{16}$$

A value of 0.032 for this quantity produces $20 \log(0.032) = -30$ dB peak sidelobe ratio (PSLR). At the 0.1-ft wavelength with $A > 0.01\ g$, this implies a quantization of 0.01 ft/sec. It is acknowledged that a specific force above the allowable level of each bias contributor ($0.0002\ g$ in the current example) but below the ratio (quantization level)/t_c, held steady to within $0.0002\ g$ during t_c at that low level, would interact with this quantization to produce unacceptable quadratic phase shift. These conditions are unlikely to hold, however, even in cruise. Velocity quantization is best characterized as an error source that distributes itself among a wide span of SAR image cells, and thus fails to produce the severe degradation that would have resulted from concentration within a narrow spectral region.

D. GYRO DRIFT

Cubic phase shift arises from a rate of change of acceleration error

$$\tilde{J} = gn_{\omega} \tag{17}$$

triply integrated over each half of the coherent integration interval [recall the discussion following (4)]. For a maximum allowable shift of $\pi/8$ rad,

$$(4\pi/\lambda)(1/6)gn_{\omega}(t_c/2)^3 < \pi/8 \tag{18}$$

or

$$n_{\omega} < 3\lambda/(2\,gt_c^3), \tag{19}$$

which is superseded by (12), since $K^2T_M \gg t_c$. Gyro coefficients will now be determined by the methods used in Section III,B.

1. *Null bias.* With the null defined as the motion-insensitive component of drift bias, a properly conducted calibration can reduce this effect to the dynamic drift level, and a steady bias is not critical. A practical specification might reasonably call for reduction of this effect by an order of magnitude, but to do so requires separation of null bias from other drifts during calibration.
2. *Turn-on bias variability.* Allowable variation is 0.1°/hr, from (12).
3. *Scale factor and misalignment.* Both of these effects are inactive in straight and level flight. Their consideration is thus deferred to Section IV.
4. *OA Sensitivity.* The coefficient, denoted j/H, for gyro sensitivity to angular acceleration about the OA, multiplied by the 0.006 (rad/sec)^2 (parameter 4 of Section I), must not exceed 0.1°/hr, or ~0.5 μrad/sec. Thus (j/H) can be 0.00008 sec at most.
5. *Anisoinertia.* The aforementioned 0.006 (rad/sec)^2 figure, multiplied by the anisoinertia coefficient, cannot exceed the same 0.5 μrad/sec figure. Maximum allowable anisoinertia coefficient is then 0.00008 sec.
6. *Rebalancing loop response.* Since nothing above a few Hz rotational frequency was postulated in the assumed motion (Section I), typical bandwidths of available gyros do not violate any requirements here.
7. *Gyro output delay mismatch.* The previously established figure of 0.006 (rad/sec)^2, multiplied by gyro-timing mismatch, cannot exceed 0.5 μrad/sec. Allowable time-lag mismatch is then also 0.00008 sec.
8. *g-Sensitive drift.* With 1 g nominal sustained lift, allowable g-sensitive drift is simply 0.1°/hr/g. This applies to both axes (X and Z).

9. *g^2-Sensitive drift*. With vibratory translation characterized by parameter 3 in Section I, allowable g^2-sensitive drift is of order $0.1°/\text{hr}/g^2$. This applies to both anisoelastic (in-phase correlated) and cylindrical (quadrature correlated) components.
10. *Residual commutation error*. With finite gyro resolution and computational rates, some coning effects can remain at the output; in the presence of motion levels described in Section I, residual drift bias from that source cannot exceed 0.1°/hr.

E. ANGULAR INCREMENT DATA

In this application, PSLR and integrated side-lobe ratio (ISLR) are essentially determined by veolcity quantization and lever arm vector uncertainty. At typical gyro quantization levels on the order of a few arc seconds, lever arm vector uncertainty is dominated by timing effects. With the 0.08 rad/sec rms angular rate, rms angular uncertainty θ accumulated in $\tilde{t}$ sec is 0.08 $\tilde{t}$ rad and, for a 1-ft separation, d, from sensor package to antenna, PSLR corresponding to a sinusoidal error at a frequency F would be determined from the ratio

$$\tilde{v}/F\lambda = 2\pi F\theta d/F\lambda = 2\pi(0.08\tilde{t})(1 \quad \text{ft})/0.1 \quad \text{ft} \doteq 5\tilde{t}. \tag{20}$$

ISLR in this case is essentially twice the level of (20), or $10\tilde{t}$. At $t = 0.00058$, corresponding to $(1/\sqrt{12}) \times (1/500)$ sec at 500-Hz data rate for attitude information, these figures correspond to $20\log(0.0029)$ and $20\log(0.0058)$ or ~ -50 and -44 dB, respectively; thus the 30-dB figure derived in Section III,C effectively governs the side-lobe levels at 0.01-ft/sec quantization. With finer velocity quantization the system could approach the 50- and 44-dB levels just derived for the data rate used here. Actually these figures depend heavily on methods used to reprocess the short-term attitude history and on detailed assumptions regarding error waveforms (which influence the ratio of rms to peak amplitudes, error statistics, etc.); they should thus be regarded as approximations subject to a variation of a few decibels.

F. SUMMARY OF STRAPDOWN REQUIREMENTS

Results obtained for the example just cited are summarized in Table I. They are restrictive in scope as already explained; extension to more complex scenarios is discussed in the next section.

These figures should be regarded as nominal values, subject to adjustment as appropriate for design needs. There are, for example, factors to be applied such as the square root of the number of partially overlapping coherent

TABLE I. INERTIAL INSTRUMENT ERROR COEFFICIENTS (LEVEL FLIGHT)[a]

Parameter	Gyro	Accelerometer
Null bias	Calibration dependent	Calibration dependent
Turn-on bias variability	0.1°/hr	0.0002 g
Scale factor	See Section IV	See Section IV
Y-Axis misalignment	See Section IV	Not critical[b]
Z-Axis misalignment	See Section IV	Not critical[b]
OA sensitivity	0.00008 sec	0.033 g sec^2
Anisoinertia	0.00008 sec	0.033 g sec^2
Rebalancing loop response lag	Not critical	Not critical
Differential response lag	0.00008 sec	Not critical
X-Axis g-sensitive drift	0.1°/hr/g	N/A
Z-Axis g-sensitive drift	0.1°/hr/g	N/A
g^2-Sensitive drift	0.1°/hr/g^2	N/A
Residual commutation error	0.1°/hr	N/A
Residual sculling	N/A	0.0002 g
Vibropendulosity	N/A	0.0002 g/g^2
Size effect	N/A	Not critical

[a] X and Y axes denote input and output axes, respectively; Z axis denotes SRA for gyros and pendulous axis for accelerometers. Values in this table are derived from the hypothetical conditions enumerated in Section I. Velocity quantization, 0.01 ft/sec; data rate, 500 Hz.

[b] While accelerometer alignment is not critical for navigation in straight and level flight, there are other requirements (e.g., alignment between IMU and radar) that could be expressed as restrictions on allowable uncertainty in the accelerometer-based coordinate frame. In the present context, however, that frame is coincident with aircraft coordinate axes by definition, so that the total error in angular displacement between lift vector and radar beam subdivides, for straight and level flight, into (1) INS leveling error and (2) uncertainty in relative orientation between radar and IMU.

integration intervals in a SAR frame. Also, some error contributors may exceed the nominal level while others are reduced, as long as the overall rss (root sum square) does not exceed the allowable total. In any case, the methodology is available to coordinate a complete balanced design in accordance with any set of conditions and system requirements.

IV. OPERATION IN THE PRESENCE OF ACCELERATION

For aircraft experiencing speed changes, pullups or turns, static specific force is no longer restricted to the vertical direction. The basic relation governing propagation of velocity error $\tilde{\mathbf{v}}$ remains, however,

$$\dot{\tilde{\mathbf{v}}} = \boldsymbol{\psi} \times \text{specific force} + \text{accelerometer error} \tag{21}$$

as resolved along locally level reference coordinates, for strapdown as well as gimbaled platforms (Chapter 4 in [2]). In this application, however, the azimuth reference of interest is vehicle based. Unlike the straight-and-level case (wherein the constant verticality of the lift force prompts a subdivision of orientation error; see footnote in Table I), here azimuth misalignment is defined only in terms of overall uncertainty in relative orientation between radar beam and acceleration vector. This lends immediate clarification to the subject under consideration (e.g., at a nominal 45° bank the azimuth misalignment now produces time-varying velocity and position errors similar to those associated with leveling error effects). Other specific force values can be analyzed by g scaling. These generalized conditions are by turns beneficial and detrimental, as follows:

1. Horizontal acceleration components considerably enhance observability of azimuth misorientation during update–alignment [4, 6].
2. Equation (13) no longer governs allowable azimuth misalignment if these more dominating effects are active during the mapping segment; allowable radar–acceleration vector misalignment uncertainty becomes simply the ratio of static acceleration error allowed by quadratic phase shift to static acceleration.
3. Differences in static conditions often produce changes in vibratory motions that influence the strapdown inertial instrument errors.
4. Three degradations that can be ignored in straight and level flight become active. The following conditions are now added to inertial instrument requirements.
    a. *Accelerometer scale factor*. In the presence of a nominal static lift of Lg, this effect must not introduce more than ψLg error during transfer alignment, since this would not be readily distinguishable from a tilt of ψ rad in a subsequent level mapping phase. If g levels experienced during mapping deviate significantly from unity, this requirement could also be scaled accordingly. Preferably, however, the analysis should be supported by further steps guided by considerations presented at the end of this section.
    b. *Gyro scale factor*. The maximum angular orientation change to be experienced, multiplied by the scale factor error, must not exceed ψ rad (which would produce ψg in nominally level flight). Again the considerations at the end of this section are cited for further analysis.

c. *Gyro misalignments.* These are defined as departures of IA direction (or, when skewed axes are used for redundancy management, uncertainties in these departures) from the accelerometer-based coordinate frame established on p. 72 and 111 in [2]. Although the direction of this effect differs from that of gyro scale factor error, the magnitude is computed in the same way; allowable misalignment about each axis [OA and spin reference axis (SRA)] is on the order of the ratio of allowable attitude error to (rotational excursion).

Attitude changes also influence the third term of (3) directly, and the second term indirectly. As with the other effects described here, quantitative results depend on applicable scenario dynamics.

Material just presented only highlights the issues to be addressed with general motion sequences. Reasons for this are traceable to variations in correlation between initial errors at the start of a mapping segment (e.g., immediately after transfer alignment) and errors present during that segment. It is quite well known, for example, that achievable verticality accuracy during in-flight leveling is inextricably linked to accelerometer-bias effects. Typically what is obtained is a combined effect of total instrument biases algebraically summed with the interaction between attitude error and prevailing forces [See (21) for an explanation]. There are several important ramifications:

1. The common practice of including accelerometer bias remaining after in-flight calibration of Section I as augmenting states in transfer alignment algorithms is, for rapid leveling, chancy at best and flatly inappropriate at worst. Accelerometer bias is not separately observable under typical flight conditions. Except during that in-flight calibration, the accelerometer bias states should be deactivated in the transfer alignment filter.

2. The pitfall mentioned after (7) can be demonstrated as follows: A 1-mrad tilt, counteracted by 800-μrad uncorrected accelerometer error, would produce an apparent 200-μg offset in a leveling operation. If the portion of steady-state instrument error contributed by kinematic rectification subsequently changed sign while remaining at the original magnitude, SAR performance could be unacceptable (due to more than twice the allowable amount of quadratic phase shift in the example of Section III,B). This is only one simple example illustrating how short-term IMU error correlation dynamics can make or break a planned sequence of in-flight operations. In this example it would be prudent to tighten all error coefficient specifications by some factor when loss of correlation (condition 4 of Section I) appears likely; more generally a complex scenario calls for either consistently conservative design procedures or rigorous simulation.

V. MODIFICATION TO PERMIT UPDATE/ALIGNMENT DURING A FRAME

The original (1985) presentation of this analysis, in *AIAA Journal of Guidance, Control and Dynamics* left open the subject of unacceptable performance prohibitive cost. A condition could nevertheless be postulated that, after all the design analysis, no acceptable cost–performance trade-off is found. The purpose of this addition is to provide a brief mention of an alternate approach, permitting transfer alignment adjustments between (and even within) SAR frames.

In applications where SAR motion compensation is the sole raison d'être for the strapdown IMU, attitude information is not inherently critical in its own right; its function is to produce dependable short-term translational information. Modest distortions of attitude data can be allowed if they result in better translational information. With the nine-state transfer alignment formulation [6] assumed in this article, for example, a 100-μg accelerometer bias will typically be "blamed" on a nonexistent 100-μrad leveling offset.[2] In [7] the willful distortion of data is carried further, allowing fictitious high-frequency motion, translational as well as rotational. These repetitive brief low-amplitude transients come from transfer alignment Kalman estimator adjustments, which are now allowed to take place between SAR frames and, if necessary, even within the coherent integration time. All of the quadratic and cubic phase errors analyzed in this article are counteracted, with a potential penalty of degraded ISLR and PSLR. Acceptable performance, as always, lies in finding the appropriate balance.

VI. CONCLUSIONS

Procedures have been indicated for establishing strapdown IMU requirements to meet the needs of synthetic aperture radar. Common deficiencies in strapdown IMU specifications as normally stated are identified and analyzed. Quantitative results do not represent any existing system, and are not intended to (some may in fact look strange); nevertheless the procedure followed in this

[2] Note that this procedure might well be unacceptable for precision pointing applications; transfer alignment for precise pointing will be similar, but not necessarily identical, to transfer alignment for SAR.

article provides an illustrative rationale to be used. Results obtained in any application will depend heavily on coherent integration time, duration of update–alignment phase and mapping segments, and applicable dynamics. Of these, the vibration environments (including rotational effects, with cross-axis correlations—both in-phase and quadrature) are generally the least available. Absence of this information will force designers to make assumptions, with the inevitable risks of degraded performance versus overdesign.

APPENDIX A. BACKGROUND FOR SYNTHETIC APERTURE RADAR DEGRADATION ANALYSIS

The following brief descriptive analysis is provided for those unfamiliar with synthetic aperture radar. Concepts presented here do not rigorously conform to mechanization, but are simplified to demonstrate only susceptibility to degradations. For that purpose, SAR imaging is characterized as the determination of relative amplitudes for radar reflections from each range–Doppler cell (analogous to a television picture element) in a swath. Geometric relations that define intersections of range and Doppler loci are combined with the collection of amplitude information, which may be obtained from magnitudes of complex numbers C_{nk} as follows. After centering to maintain the origin of a frame at a fixed range–Doppler cell index designation, the response history of the nth range gate to a sequence of transmitted radar pulses may be expressed as

$$G_n(t) = \sum_k C_{nk} \exp\{-i\omega_k t\}, \qquad 0 \le t \le t_c, \tag{22}$$

where $i = \sqrt{-1}$ and ω_k denotes the kth Doppler frequency in rads/second. The contents of this gate can then be decomposed into spectral components through repeated operations with correlating functions of the form

$$E_j(t) = \exp\{i(\omega_j t + \tilde{\phi})\}, \tag{23}$$

where $\tilde{\phi}$ represents the phase error from (2), which, if zero, would have produced just the complex amplitude C_{nk} when $j = k$ in the correlating operation,

$$\frac{1}{t_c} \int_0^{t_c} G_n(t) E_j(t)\, dt. \tag{24}$$

In general, however, correlator outputs include extraneous information exemplified as follows: Suppose a velocity error history $\tilde{v}(t)$ produces a range error history in accordance with

$$\delta|\mathbf{R}_{\mathrm{R}}| = \int \tilde{v}(t)\,dt \tag{25}$$

and, if

$$\tilde{v}(t) = \tilde{v}_0 \sin \omega_0 t, \tag{26}$$

then from (2), (25), and (26),

$$\tilde{\phi} = -[4\pi\tilde{v}_0/\lambda\omega_0]\cos \omega_0 t. \tag{27}$$

When this amplitude is $\ll 1$, (23) reduces to

$$E_j(t) \doteq [1 - i(4\pi\tilde{v}_0/\lambda\omega_0)\cos \omega_0 t]\exp\{i\omega_j t\}, \tag{28}$$

and, superimposed on the desired correlator output, there will be attenuated responses (consistent with the sideband amplitude just established) from reflectors in the nth range gate but displaced in Doppler by ω_0 rad/sec above and below the appropriate frequency. As another analytical step, (26) can be replaced by a sum of sinusoids producing attenuated responses from several remote Doppler cells. The amplitude of the largest of these, normalized by $|C_{nk}|$, is a dimensionless ratio p, which defines the PSLR as $-20\log(p)$ or $-10\log(p^2)$. The sum of squares of amplitudes in all remote Doppler cells, divided by $|C_{nk}|^2$, is a dimensionless ratio Q, which defines the ISLR as $-10\log(Q)$.

When ω_0 is small, (28) is not valid and the correlator output contains extraneous responses from nearby, rather than remote, cells. The image is then smeared rather than speckled. Suppose that

$$\cos \omega_0 t \doteq 1 - \tfrac{1}{2}(\omega_0 t)^2, \qquad 0 \le t \le t_{\mathrm{c}}. \tag{29}$$

From (23) and (27) this introduces a quadratic phase shift into the correlator, just as if an acceleration bias had been present as in (4). Superposition of a cubic term, (17) and (18), will make the correlator output still more responsive to the contents of nearby Doppler regions, thus further smearing the image. There is no abrupt separation between acceptable and unacceptable amounts, but the limits imposed by (5) and (18) are fairly representative.

APPENDIX B. CHECKLIST OF ISSUES TO BE RAISED

In extending the scope of inertial instrument-supported operations to include SAR motion compensation (or sensor stabilization in general), attention is drawn to several items beyond the standard list of cost, size, weight, power, reliability, maintainability, etc. This Appendix gives primary attention to issues calling for modified emphasis or interpretation as a result of adding sensor stabilization to the application of inertial instruments. For brevity they are categorized and tabulated in Table II.

TABLE II. IMU DESIGN ISSUES RELEVANT TO SENSOR STABILIZATION

Performance issues

Top priority to short-term accuracy stability—translational and rotational.

Inclusion of short-term translational accuracy in IMU specifications.

Inclusion of short-term rotational accuracy in IMU specifications.

Loss of short-term misorientation stability from gyro scale factor and misalignment effects.

Errors introduced through shock mounting; difference between motions experienced by instruments themselves and by their mounting base.

Rectification of motion-sensitive errors.

System configuration and function issues

Role of precise rotational information in its own right (e.g., for precise pointing) versus merely a means of ensuring accurate translational data.

Hand over from one tracking sensor to another.

Multiple or interleaved operations: navigation over geoid (latitude–longitude coordinates) and over "flat Earth" [Cartesian coordinates w.r.t. (with respect to) designated point] while performing concurrent radar modes (e.g., terrain follow, digital beamforming, tracking multiple airborne and moving or stationary ground targets, with multiple or time-shared sensors), all supported by IMU data and all interleaved in SAR.

Practicality of three or four standard IMU classes, based on performance.

Usage of three or four standard gyro specifications tagged to performance class.

One high-quality accelerometer standard.

Algorithm design issues

Multiplicity of individual contributors to overall accelerometer error.

Multiplicity of individual contributors to overall drift.

Independent changes in individual error contributors as motion is changed.

Extent of distinguishability among error contributions of different origins.

Extent of distinguishability of accelerometer bias from tilt effects.

Characterization of overall inertial instrument errors in terms of bias and random components—"dos and don'ts," ramifications.

Calibration via state augmentation—benefits and limitations.

Updating of master INS, slave INS, master and slave INS in tandem, and relative (slave INS w.r.t master INS) transfer alignment—transfer alignment algorithms, transfer alignment observables, and transfer alignment observability with large rotational deformations (i.e., greater than allowable misorientation); azimuth observability.

Requirements imposed by mission (e.g., requisite speed of transfer alignment response due to terrain masking).

TABLE II. *(Continued)*

Mechanization issues

Need for strapdown IMU (slave) colocated with sensor when structural connection from master INS to sensor is either long or nonrigid.

Physical separability of sensor package from the remainder of the strapdown inertial unit (to minimize structural impact of added weight on antenna; implications on RLG power supply).

Compensation of inertial instrument errors ahead of attitude and velocity computation; stability of calibration coefficients.

Sensor-to-processor electronics—input/output and algorithm speed.

Timing, applied to both attitude and velocity—(1) lags (buffering, degradation of real-time control operations); (2) data rates, which at low values can compromise the opportunity to exploit very fine instrument resolution; (3) data reprocessing (e.g., motion compensation calls for correction of phase at the time corresponding to each transmitted pulse, not at the time of navigation computer updates); (4) uncertainty in time tags; (5) necessity of synchronizing clocks from different units (via commonality of mechanized timing reference or by computational means; adverse effects when the computational approach must be adopted due to asynchronous CPUs); (6) extreme desirability for strapdown IMU to accept interrogations for gyro and accelerometer increments, at precise instants of time chosen by the sensor controls.

Possible standardization of interface.

Protocols.

Alignment—mechanical (mounting), electrical (sensor boresight), and optical.

Turntable—benefit (tumble test capability) versus introduction of movable parts.

Radar beam pointing: mechanical slewing (rates and angular accelerations) versus electronic steering via phase controls (attitude data latency).

Verification issues

Validation of performance—coordination of simulation with lab test and flight test.

Testability for short-term accuracy in the presence of velocity quantization, with attention given to separate short-term effects of each individual sensor degradation.

Environment issues

Vibration data including (1) rotational as well as translational, (2) cross-correlations as well as rms values, and (3) phase quadrature as well as in-phase components.

Mission constraints on allowable flight paths and motions.

Shock mounting—instrument protection versus vibration-sensitive gyro and accelerometer errors.

Range of stability for temperature compensation coefficients.

ACKNOWLEDGMENT

The analytical portion of this chapter was presented in the *AIAA Journal of Guidance & Control* **8**, 433–439, published by the American Institute of Aeronautics and Astronautics. The present chapter extends that material somewhat, and clarifies various issues that the journal manuscript failed to state.

REFERENCES

1. D. E. NEWLAND, "An Introduction to Randon Vibrations and Spectral Analysis." Longman, London, 1978.
2. J. L. FARRELL, "Integrated Aircraft Navigation." Academic Press, New York, 1976.
3. J. BESER and B. W. PARKINSON, "The Application of NAVSTAR Differential GPS in the Civilian Community," *Ion J.* **29**, 107–135 (1982).
4. A. E. BRYSON, "Rapid In-Flight Estimation of IMU Platform Misalignments Using External Position and Velocity Data," AFAL-TR-73-288, pp. 19–24 (1973).
5. F. G. UNGER, "Vector and Matrix Representation of Inertial Instruments," *IEEE Aerosp. Navig. Electron., 11th Annu. East Coast Conf.* (1964).
6. J. FARRELL, "Transfer Alignment for Precision Pointing Applications," *Naecon Symp., Dayton, Ohio* (1979).
7. J. FARRELL, "Airborne Transfer Alignment Simulation Results," *Plans, 88 Symp., Orlando, Fla.* (1988).

TRAJECTORY ESTIMATION ALGORITHM USING ANGLES-ONLY, MULTISENSOR TRACKING TECHNIQUES

TIMOTHY JOSEPH FREI

Space and Technology Group
TRW Inc.
Redondo Beach, California 90278

Department of Mechanical, Aerospace, and Nuclear Engineering
University of California
Los Angeles, California 90024

I. INTRODUCTION

This article deals with the problem of ballistic missile trajectory estimation using a pair of space-based, angles-only sensors simultaneously viewing a single target. Over the last decade, a great deal of research and development effort has been spent on formulating algorithms to process angles-only surveillance sensor data [1]. A primary focus of this research has been directed toward attempts to efficiently process observation data from a stereo pair of sensors in order to generate accurate estimates of a ballistic missile trajectory.

Multisensor trajectory estimation algorithms became important when passive, angles-only sensors were adopted into certain phases of the current strategic defense initiative (SDI) architecture. Angles-only sensors provide line-of-sight (LOS) measurements (specifically LOS elevation and azimuth) to a missile target, relative to their inertial position and boresight orientation. No range information is available with this type of passive sensor. Without ranging, a target's inertial position is not fully observable when viewed by a single sensor, specifically when the observer is located in the plane of the trajectory [2]. However, Nardon and Aidala [3] and Brown [4] have shown that properly processed observation data from two or more sensors can make the tracking problem fully observable, within geometric viewing limitations.

Copyright © 1990 by Academic Press, Inc.
All rights of reproduction in any form reserved.

The problem of target tracking is often preceded by that of target correlation. Target correlation is the process by which multiple sensors, simultaneously viewing common sets of in-flight missiles, associate identical targets against a clutter-rich background environment. After multiple sensor data has been correlated, a track file can be established for each missile and then updated throughout the trajectory, until observation data is no longer available. The track file can include a wide range of information including various tactical parameters such as launch point and time and flight heading. For today's tracking applications, such as SDI, the most important information in the track file is the target's state vector, nominally position, velocity, and possibly acceleration. Accurate state-vector estimation will provide the basis for performing boost, mid-course, and terminal-phase interdiction by SDI's sophisticated directed or kinetic energy weapons. Without accurate and timely tracking of enemy missiles, a successful SDI defense is vain hope.

In Section II, an angles-only trajectory estimation algorithm is developed. The triangulation trajectory tracker (TTT) is founded on the concept of crossing line-of-sight vectors in space. Given the approximate knowledge of sensor location in inertial space, accurate target position estimates can be obtained through analysis of the "triangle" (hence the name) formed by the base between the two sensors observing the target and the line-of-sight vectors to the target. In the absence of sensor position errors and sensor observation errors, the LOS vector from each of the sensors to the target would intersect in space. In reality, errors will always be present to prevent the vectors from precisely intersecting. The techniques utilized in the development of the TTT algorithm attempt to minimize the effects of these errors to obtain an accurate target position estimate.

After a target-position estimate has been obtained, the information is processed by a Kalman filter. The filter is used to estimate the other error corrupted trajectory parameters (velocity and acceleration). A complete development of the filter used in the TTT algorithm is included in Section II,D. Following the development of the filter, Section II,E presents the method by which the tracking process is initialized. Section II,F discusses an adaptive mechanism used in the algorithm to maintain tracking accuracy through staging events.

Sections III and IV, following the development of the TTT algorithm, are included for evaluation of the algorithm's performance. Chapter III highlights the development and presents the results of a Monte Carlo simulation built by the author to test the assumptions made in the TTT algorithm, along with its basic performance. Multiple simulations runs evaluate the algorithm for various sensor parameters (e.g., LOS accuracy and scan rate) against different types of missiles.

In Section IV, the dynamics free Kalman filter tracker (DFKFT) is presented, in a qualitative discussion, as an alternative to target triangulation methods. The DFKFT was built by members of the TRW Space and Technology Group, Military Space Systems Division, as part of the space defense scenario simulator [5]. A complete development of the DFKFT algorithm is not included here for the sake of brevity, and because of the proprietary aspects surrounding its development. Section IV,B compares the performance of the TTT algorithm against that of the DFKFT. As before, performance results for each algorithm are presented for different sensor parameters against different types of missile targets. Some conclusions about the design of the TTT algorithm, along with its overall performance, are discussed in Section V. Section V also includes some recommendations for future work and evaluation of the triangulation trajectory tracker.

The following symbols are used throughout the article:

A_i	Target azimuth as measured by ith sensor
E_i	Target elevation as measured by ith sensor
$E[u, v]$	Expectation operator, operating on quantities u, v
$\mathrm{H}(t)$	Filter measurement matrix
$\mathbf{i}, \mathbf{j}, \mathbf{k}$	Unit vectors that define the ECI coordinate system
$\mathbf{i}', \mathbf{j}', \mathbf{k}'$	Unit vectors that define a sensor coordinate system
$K(t)$	Filter gain matrix
$\mathbf{O}$	Sensor observation vector $[t, E, A]^T$
$P(t)$	Filter covariance matrix
$P_{\mathrm{m}}(t)$	Measurement covariance matrix
$\mathbf{U}_i$	Line-of-sight vector from ith sensor to the target
$v(t)$	Measurement noise (uncorrelated, white)
$w(t)$	Plant noise (uncorrelated, white)
W	Optimal position estimate weighting matrix
$\mathbf{X}(t)$	Missile state vector at time t
$\mathbf{X}_i$	ith sensor position vector (in ECI coordinates)
$\mathbf{X}_t$	Target-position estimate (in ECI coordinates)
$\mathbf{Z}(t)$	Target-position measurement
$\Delta\mathbf{Z}$	Measurement residual
Ψ	Projection of $(\mathbf{X}_1 - \mathbf{X}_2)$ onto the space spanned by $\{\mathbf{U}_1, \mathbf{U}_2\}$
Φ	State-transition matrix
∂	Partial derivative operator
ρ_i	Magnitude of the ith line-of-sight vector
Γ	Plant noise transition matrix
Σ_i	Covariance matrix of ith sensor position errors
σ_i	Covariance matrix of ith sensor line-of-sight errors
δ	Incremental change
χ^2	Chi-square statistic

II. DEVELOPMENT OF TRIANGULATION TRAJECTORY TRACKER

The triangulation trajectory tracker (TTT) algorithm processes stereo-paired sensor observation data to form an estimate of a ballistic missile's state vector over the course of its trajectory. The state vector $\mathbf{X}(t)$ includes the target's position, velocity, and acceleration:

$$\mathbf{X}(t) = \begin{bmatrix} \mathbf{x}(t) \\ \mathbf{v}(t) \\ \mathbf{a}(t) \end{bmatrix} \tag{1}$$

where $\mathbf{x}(t)$ is the target's position (x, y, z) at time t, $\mathbf{v}(t)$ is the target's velocity $(\dot{x}, \dot{y}, \dot{z})$ at time t, and $a(t)$ is the target's acceleration $(\ddot{x}, \ddot{y}, \ddot{z})$ at time t.

The six sub-sections of Section II include a complete development of the TTT algorithm. Sections III–V present the results of the computer simulation developed to analyze the TTT filter's performance.

A. COORDINATE SYSTEMS

The target state vector, defined in (1), is estimated in the traditional earth-centered inertial (ECI) reference frame, illustrated in Fig. 1. The unit vectors, $\mathbf{i}$, $\mathbf{j}$, and $\mathbf{k}$, represent the ECI directions X, Y, and $\mathbf{Z}$, respectively.

Each sensor generates a line-of-sight (LOS) vector to the target from observations of the missile's elevation and azimuth, measured in the sensor coordinate (SC) frame. The unit vectors, $\mathbf{i}'$, $\mathbf{j}'$, and $\mathbf{k}'$, that define a SC frame are illustrated in Fig. 1. These are obtained by a transformation of the sensor's position vector $\mathbf{X}_s$ measured in ECI coordinates:

$$\mathbf{k}' = -\mathbf{X}_s/|\mathbf{X}_s| = [k_x, k_y, k_z]^T \tag{2a}$$

$$\mathbf{j}' = \mathbf{k}' \times \mathbf{k}/|k' \times \mathbf{k}| = [j_x, j_y, j_z]^T \tag{2b}$$

$$\mathbf{i}' = \mathbf{j}' \times \mathbf{k}'/|\mathbf{j}' \times \mathbf{k}'| = [i_x, i_y, i_z]^T, \tag{2c}$$

where the subscripts x, y, and z identify the components of the SC frame unit vectors in ECI coordinates. In matrix notation, the local sensor coordinate transformation is given by

$$\begin{bmatrix} \mathbf{i}' \\ \mathbf{j}' \\ \mathbf{k}' \end{bmatrix} = \begin{bmatrix} -\sin(\phi)\cos(\lambda) & -\sin(\phi)\sin(\lambda) & \cos(\phi) \\ -\sin(\lambda) & \cos(\lambda) & 0 \\ -\cos(\phi)\cos(\lambda) & -\cos(\phi)\sin(\lambda) & -\sin(\phi) \end{bmatrix} \begin{bmatrix} \mathbf{i} \\ \mathbf{j} \\ \mathbf{k} \end{bmatrix}. \tag{3}$$

The angles ϕ and λ are the geodetic lattitude and longitude of the sensor, respectively, as shown in Fig. 2. The sensor is assumed to be pointing in the earth nadir direction.

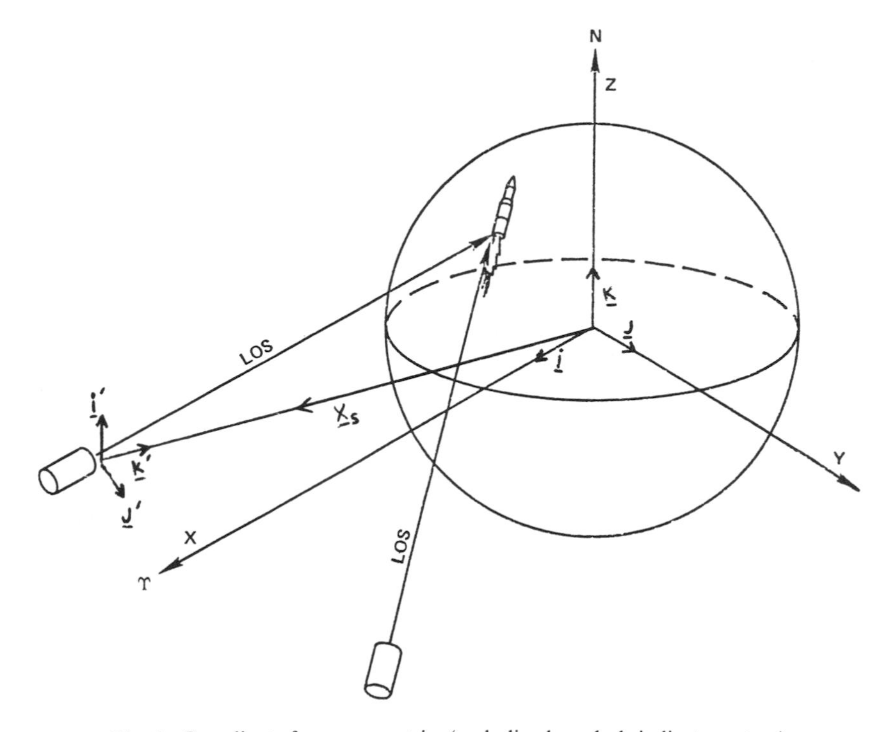

Fig. 1. Coordinate frame geometrics (underlined symbols indicate vectors).

B. SENSOR OBSERVATIONS

Each sensor measures the elevation and azimuth of the target LOS vector, in its SC frame. The measurement is conveniently held in an observation vector **O** that includes the time of the measurement and the target's elevation and azimuth:

$$\mathbf{O} = \begin{bmatrix} t \\ E \\ A \end{bmatrix} \tag{4}$$

The observation vector is transformed into a LOS vector to the target. The LOS vectors from sensors 1 and 2 to the target are

$$\mathbf{U}_1 = \cos(E_1)\cos(A_1)\mathbf{i}'_1 + \cos(E_1)\sin(A_1)\mathbf{j}'_1 + \sin(E_1)\mathbf{k}'_1 \tag{5a}$$

$$\mathbf{U}_2 = \cos(E_2)\cos(A_2)\mathbf{i}'_2 + \cos(E_2)\sin(A_2)\mathbf{j}'_2 + \sin(E_2)\mathbf{k}'_2, \tag{5b}$$

where E_i and A_i are the target's elevation and azimuth, respectively, measured in sensor 1 and sensor 2 coordinates, $[\mathbf{i}'_1, \mathbf{j}'_1, \mathbf{k}'_1]$ and $[\mathbf{i}'_2, \mathbf{j}'_2, \mathbf{k}'_2]$ respectively.

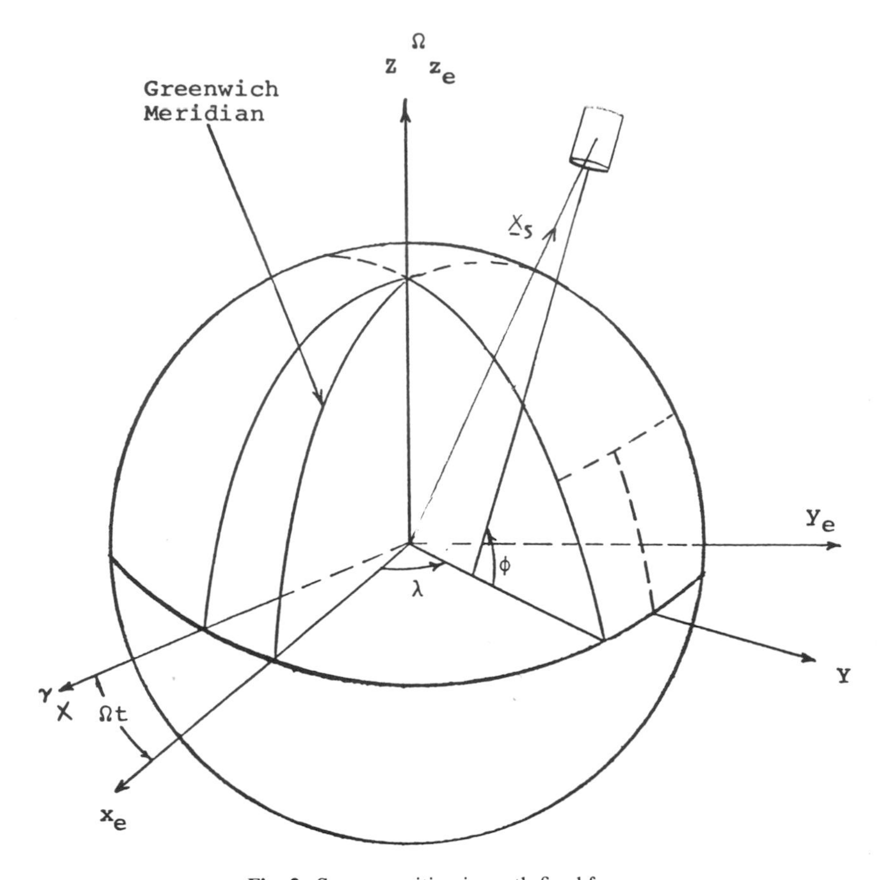

Fig. 2. Sensor position in earth-fixed frame.

C. TARGET-POSITION ESTIMATE

As discussed by Nardon and Aidala [3] and Brown [4], a missile's inertial position cannot always be accurately estimated by a single space-based, angle-only sensor due to the inherent lack of target-range information. Therefore, when this type of passive sensor is employed, at least two or more sensors are necessary to make the tracking problem fully observable.

A primary difficulty associated with the triangulation approach is the need for simultaneous observations from both sensors. In order to triangulate on the target, the sensor observations (El, Az) and sensor positions (x, y, z) must be known at a common time. Since the probability of obtaining exactly simultaneous target observations at both sensors is equal to zero, the observations must always be interpolated to a common time. The use of interpolated observation data violates the definition of the Kalman filter because it introduces correlation between the measurements. However, for the purpose of this algorithm, the measurement correlation will be neglected. The impact on performance of the neglected correlation due to measurement interpolation is presented in Section III,A.

A third-order difference interpolator has been implemented to form time-coincident target observations for both sensors. To perform the interpolation, four target measurements (for each sensor) are held in two matrices:

$$[M] = \begin{bmatrix} t_1 & El_1 & Az_1 \\ t_2 & El_2 & Az_2 \\ t_3 & El_3 & Az_3 \\ t_4 & El_4 & Az_4 \end{bmatrix}, \qquad [M]' = \begin{bmatrix} t'_1 & El'_1 & Az'_1 \\ t'_2 & El'_2 & Az'_2 \\ t'_3 & El'_3 & Az'_3 \\ t'_4 & El'_4 & Az'_4 \end{bmatrix}. \tag{6}$$

The rows of $[M]$ and $[M]'$ include the target elevation and azimuth measurements of sensors 1 and 2, respectively, at their corresponding times t_i and t'_i. The bottom row of each matrix contains the current sensor observation. The most recent observation time (either t_4 or t'_4) is the time to which the other measurement is interpolated. The scheme employs a "backwards difference" computational formula [6], weighting the most recent data more heavily than previous, older data.

After the observations have been interpolated to a common time, an estimate of the target's position can be generated, at that common time. To estimate the target position, the following problem is posed: Given the sensor position vectors, $\mathbf{X}_1$ and $\mathbf{X}_2$, and the LOS vectors to the target, $\mathbf{U}_1$ and $\mathbf{U}_2$, find ρ_1 and ρ_2 that minimize the equation:

$$\|(\mathbf{X}_1 + \rho_1\mathbf{U}_1) - (\mathbf{X}_2 + \rho_2\mathbf{U}_2)\|^2. \tag{7}$$

Equation (7) includes four vectors in the space, R^3, of which only three can be linearly independent. Therefore, the problems can be restated as the following: Given $(\mathbf{X}_1 - \mathbf{X}_2)$, $\mathbf{U}_1$ and $\mathbf{U}_2$, find a linear combination (ρ_1, ρ_2) such that

$$\|(\mathbf{X}_1 - \mathbf{X}_2) - (\rho_2\mathbf{U}_2 - \rho_1\mathbf{U}_1)\|^2 \tag{8}$$

is a minimum.

As previously noted, in the absence of observation, sensor position, and all other types of error, the LOS vectors would intersect in space, forcing the minimum value of (8) to be identically zero. However, with many sources of error corrupting the measurement, attempts must be made to minimize their effect on the position estimate.

Define a coordinate system $\mathbf{e}_1, \mathbf{e}_2, \mathbf{e}_3$, where

$$\mathbf{e}_1 = \mathbf{U}_1, \qquad \mathbf{e}_2 = \frac{(\mathbf{U}_2 - (\mathbf{U}_1 \cdot \mathbf{U}_2)\mathbf{U}_1)}{|\mathbf{U}_2 - (\mathbf{U}_1 \cdot \mathbf{U}_2)\mathbf{U}_1|}, \qquad \mathbf{e}_3 = \mathbf{e}_1 \times \mathbf{e}_2.$$

Now, define Ψ,

$$\Psi = [(\mathbf{X}_1 - \mathbf{X}_2) \cdot \mathbf{e}_1]\mathbf{e}_1 + [(\mathbf{X}_1 - \mathbf{X}_2) \cdot \mathbf{e}_2]\mathbf{e}_2 \tag{9}$$

to be the orthogonal projection of $(\mathbf{X}_1 - \mathbf{X}_2)$ onto the space spanned by $\{\mathbf{U}_1, \mathbf{U}_2\}$. The expressions for $\mathbf{e}_1$ and $\mathbf{e}_2$ are substituted into (9) to obtain

$$\Psi = \frac{(\mathbf{X}_1 - \mathbf{X}_2)^T(\mathbf{U}_1 - \mathbf{U}_2\mathbf{U}_2^T U_1)\mathbf{U}_1}{[1 - (\mathbf{U}_1^T\mathbf{U}_2)^2]} + \frac{(\mathbf{X}_1 - \mathbf{X}_2)^T(\mathbf{U}_2 - \mathbf{U}_1\mathbf{U}_1^T\mathbf{U}_2)\mathbf{U}_2}{[1 - (\mathbf{U}_1^T\mathbf{U}_2)^2]}. \tag{10}$$

From (8) it is clear that $-\rho_1$ is the coefficient of $\mathbf{U}_1$ and ρ_2 is the coefficient of $\mathbf{U}_2$ in (10), that is,

$$\rho_1 = -P(\mathbf{X}_1, \mathbf{X}_2, \mathbf{U}_1, \mathbf{U}_2) \cdot \mathbf{U}_1 = \frac{(\mathbf{X}_2 - \mathbf{X}_1)^T(I - \mathbf{U}_2\mathbf{U}_2^T)\mathbf{U}_1}{[1 - (\mathbf{U}_1^T\mathbf{U}_2)^2]} \tag{11a}$$

$$\rho_2 = P(\mathbf{X}_1, \mathbf{X}_2, \mathbf{U}_1, \mathbf{U}_2) \cdot \mathbf{U}_2 = \frac{(\mathbf{X}_1 - \mathbf{X}_2)^T(I - \mathbf{U}_1\mathbf{U}_1^T)\mathbf{U}_2}{[1 - (\mathbf{U}_1^T\mathbf{U}_2)^2]}. \tag{11b}$$

After the LOS vector magnitudes have been determined, a number of methods exist by which the target position can be estimated. Either of the LOS vectors could be combined separately with its sensor position to estimate the target position:

$$\mathbf{X}_t = \mathbf{X}_1 + \rho_1\mathbf{U}_1 \tag{12a}$$

$$\mathbf{X}_t' = \mathbf{X}_2 + \rho_1\mathbf{U}_2. \tag{12b}$$

However, a more accurate target position can be obtained by forming a weighted estimate, based on both of the individual estimates:

$$\mathbf{X}_t = W(\mathbf{X}_1 + \rho_1\mathbf{U}_1) + (I - W)(\mathbf{X}_2 + \rho_2\mathbf{U}_2), \tag{13}$$

where W is some desired weighting matrix and I is the 3×3 identity. In designing the algorithm with maximum accuracy in mind, the selection of W represents an optimization problem. In this algorithm, a matrix W will be found that minimizes the variance of the position estimate given by (13). To find such an optimal W, an expression for the measurement covariance matrix must first developed. The observation error in the position estimate can then be minimized by minimizing the trace of the measurement covariance matrix.

Let $\mathbf{X}_t$ be the measured estimate of the target position obtained by using $\mathbf{X}_1$, ρ_1, and $\mathbf{U}_1$. $\mathbf{X}_t$ is given by the expression

$$\mathbf{X}_t = \mathbf{X}_1 + \rho_1(\mathbf{X}_1, \mathbf{X}_2, \mathbf{U}_1, \mathbf{U}_2)\mathbf{U}_1. \tag{14}$$

As discussed by Lerner [7], if the measurement errors are sufficiently small, the error in $\mathbf{X}_t$ may be estimated by a first-order Taylor series:

$$\begin{aligned}\delta \mathbf{X}_t = \delta \mathbf{X}_1 &+ \mathbf{U}_1\left(\frac{\partial \rho_1}{\partial \mathbf{X}_1}\right)^T \delta \mathbf{X}_1 + \mathbf{U}_1\left(\frac{\partial \rho_1}{\partial \mathbf{U}_1}\right)^T\left(\frac{\partial \mathbf{U}_1}{\partial \mathbf{X}_1}\right)\delta \mathbf{X}_1 \\ &+ \mathbf{U}_1\left(\frac{\partial \rho_1}{\partial \mathbf{U}_1}\right)^T \delta \mathbf{U}_1 + \mathbf{U}_1\left(\frac{\partial \rho_1}{\partial \mathbf{X}_2}\right)^T \delta \mathbf{X}_2 \\ &+ \mathbf{U}_1\left(\frac{\partial \rho_1}{\partial \mathbf{U}_2}\right)^T\left(\frac{\partial \mathbf{U}_2}{\partial \mathbf{X}_2}\right)\delta \mathbf{X}_2 \\ &+ \mathbf{U}_1\left(\frac{\partial \rho_1}{\partial \mathbf{U}_2}\right)^T \delta \mathbf{U}_2 + \rho_1\, \delta \mathbf{U}_1 + \rho_1\left(\frac{\partial \mathbf{U}_1}{\partial \mathbf{X}_1}\right)\delta \mathbf{X}_1. \end{aligned} \tag{15}$$

The terms of (15) can be grouped to form the equation:

$$\delta \mathbf{X}_t = a_{11}\, \delta \mathbf{X}_1 + a_{12}\, \delta \mathbf{X}_2 + a_{13}\, \delta \mathbf{U}_1 + a_{14}\, \delta \mathbf{U}_2. \tag{16}$$

By noting that $\delta \mathbf{U}_i = [\partial \mathbf{U}_i/\partial(E_i, A_i)][\delta E_i, \delta A_i]^T$, (16) can be rewritten as

$$\delta \mathbf{X}_t = A_{11}\, \delta \mathbf{X}_1 + A_{12}\, \delta \mathbf{X}_2 + A_{13}[\delta E_1, \delta A_1]^T + A_{14}[\delta E_2, \delta A_2]^T \tag{17}$$

where

$$A_{11} = I + \mathbf{U}_1[(\partial \rho_1/\partial \mathbf{X}_1)^T + (\partial \rho_1/\partial \mathbf{U}_1)^T(\partial \mathbf{U}_1/\partial \mathbf{X}_1)] + \rho_1(\partial \mathbf{U}_1/\partial \mathbf{X}_1) \tag{18a}$$

$$A_{12} = \mathbf{U}_1[(\partial \rho_1/\partial \mathbf{X}_2)^T + (\partial \rho_1/\partial \mathbf{U}_2)^T(\partial \mathbf{U}_2/\partial \mathbf{X}_2)] \tag{18b}$$

$$A_{13} = [\rho_1 I + \mathbf{U}_1(\partial \rho_1/\partial \mathbf{U}_1)^T][\partial \mathbf{U}_1/\partial(E_1, A_1)] \tag{18c}$$

$$A_{14} = [\mathbf{U}_1(\partial \rho_1/\partial \mathbf{U}_2)^T][\partial \mathbf{U}_2/\partial(E_2, A_2)]. \tag{18d}$$

Similarly, let $\mathbf{X}_t'$ be the estimate of the target position using $\mathbf{X}_2$, ρ_2, and $\mathbf{U}_2$, given by the expression:

$$\mathbf{X}_t' = \mathbf{X}_2 + \rho_2(\mathbf{X}_1, \mathbf{X}_2, \mathbf{U}_1, \mathbf{U}_2)\mathbf{U}_2. \tag{19}$$

By expanding (19) in a Taylor series as before, we obtain

$$\delta \mathbf{X}_t' = A_{21}\, \delta \mathbf{X}_1 + A_{22}\, \delta \mathbf{X}_2 + A_{23}[\delta E_1, \delta A_1]^T + A_{24}[\delta E_2, \delta A_2]^T, \tag{20}$$

where

$$A_{21} = \mathbf{U}_2[(\partial \rho_2/\partial \mathbf{X}_1)^T + (\partial \rho_1/\partial \mathbf{U}_1)^T(\partial \mathbf{U}_1/\partial \mathbf{X}_1)] \tag{21a}$$

$$A_{22} = I + \mathbf{U}_2[(\partial \rho_2/\partial \mathbf{X}_2)^T + (\partial \rho_2/\partial \mathbf{U}_2)^T(\partial \mathbf{U}_2/\partial \mathbf{X}_2)] + \rho_2(\partial \mathbf{U}_2/\partial \mathbf{X}_2) \tag{21b}$$

$$A_{23} = [\mathbf{U}_2(\partial \rho_2/\partial \mathbf{U}_1)^T][\partial \mathbf{U}_1/\partial(E_1, A_1)] \tag{21c}$$

$$A_{24} = [\rho_2 I + \mathbf{U}_2(\partial \rho_2/\partial \mathbf{U}_2)^T][\partial \mathbf{U}_2/\partial(E_2, A_2)]. \tag{21d}$$

To compute the coefficients A_{ij}, a number of partial derivatives must be determined. Without formal proof, the necessary partials are given as follows.

The partials of the LOS vectors, $\mathbf{U}_n (n = 1, 2)$, with respect to the sensor position vectors $\mathbf{X}_n$ and the observation vectors (E_n, A_n) are given by

$$\frac{\partial \mathbf{U}_n}{\partial \mathbf{X}_n} = \cos(E_n)\cos(A_n)\left(\frac{\partial \mathbf{i}'}{\partial \mathbf{X}_n}\right) + \cos(E_n)\sin(A_n)\left(\frac{\partial \mathbf{i}'}{\partial \mathbf{X}_n}\right) + \sin(E_n)\left(\frac{\partial \mathbf{k}'}{\partial \mathbf{X}_n}\right), \quad (22)$$

where

$$\mathbf{k}' = -\mathbf{X}_n/|\mathbf{X}_{s_n}| = [k_x, k_y, k_z]^T$$

$$\mathbf{j}' = \mathbf{k}' \times \mathbf{k}/|\mathbf{k}' \times \mathbf{k}| = [j_x, j_y, j_z]^T$$

$$\mathbf{i}' = \mathbf{j}' \times \mathbf{k}'/|\mathbf{j}' \times \mathbf{k}'| = [i_x, i_y, i_z]^T$$

as before, and

$$(\partial \mathbf{k}'/\partial \mathbf{X}_n) = -[I - \mathbf{k}'\mathbf{k}'^T]/|\mathbf{X}_n| \quad (23a)$$

$$(\partial \mathbf{j}'/\partial \mathbf{X}_n) = \begin{bmatrix} 0 & 1 & 0 \\ -1 & 0 & 0 \\ 0 & 0 & 0 \end{bmatrix} (\partial \mathbf{k}'/\partial \mathbf{X}_n) \quad (23b)$$

$$(\partial \mathbf{i}'/\partial \mathbf{X}_n) = \begin{bmatrix} 0 & k_z & -k_y \\ -k_z & 0 & k_x \\ k_x & k_y & 0 \end{bmatrix} (\partial \mathbf{j}'/\partial \mathbf{X}_n) + \begin{bmatrix} 0 & -k_z & j_y \\ j_z & 0 & -j_x \\ -j_y & j_x & 0 \end{bmatrix} (\partial \mathbf{k}'/\partial \mathbf{X}_n). \quad (23c)$$

The partials of ρ_1 and ρ_2 with respect to $\mathbf{X}_1$, $\mathbf{X}_2$, $\mathbf{U}_1$, and $\mathbf{U}_2$ are

$$\frac{\partial \rho_1}{\partial \mathbf{X}_1} = \frac{-[(I - \mathbf{U}_2\mathbf{U}_2^T)\mathbf{U}_1]}{1 - (\mathbf{U}_1^T\mathbf{U}_2)^2} \quad (24a)$$

$$\frac{\partial \rho_1}{\partial \mathbf{X}_2} = \frac{-\partial \rho_1}{\partial \mathbf{X}_1} \quad (24b)$$

$$\frac{\partial \rho_1}{\partial \mathbf{U}_1} = \frac{I - (\mathbf{U}_2\mathbf{U}_2^T)(\mathbf{X}_2 - \mathbf{X}_1) + 2\rho_1(\mathbf{U}_1^T\mathbf{U}_2)\mathbf{U}_2}{1 - (\mathbf{U}_1^T\mathbf{U}_2)^2} \quad (24c)$$

$$\frac{\partial \rho_1}{\partial \mathbf{U}_2} = \frac{-[(\mathbf{U}_1^T\mathbf{U}_2)I + \mathbf{U}_1\mathbf{U}_2^T](\mathbf{X}_2 - \mathbf{X}_1) + 2\rho_1(\mathbf{U}_1^T\mathbf{U}_2)\mathbf{U}_1}{1 - (\mathbf{U}_1^T\mathbf{U}_2)^2} \quad (24d)$$

$$\frac{\partial \rho_2}{\partial \mathbf{U}_1} = \frac{[(\mathbf{U}_1^T\mathbf{U}_2)I + \mathbf{U}_2\mathbf{U}_1^T](\mathbf{X}_1 - \mathbf{X}_2) - 2\rho_2(\mathbf{U}_1^T\mathbf{U}_2)\mathbf{U}_2}{1 - (\mathbf{U}_1^T\mathbf{U}_2)^2} \quad (24e)$$

$$\frac{\partial \rho_2}{\partial \mathbf{U}_2} = \frac{I - (\mathbf{U}_1\mathbf{U}_1^T)(\mathbf{X}_1 - \mathbf{X}_2) + 2\rho_2(\mathbf{U}_1^T\mathbf{U}_2)\mathbf{U}_1}{1 - (\mathbf{U}_1^T\mathbf{U}_2)^2} \quad (24f)$$

$$\frac{\partial \rho_2}{\partial \mathbf{X}_1} = \frac{(I - \mathbf{U}_1\mathbf{U}_1^T)\mathbf{U}_2}{1 - (\mathbf{U}_1^T\mathbf{U}_2)^2} \quad (24g)$$

$$\frac{\partial \rho_2}{\partial \mathbf{X}_2} = \frac{-\partial \rho_2}{\partial \mathbf{X}_1}. \tag{24h}$$

Given (17)–(24), the expression for the covariance of the measurement is given by

$$P_{\mathrm{m}} = E\{[W\delta\mathbf{X}_{\mathrm{t}} + (I - W)\delta\mathbf{X}_{\mathrm{t}}'][\delta\mathbf{X}_{\mathrm{t}}^T W^T + \delta\mathbf{X}_{\mathrm{t}}'^T(I - W^T)]\}. \tag{25}$$

Expanding (25) yields

$$\begin{aligned} P_{\mathrm{m}} = {} & W\{E[\delta\mathbf{X}_{\mathrm{t}}\delta\mathbf{X}_{\mathrm{t}}^T] - E[\delta\mathbf{X}_{\mathrm{t}}\delta\mathbf{X}_{\mathrm{t}}'^T] - E[\delta\mathbf{X}_{\mathrm{t}}'\delta\mathbf{X}_{\mathrm{t}}^T] + E[\delta\mathbf{X}_{\mathrm{t}}'\delta\mathbf{X}_{\mathrm{t}}'^T]\}W^T \\ & + W\{E[\delta\mathbf{X}_{\mathrm{t}}\delta\mathbf{X}_{\mathrm{t}}'^T] - E[\delta\mathbf{X}_{\mathrm{t}}'\delta\mathbf{X}_{\mathrm{t}}'^T]\} + \{E[\delta\mathbf{X}_{\mathrm{t}}'\delta\mathbf{X}_{\mathrm{t}}^T] \\ & - E[\delta\mathbf{X}_{\mathrm{t}}'\delta\mathbf{X}_{\mathrm{t}}'^T]\}W^T + \{E[\delta\mathbf{X}_{\mathrm{t}}'\delta\mathbf{X}_{\mathrm{t}}'^T]\} \end{aligned} \tag{26}$$

or

$$P_{\mathrm{m}} = WAW^T + WB^T + BW^T + C \tag{27}$$

where A, B^T, B, and C are the bracketed terms, respectively, in (26). The linearized expectations in (26) are computed from the following relations:

$$E[\delta\mathbf{X}_{\mathrm{t}}\delta\mathbf{X}_{\mathrm{t}}^T] = A_{11}\Sigma_1 A_{11}^T + A_{12}\Sigma_2 A_{12}^T + A_{13}\sigma_1 A_{13}^T + A_{14}\sigma_2 A_{14}^T \tag{28a}$$

$$E[\delta\mathbf{X}_{\mathrm{t}}\delta\mathbf{X}_{\mathrm{t}}'^T] = A_{11}\Sigma_1 A_{21}^T + A_{12}\Sigma_2 A_{22}^T + A_{13}\sigma_1 A_{23}^T + A_{14}\sigma_2 A_{24}^T \tag{28b}$$

$$E[\delta\mathbf{X}_{\mathrm{t}}'\delta\mathbf{X}_{\mathrm{t}}^T] = A_{21}\Sigma_1 A_{11}^T + A_{22}\Sigma_2 A_{12}^T + A_{23}\sigma_1 A_{13}^T + A_{24}\sigma_2 A_{14}^T \tag{28c}$$

$$E[\delta\mathbf{X}_{\mathrm{t}}'\delta\mathbf{X}_{\mathrm{t}}'^T] = A_{21}\Sigma_1 A_{21}^T + A_{22}\Sigma_2 A_{22}^T + A_{23}\sigma_1 A_{23}^T + A_{24}\sigma_2 A_{24}^T, \tag{28d}$$

where Σ_1 and Σ_2 are the covariance matrices of position errors for sensors 1 and 2, respectively, and σ_1 and σ_2 are the covariance matrices of sensor LOS errors for sensors 1 and 2, respectively. To find an optimal weighting matrix W, the trace of P_{m} must be minimized.

Find w_{ij} that minimizes $\mathrm{Tr}(WAW^T + WB^T + BW^T + C)$, $i = 1, 2, 3$ and $j = 1, 2, 3$. In tensor notation,

$$\mathrm{Tr}(WAW^T + WB^T + BW^T + C) = w_{ij}a_{jk}w_{ik} + w_{il}b_{il} + b_{ip'}w_{ip'} + c_{ii}. \tag{29}$$

Taking partials w.r.t. w_{st} yields

$$\delta_{is}\delta_{jt}a_{jk}w_{ik} + w_{ij}a_{jk}\delta_{is}\delta_{kt} + \delta_{is}\delta_{lt}b_{il} + b_{ip'}\delta_{is}\delta_{p't} = 0 \tag{30a}$$

$$a_{tk}w_{sk} + w_{sj}a_{jt} + b_{st} + b_{st} = 0 \tag{30b}$$

$$w_{sk}a_{tk} + w_{sj}a_{jt} + 2b_{st} = 0. \tag{30c}$$

By exploiting the symmetry of A, (30a)–(30c) can be combined to obtain

$$2WA + 2B = 0 \Rightarrow W = -BA^{-1}. \tag{31}$$

Now that an expression for the weighting matrix W has been obtained, the optimal target-position estimate and its associated measurement covariance matrix can be computed.

D. FILTER EQUATIONS

The form of the Kalman filter equations used in the triangulation trajectory tracker algorithm are taken from Meditch [8]. The state model is given by the relation

$$\mathbf{X}(t_{k+1}) = \Phi(t_{k+1}, t_k)\mathbf{X}(t_k) + \Gamma(t_{k+1}, t_k)w(t_k), \tag{32}$$

where $\mathbf{X}(t_{k+1})$ is the target state at time t_{k+1}; $\mathbf{X}(t_k)$ is the target state at a previous time t_k; $\Phi(t_{k+1}, t_k)$ is the state transition matrix; $\Gamma(t_{k+1}, t_k)$ is the noise transition matrix; and $w(t_{k+1})$ is the plant noise (uncorrelated, white).

The state-transition matrix chosen for the TTT algorithm is given by

$$\Phi(t_{k+1}, t_k) = \begin{bmatrix} I & \Delta t_k I & (\Delta t_k^2/2)I \\ 0 & I & \Delta t_k I \\ 0 & 0 & I \end{bmatrix}, \tag{33}$$

where I is the 3×3 identity matrix; 0 is the 3×3 zero matrix; and Δt_k is the time difference $(t_{k+1} - t_k)$. This state-transition matrix was shown by Danchick [5] to perform reasonably well in both the boost and ballistic trajectory phases. It is attractive, not only because of its linearity, but also for its simplicity. It is considered to be "dynamics free" because it models the target motion in purely kinematic terms.

The plant noise used in the TTT is taken directly from that used in the DFKFT [5]. The plant noise model is built on the typically neglected third derivatives (i.e., the time rate of change of the missile acceleration). For simulation purposes, the plant noise is computed by an off-line numerical integration of the target dynamical equations of motion and then read into the algorithm by the main program. The noise-transition matrix then scales the plant noise according to the time-step size Δt_k. The DFKFT algorithm was developed under Independent Research and Development funds, and the plant noise model has been deemed proprietary. For this reason, it cannot be explicitly included here. It should be noted that because the plant noise involves a statistical treatment of the dynamics, the filter can only be claimed to be essentially dynamics free.

The measurement equation in the TTT algorithm is given by

$$\mathbf{Z}(t_{k+1}) = H(t_{k+1})\mathbf{X}(t_{k+1}) + v(t_{k+1}), \tag{34}$$

where $\mathbf{Z}(t_{k+1})$ is the observation at time t_{k+1}; $H(t_{k+1})$ is the measurement matrix at time t_{k+1}; $\mathbf{X}(t_{k+1})$ is the state at time t_{k+1}; and $v(t_{k+1})$ is the measurement noise at time t_{k+1}.

In the TTT algorithm, the target position is said to be measured directly if one includes the linearization process by which the measurement covariance matrix was obtained. The filter measurement matrix H is defined by the constant matrix:

$$H = \begin{bmatrix} 1 & 0 & 0 & 0 & 0 & 0 & 0 & 0 & 0 \\ 0 & 1 & 0 & 0 & 0 & 0 & 0 & 0 & 0 \\ 0 & 0 & 1 & 0 & 0 & 0 & 0 & 0 & 0 \end{bmatrix}. \tag{35}$$

The measurements are assumed to be disturbed by uncorrelated, white noise that corrupts the sensor observations of a target's elevation and azimuth. The disturbance results in erroneous LOS vectors which in turn corrupt the target-position estimate. In the absence of any measurement errors, the LOS vectors would intersect, resulting in a completely accurate triangulated target-position estimate.

The filtered state vector estimate at time t_{k+1} is given by the relation

$$\mathbf{X}(t_{k+1} \mid t_{k+1}) = \mathbf{X}(t_{k+1} \mid t_k) + K(t_{k+1})\Delta\mathbf{Z}(t_{k+1}) \tag{36}$$

where $\mathbf{X}(t_{k+1} \mid t_k)$ is an estimate of the target state vector at t_{k+1} given the filtered state vector estimate at time t_k; $K(t_{k+1})$ is the filter gain matrix at time t_{k+1}; and $\Delta\mathbf{Z}(t_{k+1})$ is the residual error in the measurement at time t_{k+1}.

The Kalman filter operates in a predictor–corrector fashion. First, the state from the preceding update is propagated forward to produce a predicted state at the current time. A correction term is then added to the extrapolated state, based on the residual of the observed state and the predicted state.

The extrapolated prediction $\mathbf{X}(t_{k+1} \mid t_k)$ is obtained by propagating the $\mathbf{X}(t_k)$ estimate forward in time using the state-transition matrix Φ:

$$\mathbf{X}(t_{k+1} \mid t_k) = \Phi(t_{k+1}, t_k)\mathbf{X}(t_k). \tag{37}$$

The gain matrix $K(t_{k+1})$ is computed using relation

$$K(t_{k+1}) = P(t_{k+1} \mid t_k)H^T(t_{k+1})[H(t_{k+1})P(t_{k+1} \mid t_k)H^T(t_{k+1}) + P_{\mathrm{m}}(t_{k+1})]^{-1} \tag{38}$$

where $H(t_{k+1})$ is given in (35); $P_{\mathrm{m}}(t_{k+1})$ is the covariance matrix of the measurement at time t_{k+1}; $P(t_{k+1} \mid t_k)$ is the filter covariance matrix extrapolated to time t_{k+1} based on measurements up to and including that at time t_k. It is computed from

$$P(t_{k+1} \mid t_k) = \Phi(t_{k+1}, t_k)P(t_k)\Phi^T(t_{k+1}, t_k) + \Gamma(t_{k+1}, t_k)w(t_k)\Gamma^T(t_{k+1}, t_k) \tag{39}$$

where $\Gamma(t_{k+1}, t_k)$ is the noise-transition matrix, $w(t_k)$ is the plant noise at time t_k, $P(t_k)$ is the filter covariance matrix at time t_k, based on all previous measurements up to and including the current measurement. $P(t_k)$ is obtained from the relation

$$P(t_k) = [I - K(t_{k+1})H(t_{k+1})]P(t_{k+1} \mid t_k) \tag{40}$$

where I is the 9×9 identity matrix, and $K(t_{k+1})$, $H(t_{k+1})$, and $P(t_{k+1} \mid t_k)$ are as defined previously. The measurement residual, $\Delta\mathbf{Z}(t_{k+1})$ is computed from the relation

$$\Delta\mathbf{Z}(t_{k+1}) = \mathbf{Z}(t_{k+1}) - H(t_{k+1})\mathbf{X}(t_{k+1} \mid t_k), \tag{41}$$

where $\mathbf{Z}(t_{k+1})$ is the measured target position vector at t_{k+1}, and $H(t_{k+1})$ and $\mathbf{X}(t_{k+1} \mid t_k)$ are as defined previously.

E. TRACKING FILTER INITIALIZATION

A very important part of the tracking filter process is the means by which the state vector and the filter covariance matrix are initialized. In this algorithm, a three-point initialization scheme has been implemented to provide an initial-state vector and covariance matrix following the third target observation.

Let $\mathbf{X}_{t_i} = [x_i, y_i, z_i]^T$ be the ith target-position estimate at time $t_i (i = 1, 2, 3)$. Suppose $x(t)$ is the component of the target position in the direction of $\mathbf{i}$, at time t. $x(t)$ can be expressed by the relation

$$x(t) = a_0 + a_1(t - t_3) + (a_2/2)(t - t_3)^2. \tag{42}$$

Then,

$$x(t_3) = a_0 \tag{43}$$

$$x(t_1) - x(t_3) = a_1(t_1 - t_3) + (a_2/2)(t_1 - t_3)^2 \tag{44}$$

$$x(t_2) - x(t_3) = a_1(t_2 - t_3) + (a_2/2)(t_2 - t_3)^2. \tag{45}$$

Equations (43)–(45) can be solved for a_0, a_1, and a_2:

$$\begin{bmatrix} a_0 \\ a_1 \\ a_2 \end{bmatrix} = [D][E] \begin{bmatrix} x(t_1) \\ x(t_2) \\ x(t_3) \end{bmatrix}, \tag{46}$$

where

$$[D] = \begin{bmatrix} 1 & 0 & 0 \\ 0 & \dfrac{(t_2 - t_3)}{(t_2 - t_1)} & \dfrac{(t_3 - t_1)}{(t_2 - t_1)} \\ 0 & \dfrac{-2}{(t_2 - t_1)} & \dfrac{2}{(t_2 - t_1)} \end{bmatrix} \tag{47a}$$

and

$$[E] = \begin{bmatrix} 0 & 0 & 1 \\ \dfrac{1}{(t_1 + t_3)} & 0 & \dfrac{-1}{(t_1 - t_3)} \\ 0 & \dfrac{1}{(t_2 - t_3)} & \dfrac{-1}{(t_2 - t_3)} \end{bmatrix}. \tag{47b}$$

Now, define $[C] = [D][E]$, and partition $[C]$ into 1×3 row vectors:

$$C = \begin{bmatrix} C_1 \\ \hline C_2 \\ \hline C_3 \end{bmatrix}.$$

It can be shown that the initial-state vector (at time t_3), based on the first three observations, is given by

$$X(t_3 \mid t_3) = \begin{bmatrix} x \\ y \\ z \\ x \\ y \\ z \\ x \\ y \\ z \end{bmatrix} = \begin{bmatrix} C_1 & \mathbf{0} & \mathbf{0} \\ \mathbf{0} & C_1 & \mathbf{0} \\ \mathbf{0} & \mathbf{0} & C_1 \\ C_2 & \mathbf{0} & \mathbf{0} \\ \mathbf{0} & C_2 & \mathbf{0} \\ \mathbf{0} & \mathbf{0} & C_2 \\ C_3 & \mathbf{0} & \mathbf{0} \\ \mathbf{0} & C_3 & \mathbf{0} \\ \mathbf{0} & \mathbf{0} & C_3 \end{bmatrix} \begin{bmatrix} x_1 \\ x_2 \\ x_3 \\ y_1 \\ y_2 \\ y_3 \\ z_1 \\ z_2 \\ z_3 \end{bmatrix} \tag{48}$$

where $\mathbf{O}$ is the 1×3 row matrix $[0 \quad 0 \quad 0]$.

To generate the initial filter covariance matrix, several definitions are required. Let $\sigma_{ij}^{(k)}$ be the ijth element of the kth measurement covariance matrix, and define

$$[\sigma_{ij}] = \begin{bmatrix} \sigma_{ij}^{(1)} & 0 & 0 \\ 0 & \sigma_{ij}^{(2)} & 0 \\ 0 & 0 & \sigma_{ij}^{(3)} \end{bmatrix}. \tag{49}$$

Also, let $\mathbf{x} = (x - \bar{x}, y - \bar{y}, z - \bar{z})^T$, $\mathbf{v} = (\dot{x} - \bar{\dot{x}}, \dot{y} - \bar{\dot{y}}, \dot{z} - \bar{\dot{z}})^T$, and $\mathbf{a} =$ some column vectors $\mathbf{u}$ and $\mathbf{v}$.

The initial covariance matrix (at time t_3), based on the first three observations, is given by

$$P(t_3 \mid t_3) = \begin{bmatrix} E[\mathbf{x} \quad \mathbf{x}^T] & E[\mathbf{x} \quad \mathbf{v}^T] & E[\mathbf{x} \quad \mathbf{a}^T] \\ E[\mathbf{v} \quad \mathbf{x}^T] & E[\mathbf{v} \quad \mathbf{v}^T] & E[\mathbf{v} \quad \mathbf{a}^T] \\ E[\mathbf{a} \quad \mathbf{x}^T] & E[\mathbf{a} \quad \mathbf{v}^T] & E[\mathbf{a} \quad \mathbf{a}^T] \end{bmatrix}. \tag{50}$$

Each of the nine blocks of P is a 3×3 matrix that must be evaluated. The upper left block, $P(1,1)$, is given by the expression

$$E[\mathbf{x}\quad \mathbf{x}^T] = \begin{bmatrix} \mathbf{C}_1[\sigma_{11}]\mathbf{C}_1^T & \mathbf{C}_1[\sigma_{12}]\mathbf{C}_1^T & \mathbf{C}_1[\sigma_{13}]\mathbf{C}_1^T \\ \mathbf{C}_1[\sigma_{12}]\mathbf{C}_1^T & \mathbf{C}_1[\sigma_{22}]\mathbf{C}_1^T & \mathbf{C}_1[\sigma_{23}]\mathbf{C}_1^T \\ \mathbf{C}_1[\sigma_{13}]\mathbf{C}_1^T & \mathbf{C}_1[\sigma_{23}]\mathbf{C}_1^T & \mathbf{C}_1[\sigma_{33}]\mathbf{C}_1^T \end{bmatrix}. \tag{51a}$$

Noting that P is a symmetric matrix, the other eight blocks are given by the following expressions:

$$E[\mathbf{x}\quad \mathbf{v}^T] = \begin{bmatrix} \mathbf{C}_1[\sigma_{11}]\mathbf{C}_2^T & \mathbf{C}_1[\sigma_{12}]\mathbf{C}_2^T & \mathbf{C}_1[\sigma_{13}]\mathbf{C}_2^T \\ \mathbf{C}_1[\sigma_{12}]\mathbf{C}_2^T & \mathbf{C}_1[\sigma_{22}]\mathbf{C}_2^T & \mathbf{C}_1[\sigma_{23}]\mathbf{C}_2^T \\ \mathbf{C}_1[\sigma_{13}]\mathbf{C}_2^T & \mathbf{C}_1[\sigma_{23}]\mathbf{C}_2^T & \mathbf{C}_1[\sigma_{33}]\mathbf{C}_2^T \end{bmatrix} \tag{51b}$$

$$E[\mathbf{x}\quad \mathbf{a}^T] = \begin{bmatrix} \mathbf{C}_1[\sigma_{11}]\mathbf{C}_3^T & \mathbf{C}_1[\sigma_{12}]\mathbf{C}_3^T & \mathbf{C}_1[\sigma_{13}]\mathbf{C}_3^T \\ \mathbf{C}_1[\sigma_{12}]\mathbf{C}_3^T & \mathbf{C}_1[\sigma_{22}]\mathbf{C}_3^T & \mathbf{C}_1[\sigma_{23}]\mathbf{C}_3^T \\ \mathbf{C}_1[\sigma_{13}]\mathbf{C}_3^T & \mathbf{C}_1[\sigma_{23}]\mathbf{C}_3^T & \mathbf{C}_1[\sigma_{33}]\mathbf{C}_3^T \end{bmatrix} \tag{51c}$$

$$E[\mathbf{v}\quad \mathbf{x}^T] = E[\mathbf{x}\quad \mathbf{v}^T]^T \tag{51d}$$

$$E[\mathbf{v}\quad \mathbf{v}^T] = \begin{bmatrix} \mathbf{C}_2[\sigma_{11}]\mathbf{C}_2^T & \mathbf{C}_2[\sigma_{12}]\mathbf{C}_2^T & \mathbf{C}_2[\sigma_{13}]\mathbf{C}_2^T \\ \mathbf{C}_2[\sigma_{12}]\mathbf{C}_2^T & \mathbf{C}_2[\sigma_{22}]\mathbf{C}_2^T & \mathbf{C}_2[\sigma_{23}]\mathbf{C}_2^T \\ \mathbf{C}_2[\sigma_{13}]\mathbf{C}_2^T & \mathbf{C}_2[\sigma_{23}]\mathbf{C}_2^T & \mathbf{C}_2[\sigma_{33}]\mathbf{C}_2^T \end{bmatrix} \tag{51e}$$

$$E[\mathbf{v}\quad \mathbf{a}^T] = \begin{bmatrix} \mathbf{C}_2[\sigma_{11}]\mathbf{C}_3^T & \mathbf{C}_2[\sigma_{12}]\mathbf{C}_3^T & \mathbf{C}_2[\sigma_{13}]\mathbf{C}_3^T \\ \mathbf{C}_2[\sigma_{12}]\mathbf{C}_3^T & \mathbf{C}_2[\sigma_{22}]\mathbf{C}_3^T & \mathbf{C}_2[\sigma_{23}]\mathbf{C}_3^T \\ \mathbf{C}_2[\sigma_{13}]\mathbf{C}_3^T & \mathbf{C}_2[\sigma_{23}]\mathbf{C}_3^T & \mathbf{C}_2[\sigma_{33}]\mathbf{C}_3^T \end{bmatrix} \tag{51f}$$

$$E[\mathbf{a}\quad \mathbf{x}^T] = E[\mathbf{x}\quad \mathbf{a}^T]^T \tag{51g}$$

$$E[\mathbf{a}\quad \mathbf{v}^T] = E[\mathbf{v}\quad \mathbf{a}^T]^T \tag{51h}$$

$$E[\mathbf{a}\quad \mathbf{a}^T] = \begin{bmatrix} \mathbf{C}_3[\sigma_{11}]\mathbf{C}_3^T & \mathbf{C}_3[\sigma_{12}]C_3^T & \mathbf{C}_3[\sigma_{13}]\mathbf{C}_3^T \\ \mathbf{C}_3[\sigma_{12}]\mathbf{C}_3^T & \mathbf{C}_3[\sigma_{22}]\mathbf{C}_3^T & \mathbf{C}_3[\sigma_{23}]\mathbf{C}_3^T \\ \mathbf{C}_3[\sigma_{13}]\mathbf{C}_3^T & \mathbf{C}_3[\sigma_{23}]\mathbf{C}_3^T & \mathbf{C}_3[\sigma_{33}]C_3^T \end{bmatrix}. \tag{51i}$$

Finally, with all nine blocks evaluated, the initial filter covariance matrix (at time t_3) can be computed by substituting the blocks into (50).

F. ADAPTIVE FILTER MECHANISM

An adaptive filter mechanism has been built into the TTT algorithm to improve overall performance, especially while tracking through staging events. Maybeck, Jenson, and Harnly [9], Moose [10], and Moose and Gholson [11] have shown both the applicability and advantage of adaptive filtering techniques applied to angle-only tracking. Danchick [5] has shown the difficulty of tracking ballistic missiles through their boost-phase staging events without the aid of an adaptive mechanism.

The mechanism implemented in the TTT algorithm is a 3-degree-of-freedom (DOF) extension of the 2-DOF mechanism used in the DFKFT algorithm [5]. The filter performance criterion that activates the mechanism is the residual chi square (χ^2) statistic, computed from the relation

$$\chi^2 = \Delta\mathbf{Z}^T[HPH^T + P_{\mathrm{m}}]^{-1}\Delta\mathbf{Z} \tag{52}$$

where $\Delta\mathbf{Z}$ is the measurement residual given in (42); H is the measurement matrix given in (36); P is the extrapolated filter covariance matrix given in (39); and P_{m} is the current measurement covariance matrix.

After computing the filter gain, $K(t_{k+1})$, the χ^2 is computed. If the value of the statistic falls outside the 99% confidence region ($\chi^2 > 11.341$) [12], a scalar multiple of the plant noise is added to the extrapolated filter covariance matrix, $P(t_{k+1} \mid t_k)$, and the gain is recomputed. The scalar multiple is determined through an eigenvalue–eigenvector analysis routine adopted from [5]. The factor provides just enough additive plant noise to the covariance matrix so as to reduce the χ^2 to a nominal value of 11.341. The explicit adaptive mechanism algorithm is not included here because of previously mentioned company proprietary reasons.

III. TRIANGULATION TRAJECTORY TRACKER ALGORITHM PERFORMANCE

A large computer simulation was developed to establish the performance of the TTT algorithm. It includes a driver program and 38 subprograms, all totaling over 3000 lines of executable FORTRAN code. The simulation is embedded within the space defense scenario simulator (SDSS) [5], which runs on a VAX 8600 computer. The SDSS fields the event trajectory and provides pointers to the target and sensor positions at program requested times. The TTT simulation uses the "truth" position data to generate noise-corrupted sensor observations at each scan time. The tracking filter uses the noisy observations to estimate the target trajectory. The estimated trajectory is then compared to the actual vehicle position to determine the filter residuals. The output of the tracking program is a file containing (1) the time of the current filter update, (2) the ECI position of each sensor, (3) the true target state vector, (4) the residuals of the filtered state vector, (5) the chi-square statistic of the filter, (6) the measurement covariance matrix, and (7) the covariance matrix of the filter.

Due to the cost of computer resources, the TTT algorithm was evaluated against only two different missile types. For security classification reasons, the missiles will only be refered to as type 1 and type 2. The initial viewing geometry was kept the same for all test cases. Sensor 1 views the target with an earth-central angle (ECA) of 55°, and sensor 2 has a 68° ECA. The random number seed sequence was also held fixed at the start of each simulation run. Although this test method does not produce a complete ensemble of trials for each test case, it does allow a fair comparison between successive runs with varied sensor parameters, against the same missile type. In all test cases, the first observation by either sensor (both for matched cases) occurs 30 sec after target launch. In all cases, tracking is terminated at $\sim$390 sec into the flight.

A. OBSERVATION INTERPOLATION EFFECTS

The first test conducted on the TTT algorithm attempts to assess the effect of interpolating the sensor observations to a common time, prior to estimating the target position. To perform this test, one trajectory was flown with matched sensor sightings (i.e., sensors 1 and 2 obtain observation data simultaneously). The same trajectory was then flown with the sightings interleaved one-half of a frame time apart. This represents the worst-case interpolation.

Four type-1 trajectories were run to generate two interpolation test cases. In the first test, both sensors have 1-arcsec LOS accuracies and 5-sec frame times. The interpolation effects on position and velocity-tracking error are illustrated in Fig. 3a and 3b, respectively. For the second case, the frame times of each sensor were increased to 10 sec, while the LOS accuracies were kept at 1 arcsec. The results for test 2 are shown is Figs. 4a and 4b.

It is evident from the results of both test cases that interpolating the observations to a common time does not significantly affect the filter's performance. This indicates that a reasonable decision was made to neglect the correlation between observations resulting from interpolation. In fact, it can be seen from Fig. 3b that interpolation acted to suppress a large error spike in velocity tracking at 375 sec into the flight.

Although only two cases were run to evaluate interpolation effects, the resulting performance appears to be consistent. However, additional cases should be run in the future to fully explore tracking errors induced by observation interpolation.

B. PERFORMANCE OF ADAPTIVE FILTER MECHANISM

To illustrate the necessity of including an adaptive filtering mechanism to maintain accurate tracking through staging events, type 1 and 2 trajectories

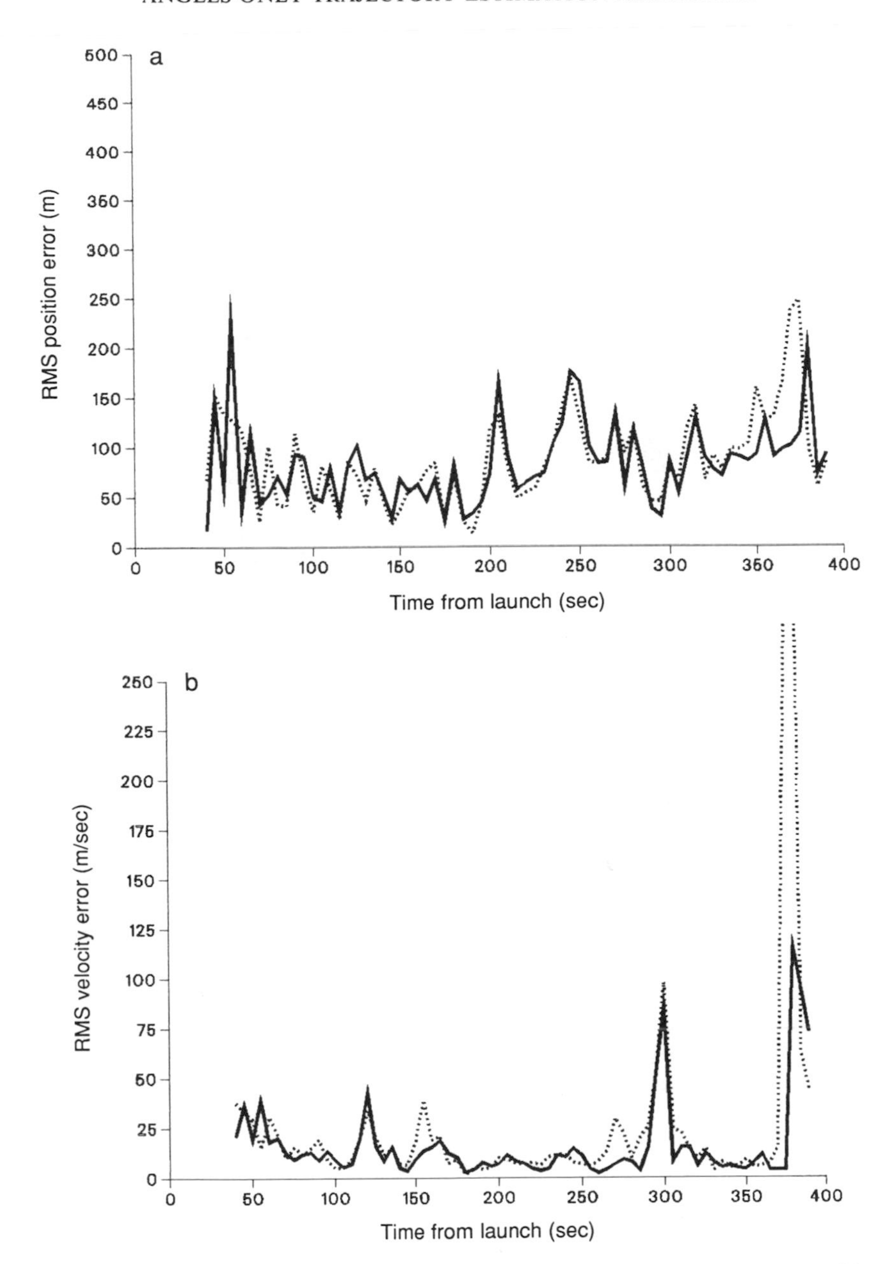

Fig. 3. TTT interpolation performance—test 1 (5-sec frame time; 1-arcsec LOS): (a) position-tracking error and (b) velocity-tracking error. Solid line, interleaved; dotted line, matched.

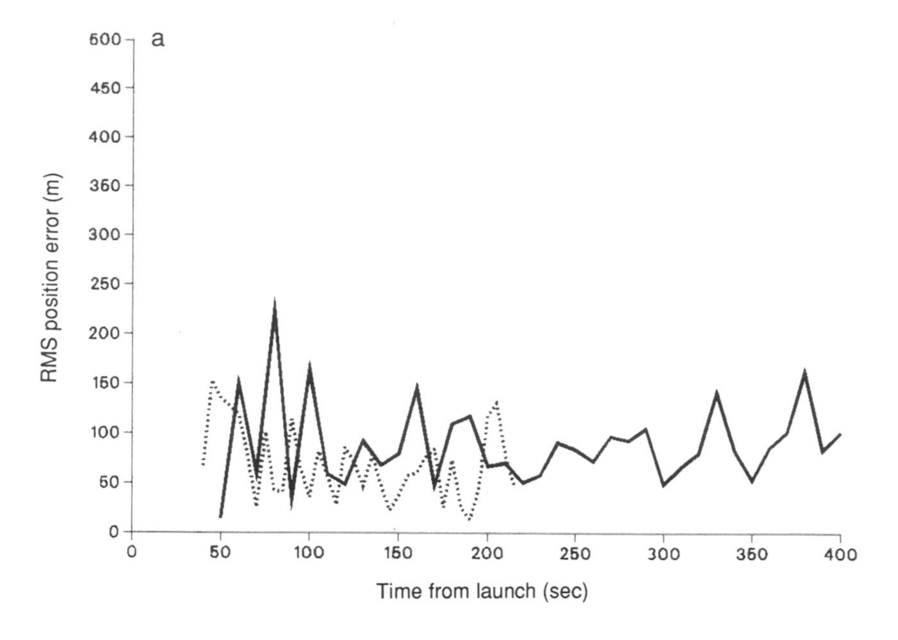

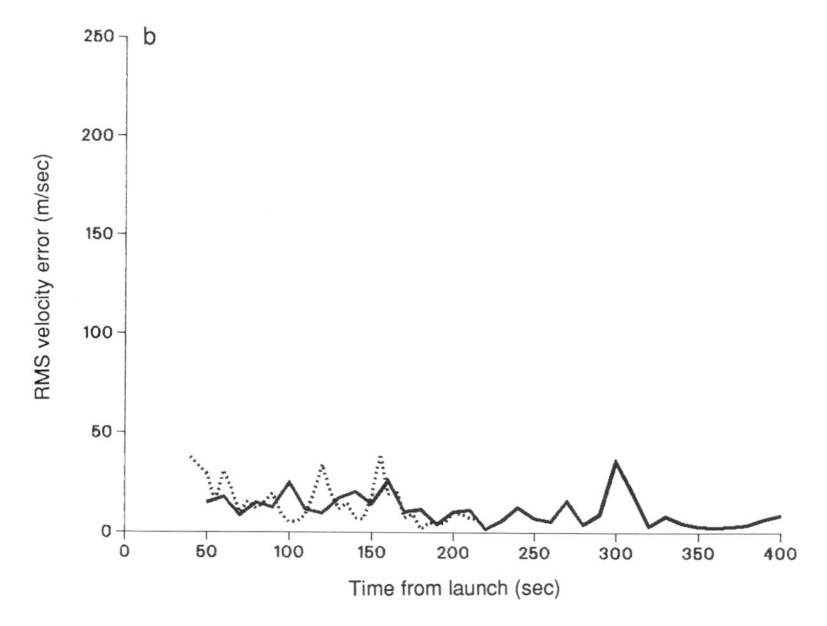

Fig. 4. TTT interpolation performance—test 2 [10-sec frame time, 1-arcsec LOS]: (a) position-tracking error and (b) velocity-tracking error. Solid line, interleaved; dotted line, matched.

were observed by the TTT filter, both with and without the mechanism enabled. In each test case, the sensors were given 1-arcsec LOS accuracies and 5-sec frame times.

Figures 5a and 5b present adaptive performance when tracking a type-1 missile. It is obvious from the results that the filter performance is greatly enhanced with the mechanism enabled. Without adaptive compensation for nonlinearities, the rms errors in both position and velocity increased substantially, immediately following each staging event (150 and 290 sec). With the mechanism enabled, the position error was unaffected by staging, while the velocity error was only slightly affected.

Figures 6a and 6b illustrate adaptive filter performance against a type-2 target, for the defined sensor parameters. In this case, the adaptive mechanism added little to the filter's tracking performance. Close examination of Figs. 6a and 6b reveals that the filter was forced to adapt only slightly (near 130 sec) throughout the entire trajectory.

The results from the two test cases make very different arguments about the need for adaptive filtering. Obviously more cases should be evaluated. It is interesting to note that, in case 1, the filter did not continue to diverge after the staging events. It is recommended that additional cases he run to test the filter's range of recoverability as well as to fully explore the overall effect of adaptive filtering.

C. SENSOR LINE-OF-SIGHT ERROR EFFECTS

Two test cases were run to establish the effect of sensor LOS errors on the overall filter performance. LOS accuracies of 1 and 5 arcsec were examined. Test case 1 consisted of a type-1 missile trajectory observed by sensors with 10-sec frame times. For the first run, the sensors were given 1-arcsec LOS accuracies. In the second run, the accuracies were degraded to 5 arcsec. The results of each case were then plotted together. The rms position- and velocity-tracking errors are shown in Figs. 7a and 7b, respectively. In this test, both the position and velocity-tracking errors are significantly larger for the 5-arcsec LOS sensors. When averaged over the entire trajectory track, the errors for the 5-arsec LOS run were, as should be expected, nearly five times larger than in the 1-arcsec run.

The second test case was also performed using a type-1 trajectory. This time, the sensor frame times were increased to 5 sec. The sensor LOS accuracies were varied as in case 1. The results for case 2 are shown in Figs. 8a and 8b. Again, the position and velocity errors increased by a factor of ~ 5 as the LOS accuracy was degraded by the same factor.

The results of the two cases just described clearly indicate a strong correlation between sensor LOS accuracy and filter performance, as expected.

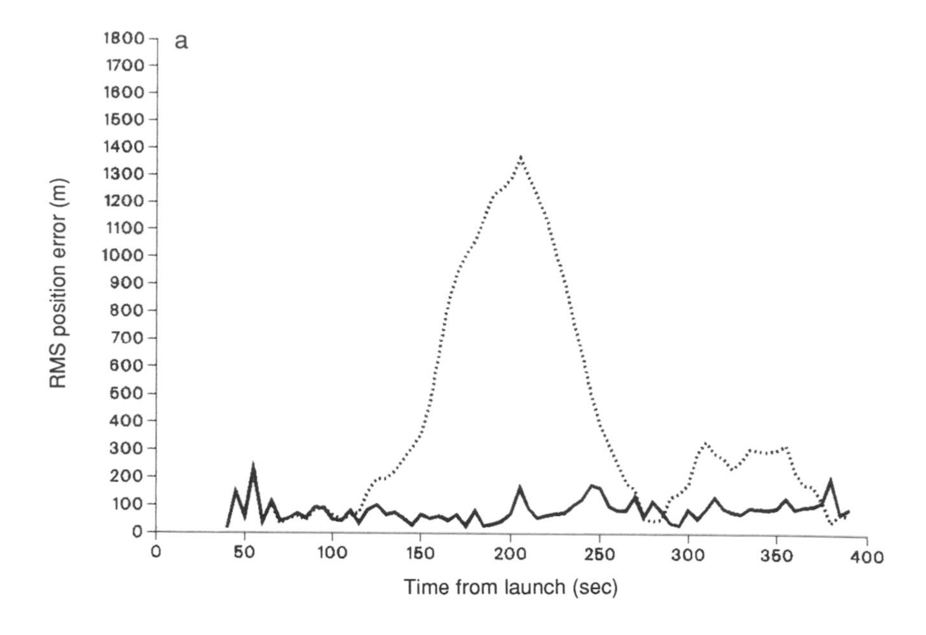

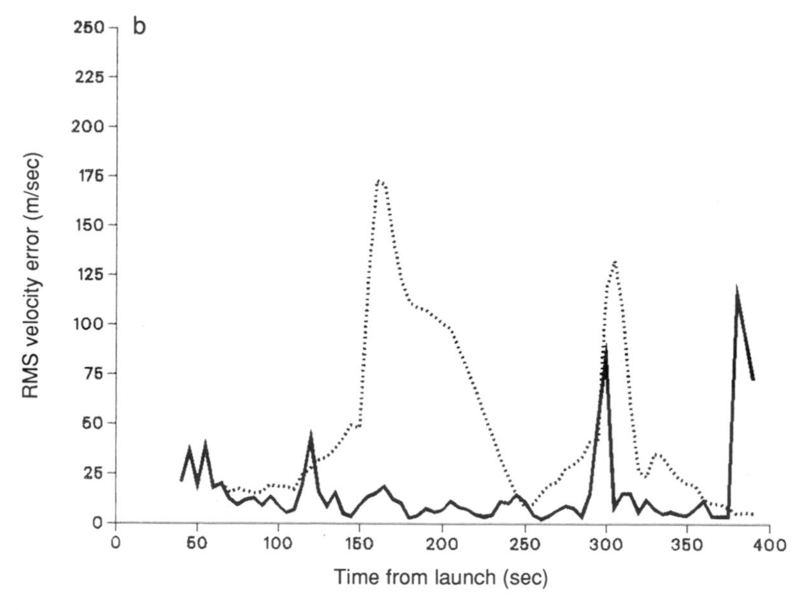

Fig. 5. TTT adaptive performance—test 1 [5-sec frame time, 1-arcsec LOS; with adaptive mechanism (solid line) and without (dotted line)]: (a) position-tracking error and (b) velocity-tracking error.

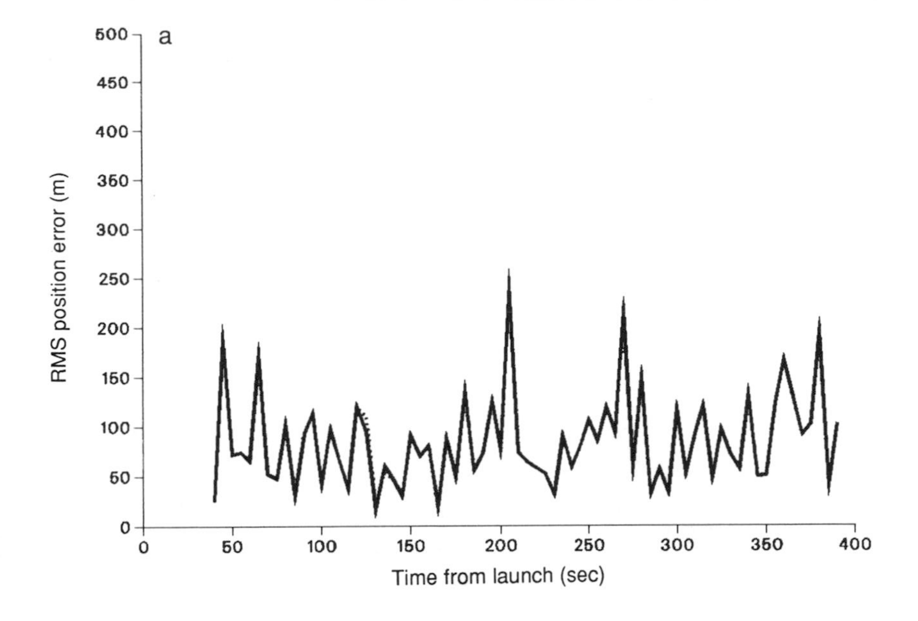

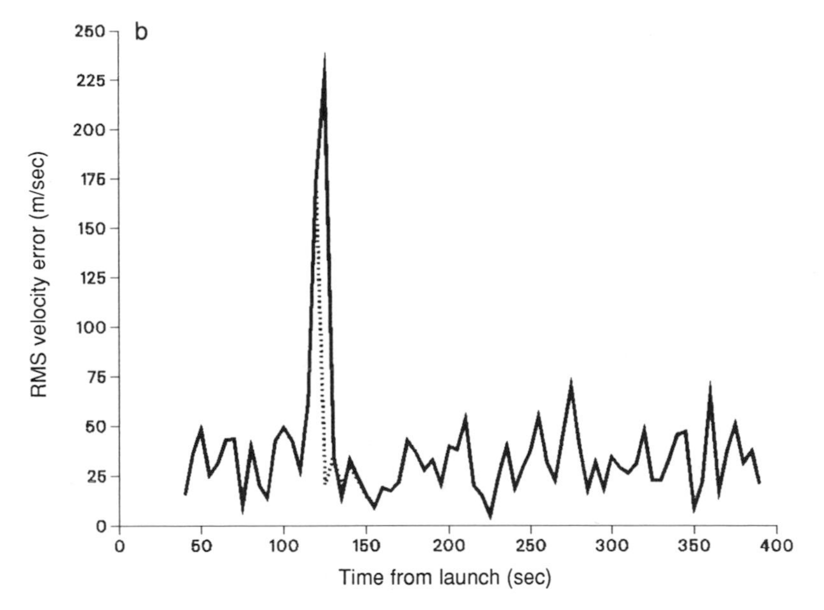

Fig. 6. TTT adaptive performance—test 2 [5-sec frame time, 1-arcsec LOS; with adaptive mechanism (solid line) and without (dotted line)] (a) position-tracking error and (b) velocity-tracking error.

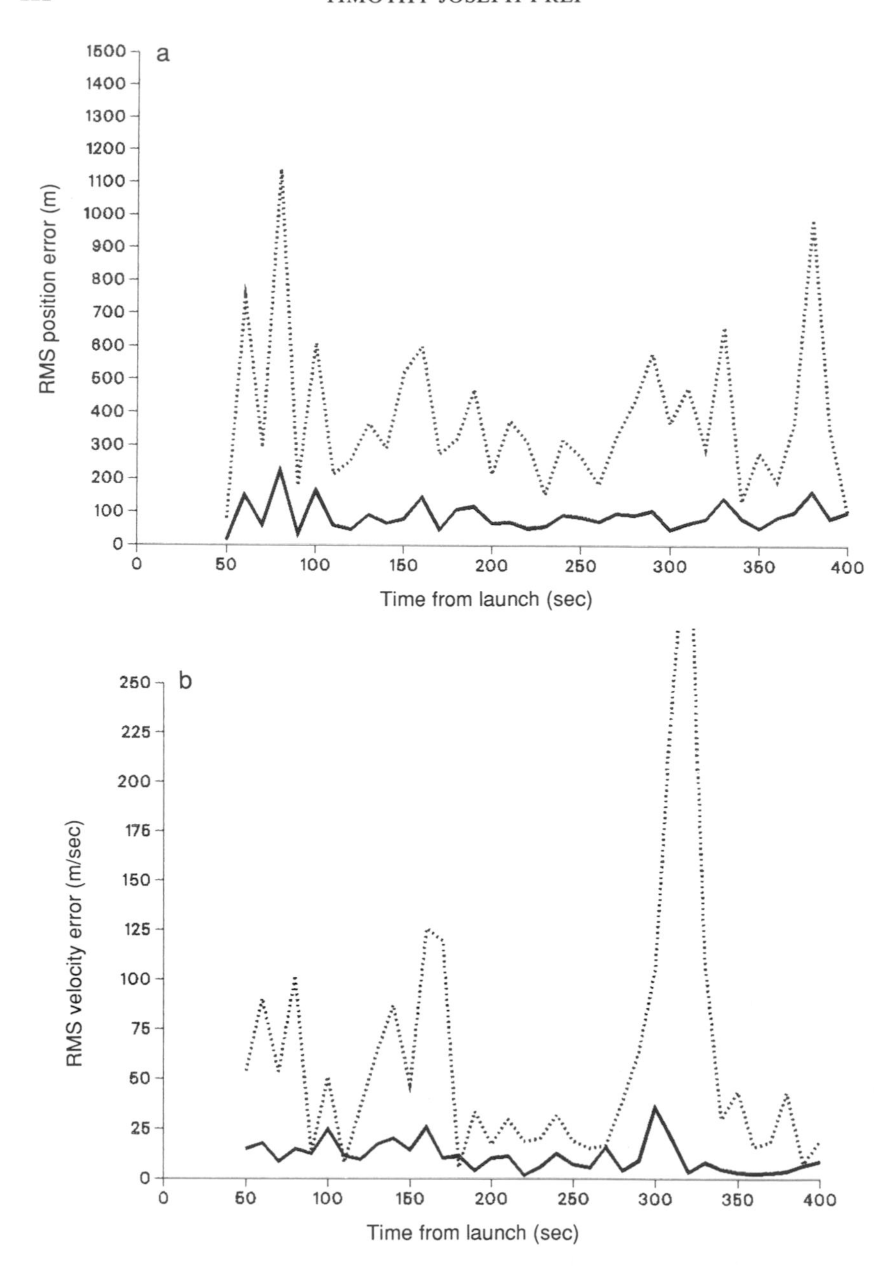

Fig. 7. Effect of LOS error on filter performance—test 1 [10-sec frame time; 1-arcsec LOS (solid line) and 5-arcsec LOS (dotted line)]: (a) position-tracking error and (b) velocity-tracking error.

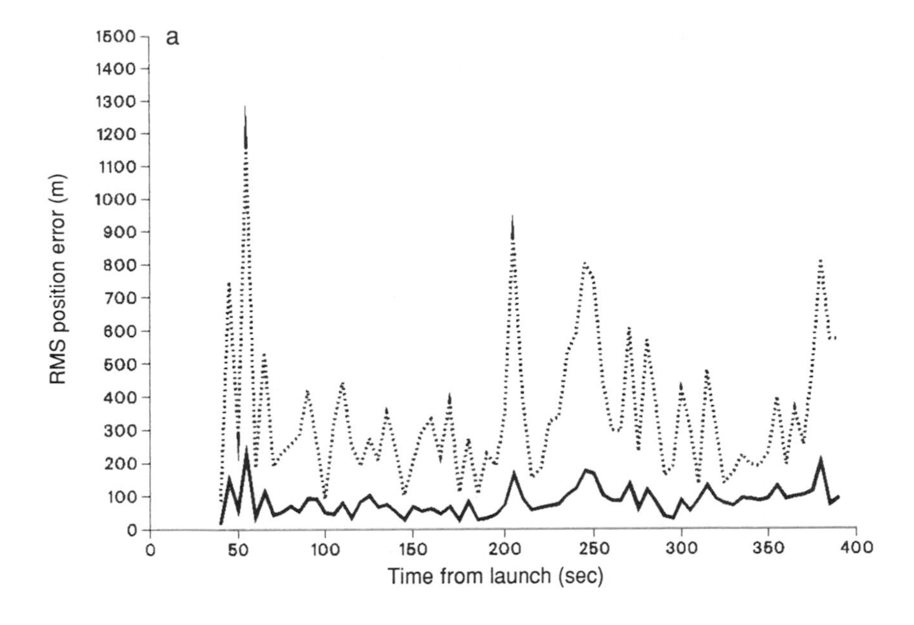

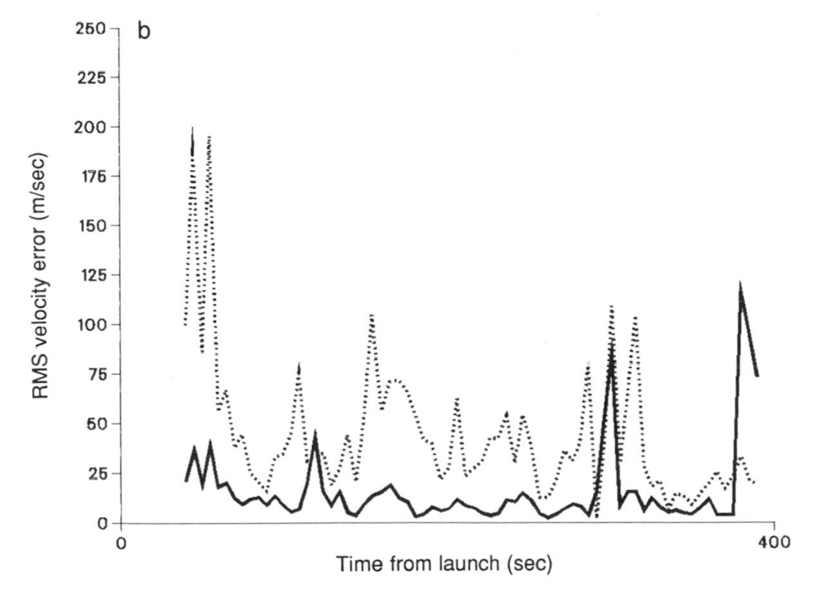

Fig. 8. Effect of LOS error on filter performance—test 2 [5-sec frame time; 1-arcsec LOS (solid line) and 5-arcsec LOS (dotted line)]: (a) position-tracking error and (b) velocity-tracking error.

As for the previous tests, a complete ensemble of test cases should be run to develop a more complete understanding of the effects of sensor LOS accuracy on the TTT tracking performance.

D. EFFECT OF SENSOR FRAME TIME ON PERFORMANCE

Four tests were conducted to determine the dependence of the TTT filter's performance on the sensor frame rate (i.e., the rate at which target observations are taken). The assumption here is that the sensor of interest is a scanning-type device. With a scanner, each target is observed once during each frame. The simulation test results are shown in the next eight figures.

In case 1, the sensors were given 5-arcsec LOS accuracies and the frame time was varied between 5 and 10 sec. A type-1 trajectory was observed. The rms position- and velocity-tracking errors for case 1 are shown in Figs. 9a and 9b. The results for this test case seem to indicate that tracking performance is not improved by decreasing the frame time from 10 to 5 sec. The average rms tracking errors are very similar in both position and velocity.

In case 2, a type-1 missile was again observed. This time, the sensors were given 1-arcsec LOS accuracies and, as the previous case, the frame times were varied between 5 and 10 sec. The case-2 results are depicted in Figs. 10a and 10b. As in the first case, the tracking accuracies were not improved by decreasing the time between successive filter updates. In fact, Fig. 10b indicates that the peak-to-peak fluctuations in the velocity error become larger with decreasing frame rates.

To further examine filter dependence on frame rate, a type-1 missile trajectory was again flown and observed by sensors with 1-arcsec LOS accuracies, and frames times varying between 2 and 5 sec. The results of this test are shown in Figs. 11a and 11b. As in the previous two cases, the filter performance again failed to increase with additional measurements. In this case, both the position and velocity errors suffered very large fluctuations when the sensors had 2-sec frame times. Most important is the fact that the velocity errors for the 2-sec scan were unacceptably large around the second staging event. This appears to have been a statistically worst-case occurrance caused by large errors in the LOS vectors, amplified by computer round-off and truncation errors.

The results from case 3 prompted an additional test case. In case 4, a type-2 trajectory was observed by sensors with identical parameters as those used in case 3. The rms position and velocity results for case 4 are shown in Figs. 12a and 12b. Again, with the exception of the spike in the velocity error curve near 130 sec, the peak-to-peak fluctuations were generally larger for the 2-sec frame time. In this case, the velocity tracking was much better than in case 3, thereby lending credibility to the worst-case error hypothesis stated previously.

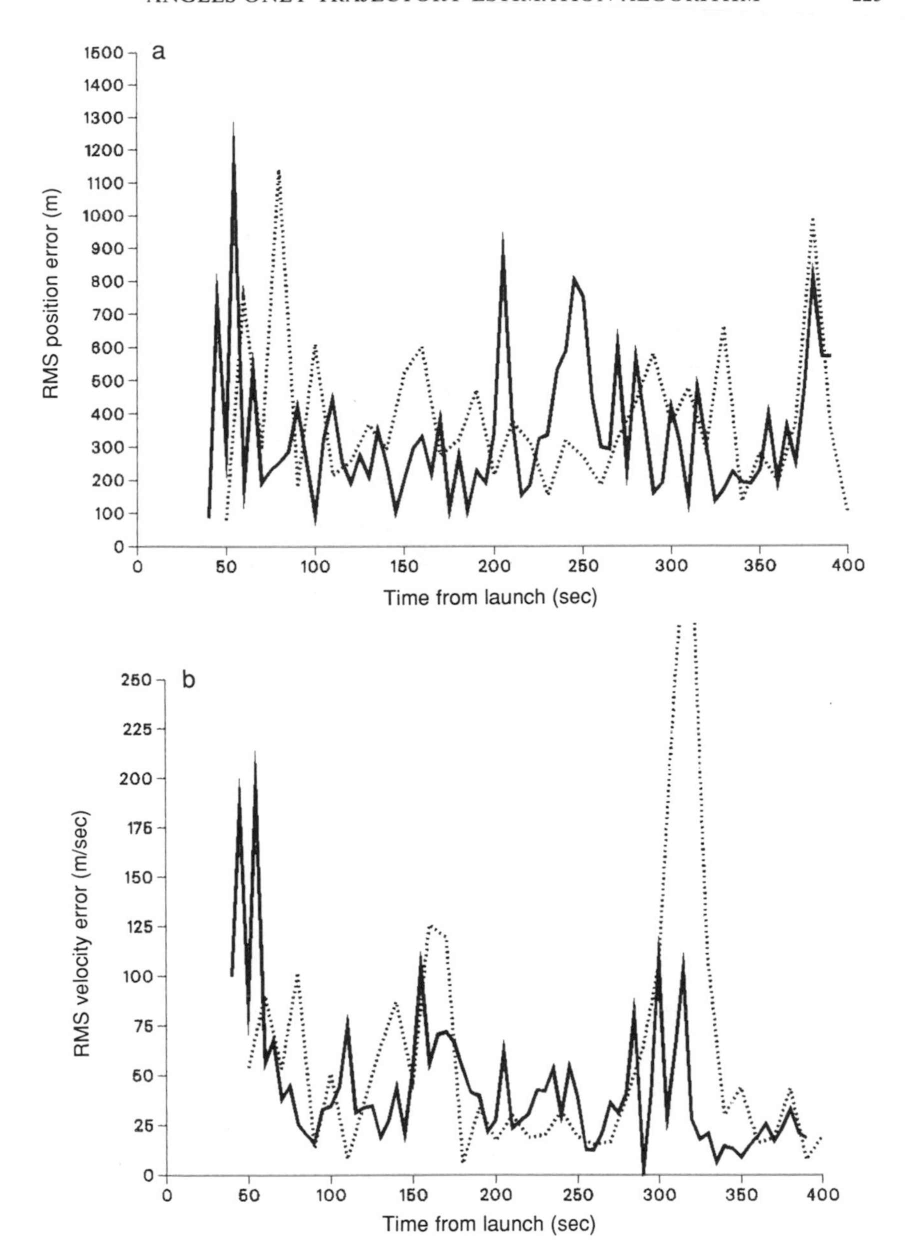

Fig. 9. Effect of frame time on filter performance—test 1 [5-arcsec LOS; 5-sec frame time (solid line) and 10-sec frame time (dotted line)]: (a) position-tracking error and (b) velocity-tracking error.

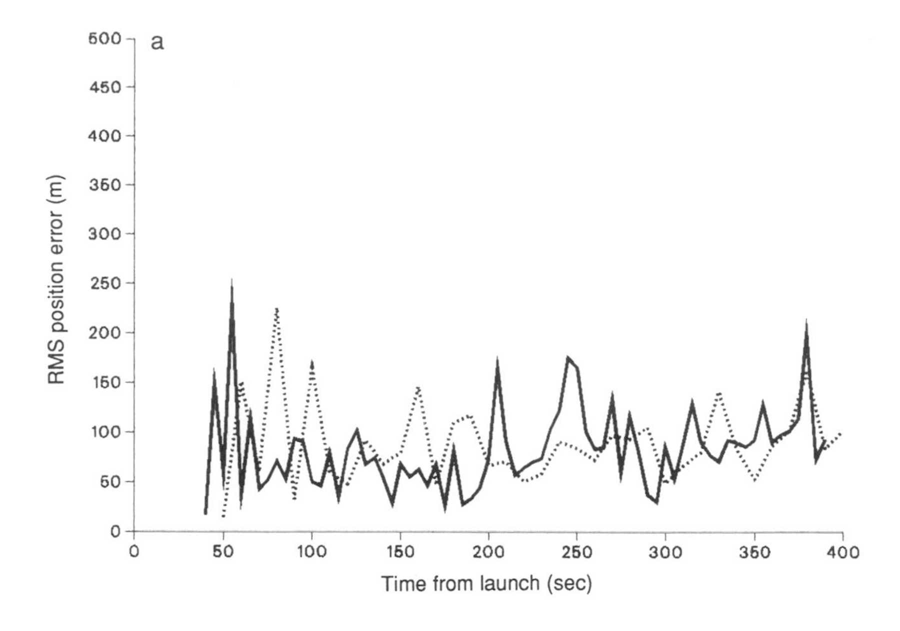

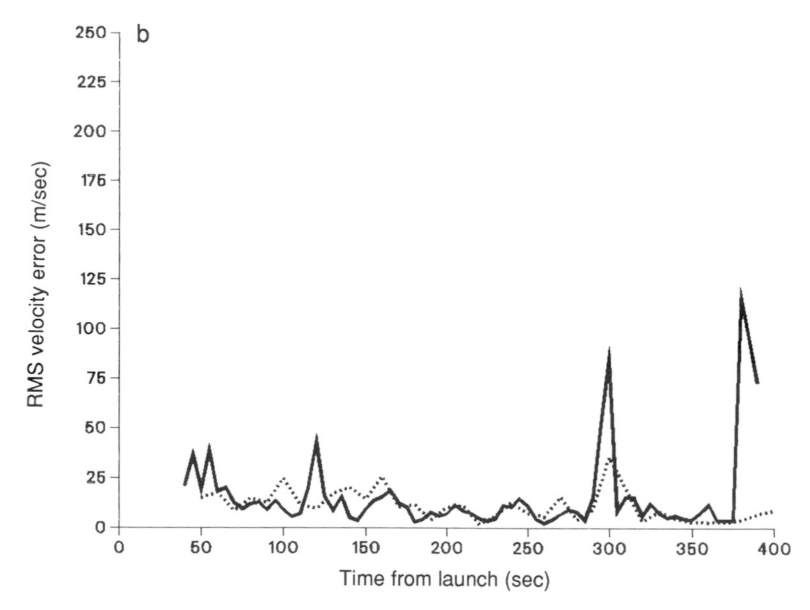

Fig. 10. Effect of frame time on filter performance—test 2 [1-arcsec LOS; 5-sec frame time (solid line) and 10-sec frame time (dotted line)]: (a) position-tracking error and (b) velocity-tracking error.

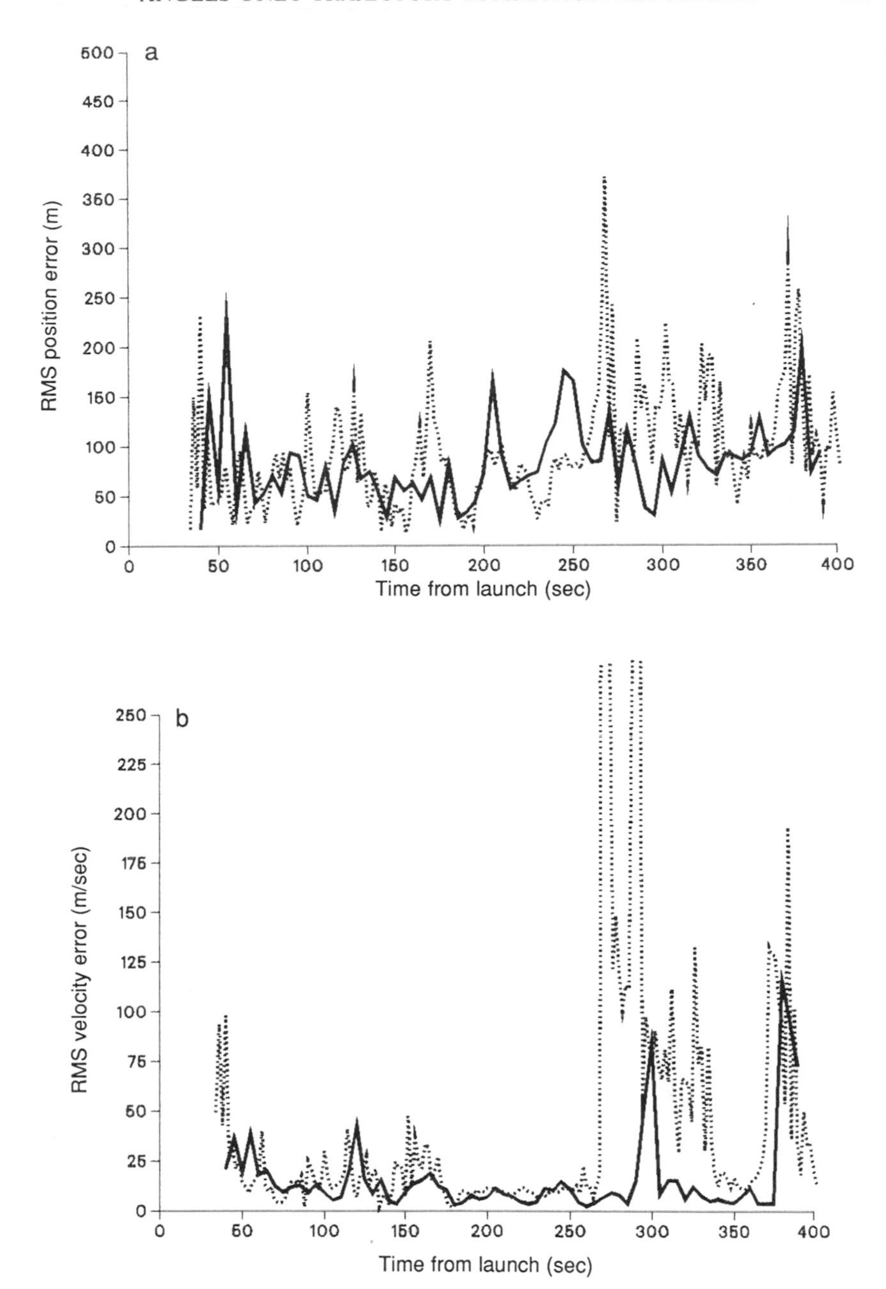

Fig. 11. Effect of frame time on filter performance—test 3 [1-arcsec LOS; 5-sec frame time (solid line) and 2-sec frame time (dotted line)]: (a) position-tracking error and (b) velocity-tracking error.

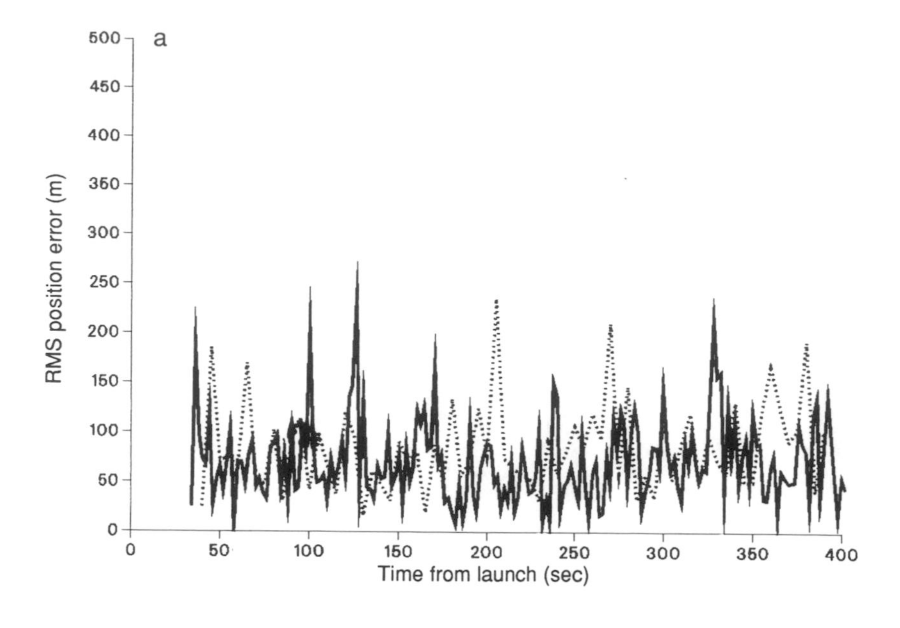

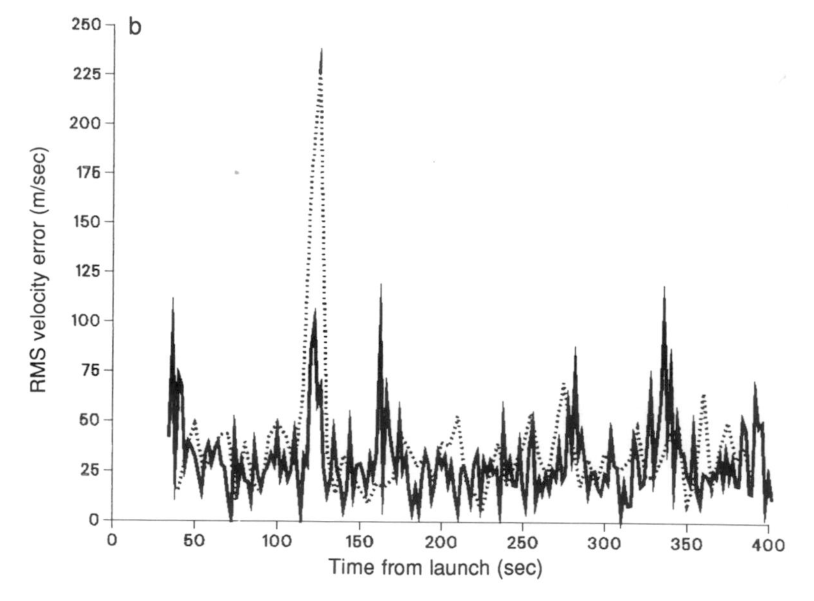

Fig. 12. Effect of frame time on filter performance—test 4 [1-arcsec LOS; 2-sec frame time (solid line) and 5-sec frame time (dotted line)]: (a) position-tracking error and (b) velocity-tracking error.

In all of the four tests just described, the average rms tracking errors in both position and velocity are nearly the same for a given missile type and LOS accuracy, regardless of the frame time used. This appears to indicate that, over the range of frame rates tested, the TTT algorithm performance is nearly independent of the frequency of observations. Obviously this hypothesis breaks down as the frame rates increase beyond the effective bandwidth of the filter. The filter bandwidth could most likely be found for a particular missile type if a number of additional test cases were examined for sufficiently varied frame times.

E. SUMMARY OF TRIANGULATION TRAJECTORY TRACKER ALGORITHM PERFORMANCE

Several additional plots were generated from previous test case data to effectively illustrate the effect on tracking of varying both the sensor LOS accuracy and frame time. This data is presented in Figs. 13a–15b. Without a detailed discussion of each plot, several generalizations can be made about the TTT algorithm and its performance. First of all, as was shown in Section III,C, the performance is significantly improved by increasing the LOS accuracies of the tracking sensors. This is no great surprise because reducing measurement error will improve the performance of any well-designed Kalman filter.

Second, relatively large fluctuations in rms position and velocity error were observed, especially with a 2-sec frame rate. These spikes are due to a combination of statistically worst-case measurement errors, addition of plant noise during the adaptive process, and an overly optimistic covariance matrix. An examination of test case data showed that the variances of the state-vector elements [diagonal entries of $P(t_k)$] periodically indicated a smaller than actual error in position, velocity, or acceleration. Excessive optimism in the filter covariance matrix tends to lead to filter divergence. The adaptive mechanism prevented divergence, but the additive noise required to reduce the chi-square statistic caused a number to relatively large rms error spikes. Additional in-depth covariance analysis is required to better understand the TTT algorithm's performance.

In most ballistic missile-tracking applications, timeliness is as important as tracking accuracy. To successfully intercept a ballistic missile, tracks must be established early in flight. This notion leads to a second generalization about the TTT algorithm. Although tracking accuracy did not improve with increased scan rates, in all cases smaller frame times led to earlier track initialization. For example, with a frame time of 2 sec, the track could be initialized as early as 4 sec after the first observation, assuming ideally matched

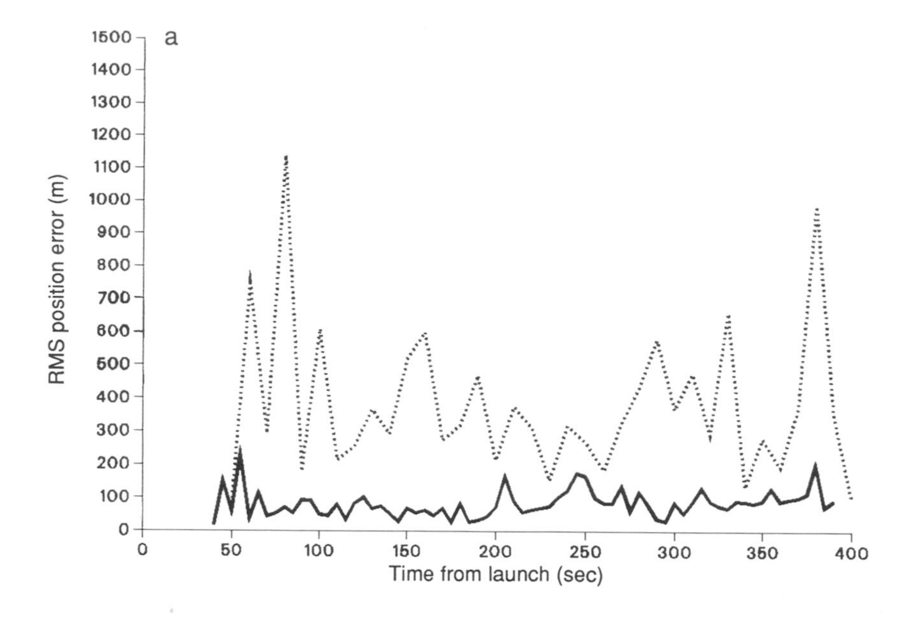

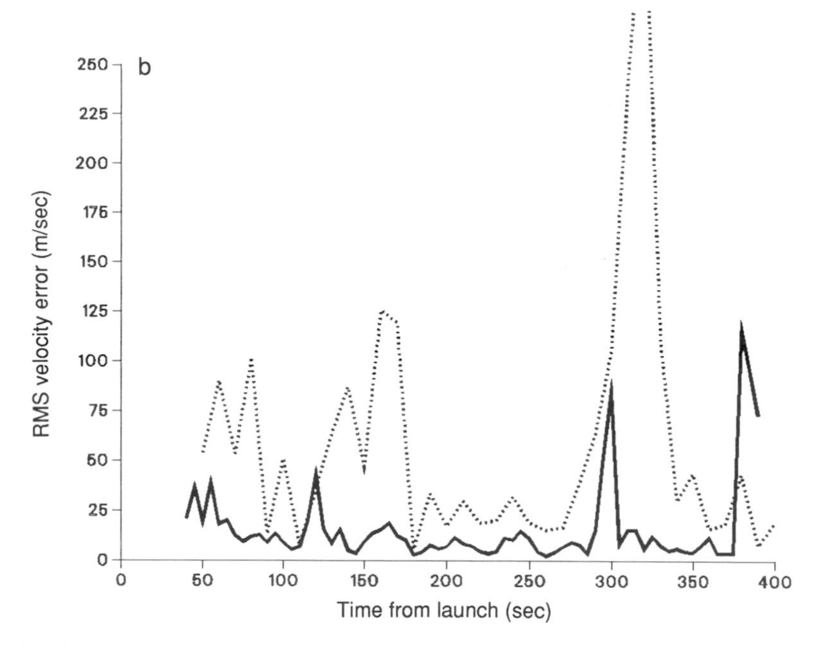

Fig. 13. TTT performance summary—test 1 [5-sec frame time, 1-arcsec LOS (solid line); 10-sec frame time, 5 arcsec LOS (dotted line)]: (a) position-tracking error and (b) velocity-tracking error.

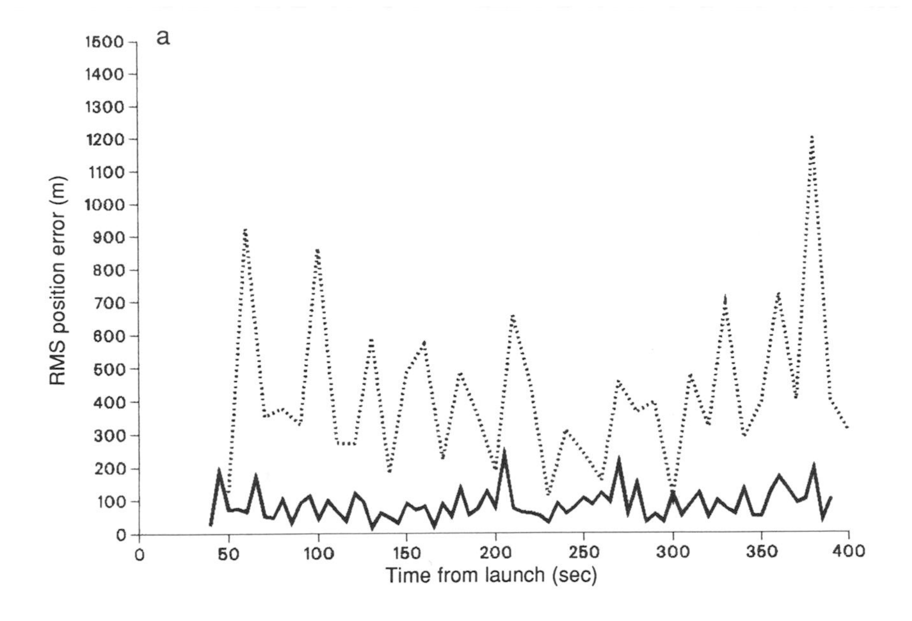

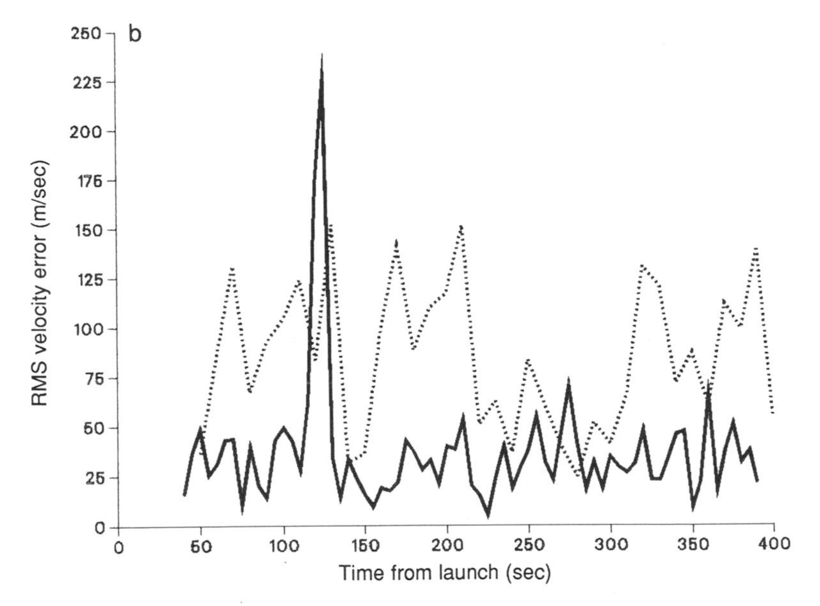

Fig. 14. TTT performance summary—test 2 [5-sec frame time, 1-arcsec LOS (solid line); 10-sec frame time, 5-arc sec LOS (dotted line)]: (a) position-tracking error and (b) velocity-tracking error.

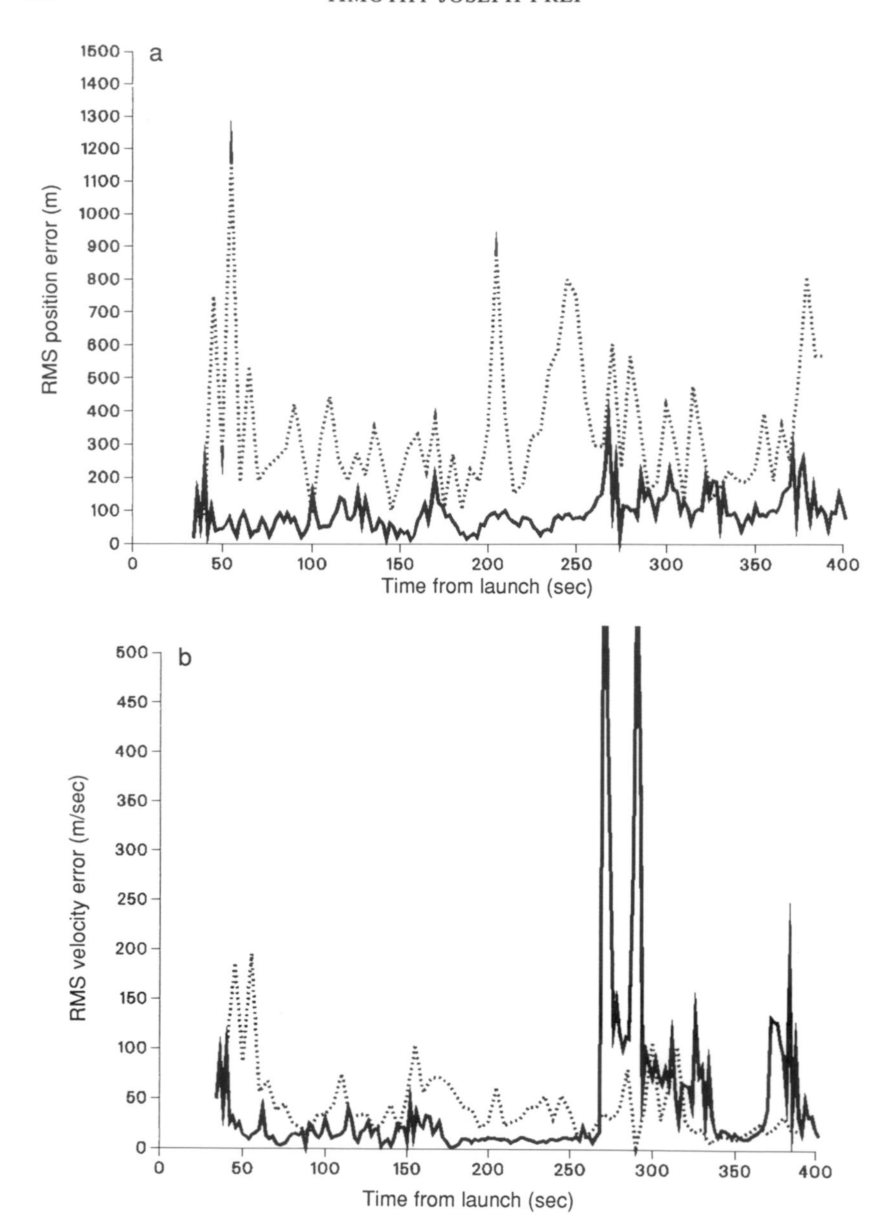

Fig. 15. TTT performance summary—test 3 [2-sec frame time, 1-arcsec LOS (solid line); 5-sec frame time, 5-arcsec LOS (dotted line)]: (a) position-tracking error and (b) velocity-tracking error.

observations and excluding electromagnetic propagation, signal and data processing delays. By comparison, with a 10-sec scan the track could be initialized no sooner that 20 sec after the first observation.

IV. ALGORITHM PERFORMANCE COMPARISON

To obtain additional insight into the performance of the triangulation trajectory tracker algorithm, head-to-head test cases were run against the fundamentally different dynamics free Kalman filter tracker [5]. These performance comparisons provide more evidence of the good tracking performance achieved by the TTT algorithm. In Section IV,A, the DFKFT is presented along with the motivations behind its formulation. In Section IV,B, the results of the comparative test cases are presented and discussed.

A. DYNAMICS FREE KALMAN FILTER TRACKER ALGORITHM

The DFKFT algorithm was built as an integral part of the space defense scenario simulator [5]. The objective of the SDSS was to develop a piece of analysis software capable of providing system-level performance evaluation for multilayered, SDI, surveillance, weapon, and communication systems. The DFKFT, embedded within the SDSS, is a launch-phase tracking algorithm for single- and dual-sensor observations. Some properties of the TTT algorithm were adopted directly or modified from the DFKFT algorithm. The kinematic-state transition matrix and the plant noise models were taken directly from the DFKFT. The adaptive mechanism used in the TTT filter is a 3-DOF extension of the adaptive mechanism implemented in the DFKFT. With these modeling similarities, the DFKFT provided an excellent candidate for comparative performance evaluations with the TTT filter.

The design of the DFKFT was based on a set of five functional requirements. First, the DFKFT had to automatically initiate the tracking process on a single LOS observation. Second, the algorithm had to track boost–ballistic-phase targets in a monoviewing (single-sensor) mode. Third, the DFKFT had to maintain monotrack through all boost-phase staging events. Fourth, it had to correlate state vectors generated by two different sensors. Last, the DFKFT algorithm had to combine two monoviewed track state vectors into one stereo-viewed state vector and then continue tracking in the stereo mode. Additionally, the DFKFT had to be both computationally efficient (short processing time and minimal memory per target) and robust against variant trajectories with minimal *a priori* kinematical or dynamical information. These requirements created a formidable challenge to formulate a high-performance tracking algorithm. By comparison, the TTT assumes *a priori* target correlation, is allowed to initiate a track on three stereo observations, and has no capability to track in a degraded monoviewing mode.

The DFKFT algorithm was developed using an extended Kalman filter. The nonlinear relationship between the LOS elevation and azimuth and the inertial target position was linearized by a first-order Taylor series approximation. In other words, the filter measurement matrix H is obtained by taking partial derivatives of the observation vector $(E1, Az)$ with respect to the target position. The filter operates in the two-dimensional elevation–azimuth space, contrary to the TTT, which forms measurements in the three-dimensional ECI space. The dimension of the DFKFT filter is, therefore, one less than that implemented in the TTT algorithm.

A key development in the DFKFT algorithm is the single-point initialization process. From one measured LOS elevation and azimuth, an initial state-vector estimate and covariance matrix is obtained for use in starting the monoview tracking process. Although the single-sighting state vectors are generally much less accurate than stereo-state vectors, they do provide for earliest possible correlation. The correlation of two monoviewed track state vectors into a single stereo-state vector is achieved by an optimal weighting of the monoviewed state estimates and associated covariance matrices. Prior to folding state vectors together, an association decision is made as to whether the candidate monoviewed states pertain to the same target. The decision is based on the chi-square statistic of the difference between the two state estimates. Association occurs if the statistic exceeds the 95% confidence region for a 9-DOF system. If association is not achieved, each sensor continues to maintain target track in the monoviewing mode until such a time when correlation is possible.

After stereo tracking has been initiated, the filter is updated after every observation by either sensor. The resulting filter frequency becomes effectively twice that of the TTT algorithm, for identical sensor frame times. This feature is very advantageous in applications where updates are required at rates greater than the revisit time of the sensors involved.

B. COMPARATIVE ALGORITHM PERFORMANCE

Seven test cases were simulated to compare the performance of the TTT algorithm with that of the DFKFT. The test parameters for all seven comparison cases are summarized in Table I. The first four comparison tests were run against the type-1 missile with various combinations of sensor frame time and LOS accuracy. Three cases were then simulated against a type-2

TABLE I. TTT VERSUS DFKFT PERFORMANCE TEST PARAMETERS

Case	Missile type	TTT sensor parameters		DFKFT sensor parameters	
		LOS (arcsec)	Frame (sec)	LOS (arcsec)	Frame (sec)
1	1	1	2	1	2
2	1	1	5	1	5
3	1	5	5	5	5
4	1	5	10	5	10
5	2	1	2	1	2
6	2	1	5	1	5
7	2	5	10	5	10

missile. In all cases and for both filters, the observations of the first and second sensor were interleaved one-half scan. The first observation by each sensor always occurred 30 sec after launch. The same initial random number seed sequence was used by both algorithms in all of the tests. In all test cases, the TTT and DFKFT tracking errors are represented by the solid and dotted line curves, respectively. As before, the tests case sensor parameters are abbreviated in the plot legend.

In the first comparison test, the sensors were given 2-sec frame times and 1-arcsec LOS accuracies. The resulting rms position- and velocity-tracking errors are shown in Figs. 16a and 16b. In this first case, the position errors for the TTT are slightly worse, by overall they are very comparable. The spikes at the beginning of the DFKFT curve correspond to the errors in the monoviewed-state estimates, prior to association and correlation. As shown in Fig. 16b, the velocity tracking by both algorithms is similar until just prior to the second staging event, at which time the TTT begins to suffer large deviations in performance throughout the remainder of the trajectory. The cause of fluctuations was discussed in Section III,E.

In the second test case, the sensor frame times were increased to 5 sec while the LOS accuracy and missile type were kept the same. The results for test 2 are summarized in Figs. 17a and 17b. For this case, both the position- and velocity-tracking errors were slightly larger for the DFKFT than for the TTT algorithm. Moderate transients in velocity errors were experienced by both trackers at the second staging event.

For the third test case, the LOS accuracy of the sensors was degraded to 5 arcsec while the frame time was held constant at 5 sec. The position and velocity results are shown in Figs. 18a and 18b. Again, the TTT performance was slightly better in both position and velocity. Peak transient errors were similar in amplitude for both algorithms, but the DFKFT experienced worst-case fluctuations at a higher frequency of occurrence.

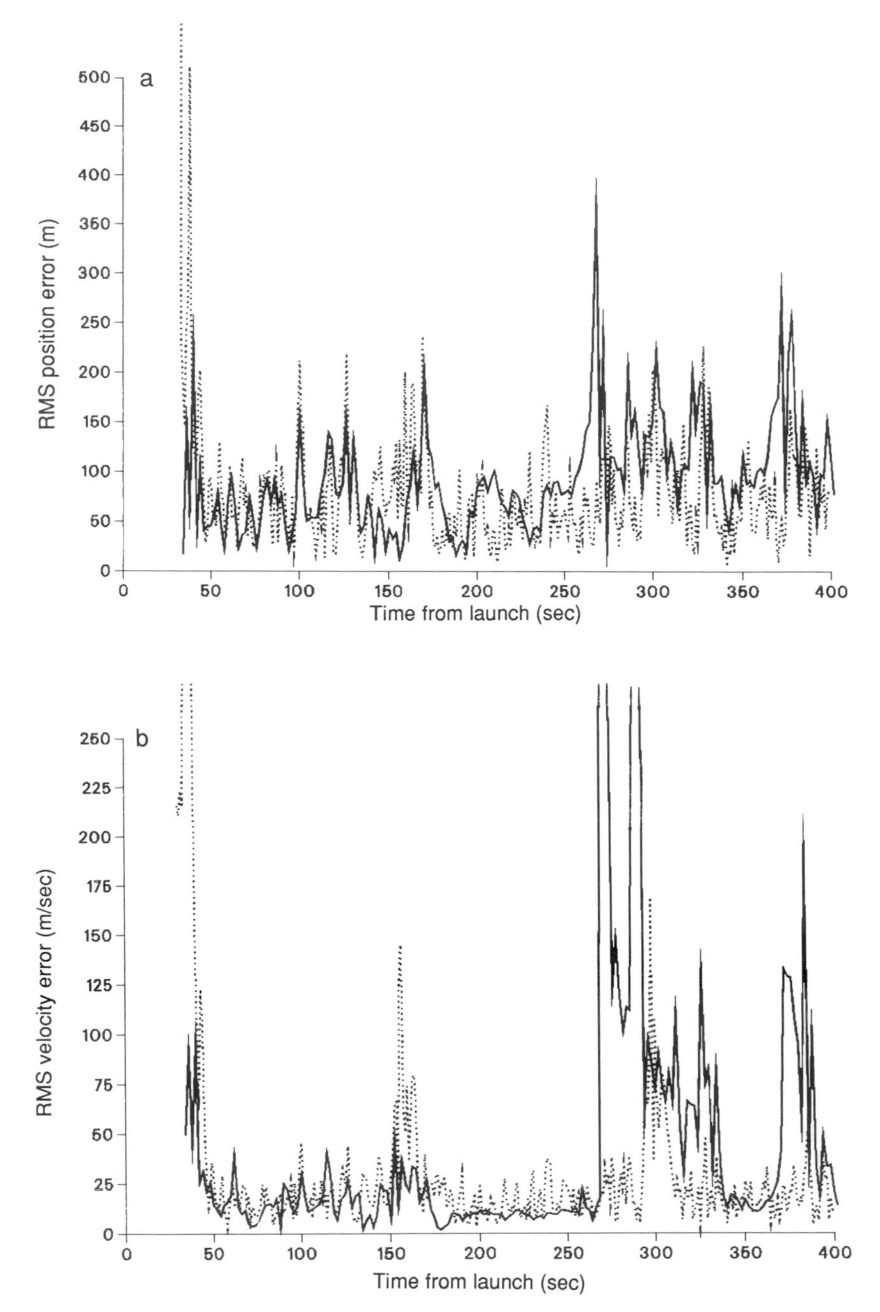

Fig. 16. TTT (solid line) versus DFKFT (dotted line) performance comparison—test 1 (2-sec frame time, 1-arcsec LOS): (a) position-tracking error and (b) velocity-tracking error.

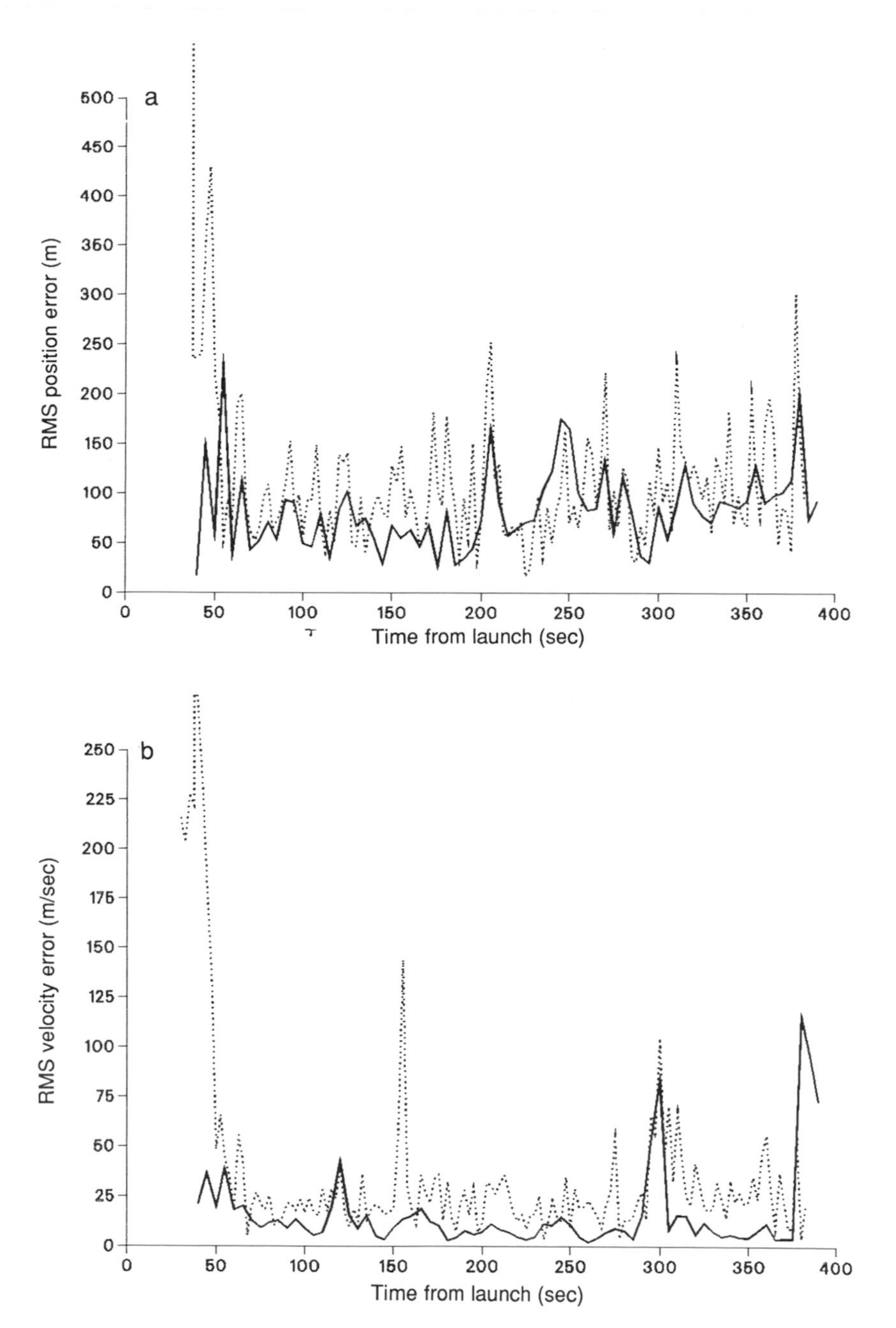

Fig. 17. TTT (solid line) versus DFKFT (dotted line) performance comparison—test 2 (5-sec frame time, 1-arcsec LOS): (a) position-tracking error and (b) velocity-tracking error.

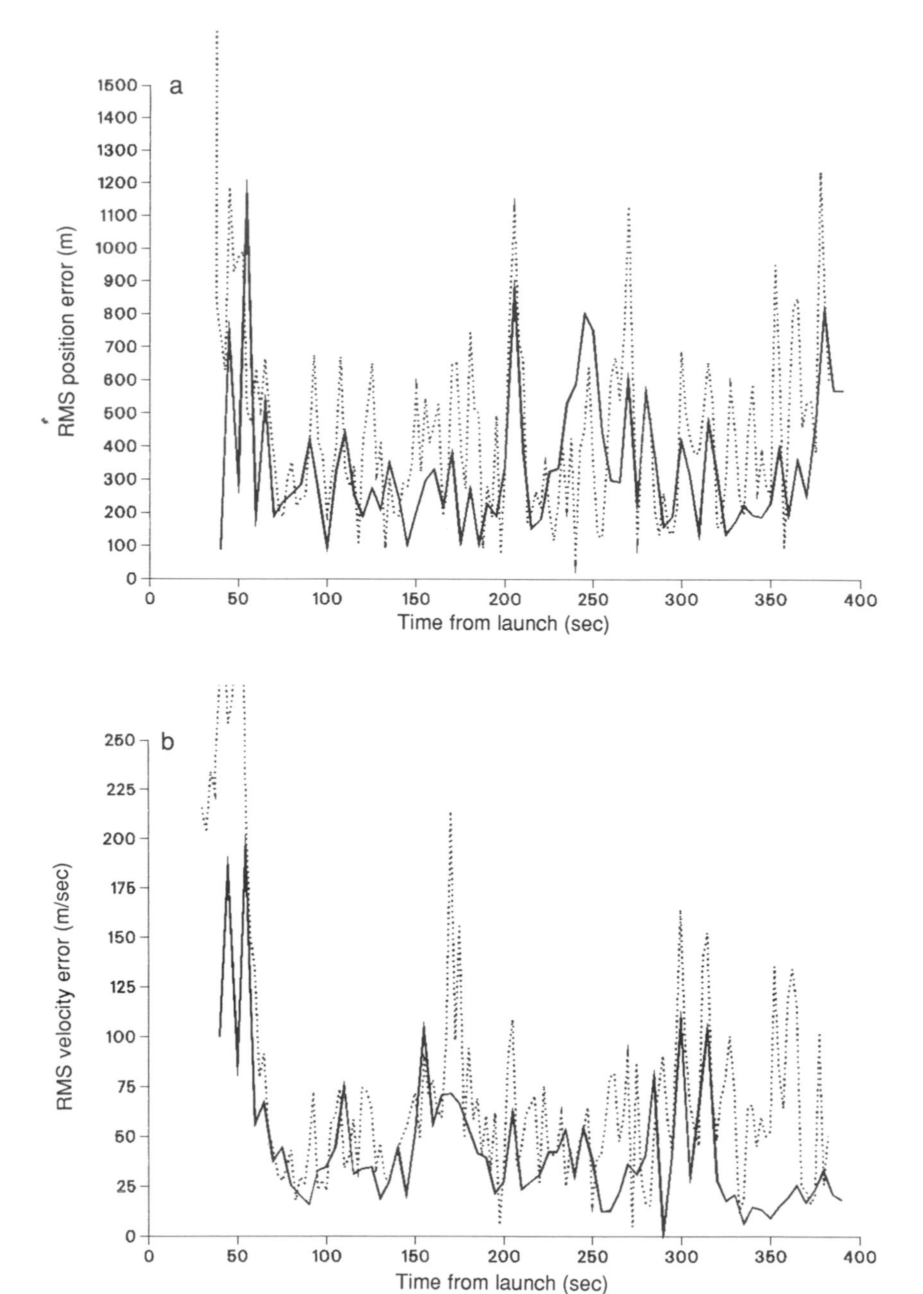

Fig. 18. TTT (solid line) versus DFKFT (dotted line) performance comparison—test 3 (5-sec frame time, 5-arcsec LOS): (a) position-tracking error and (b) velocity-tracking error.

In the fourth and last of the type-1 missile comparison tests, the sensors were given 10-sec frame times and 5-arcsec LOS accuracies. The results are plotted in Figs. 19a and 19b. In this test, the TTT demonstrated slightly better position tracking, but suffered a much larger transient error in velocity tracking following the second staging event.

For the last three performance comparison test cases, a type-2 missile trajectory was observed. In case 5, the sensors had 2-sec frame times and 1-arcsec LOS accuracies. The results for this case, Figs. 20a and 20b, indicate that both algorithms achieved virtually identical position- and velocity-tracking performance over the course of the entire trajectory (following DFKFT association and correlation). This result is very different from that seen in case 1 where the same sensors observed a type-1 trajectory.

In case 6, the sensor frame times were increased to 5 sec and the same type-2 missile trajectory was observed. The results of this case are plotted in Figs. 21a and 21b. In this test, the DFKFT filter had significant position-tracking errors following the first staging event. Both algorithms also suffered large velocity error spikes at the first event. Subsequent to the first stage spikes, both filters achieved equal performance for the remainder of the observed trajectory.

In the final comparison test case, the type-2 trajectory was observed by sensors with 10-sec frame times and 5-arcsec LOS accuracies. In this case, as is seen in Figs. 22a and 22b, the TTT filter achieved consistently better position-tracking performance and marginally better velocity tracking.

The tests described in this article represent only a very small portion of the entire spectrum of cases needed to perform a complete comparative evaluation of the TTT and DFKFT algorithms. To complete the algorithm comparison, additional tests are required. These tests should include cases for new missile types and different initial conditions such as viewing geometry and first observation time, against a greater range of sensor parameter variations. For the seven cases evaluated, however, some conclusions about relative filter performance can be drawn. In all but two of the tests (cases 1 and 5), the TTT filter demonstrated superior position-tracking performance. For frame times of 2 sec, the position errors obtained by both algorithms were nearly identical. This would indicate that the TTT filter achieves better position tracking for larger frame times, the differences becoming smaller as the revisit time is reduced.

On the other hand, the velocity-tracking results were mixed. In cases 2 and 3, the TTT filter showed better velocity results, while in cases 1 and 4 the DFKFT filter was superior in velocity tracking. In cases 5, 6, and 7, the velocity-tracking errors by both filters were nearly identical. Overall, both filters achieved acceptable velocity results with the exception of some large transient errors. The occasional fluctuations are not entirely surprising because of the method by which velocity tracking is achieved. With angles-only

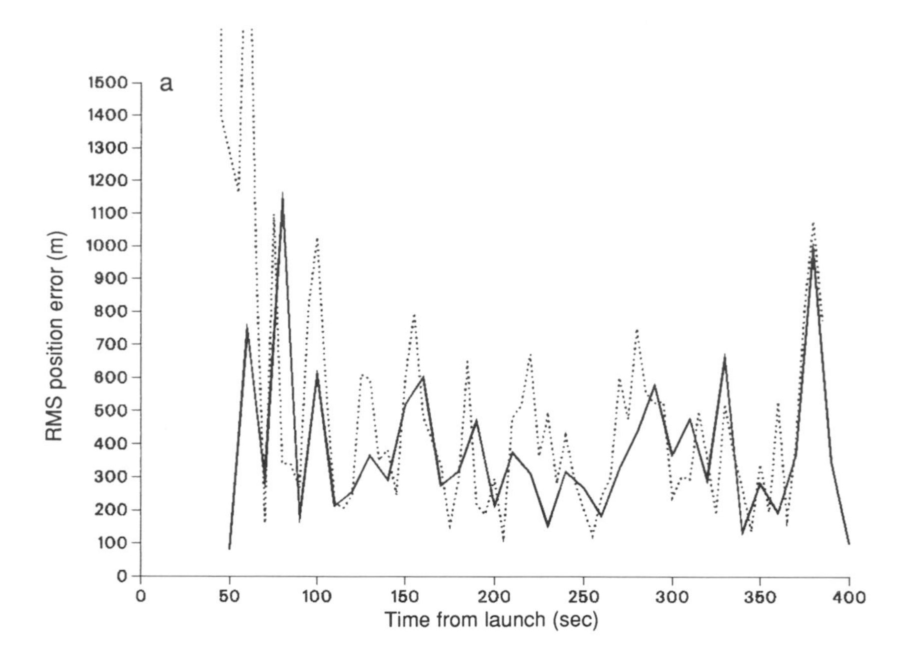

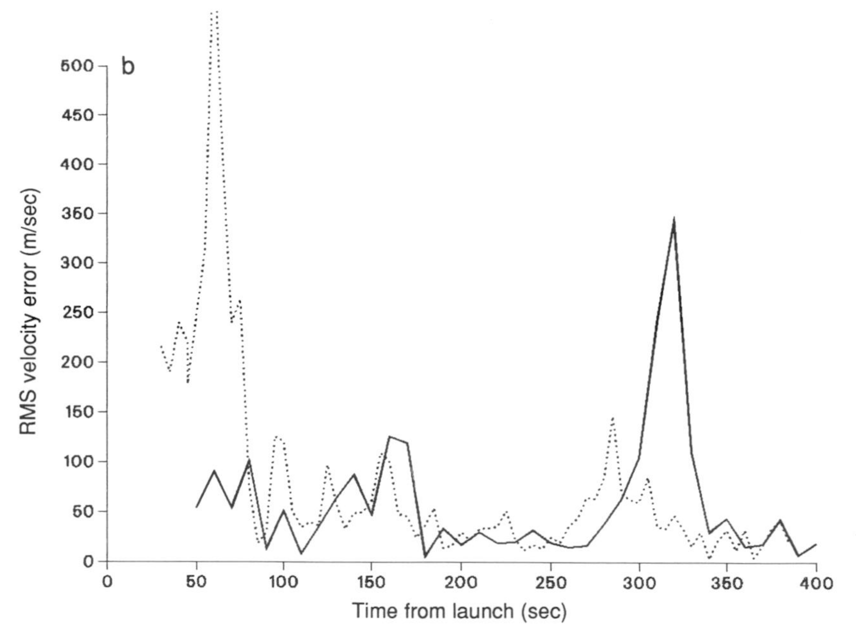

Fig. 19. TTT (solid line) versus DFKFT (dotted line) performance comparison—test 4 (10-sec frame time, 5-arcsec LOS): (a) position-tracking error and (b) velocity-tracking error.

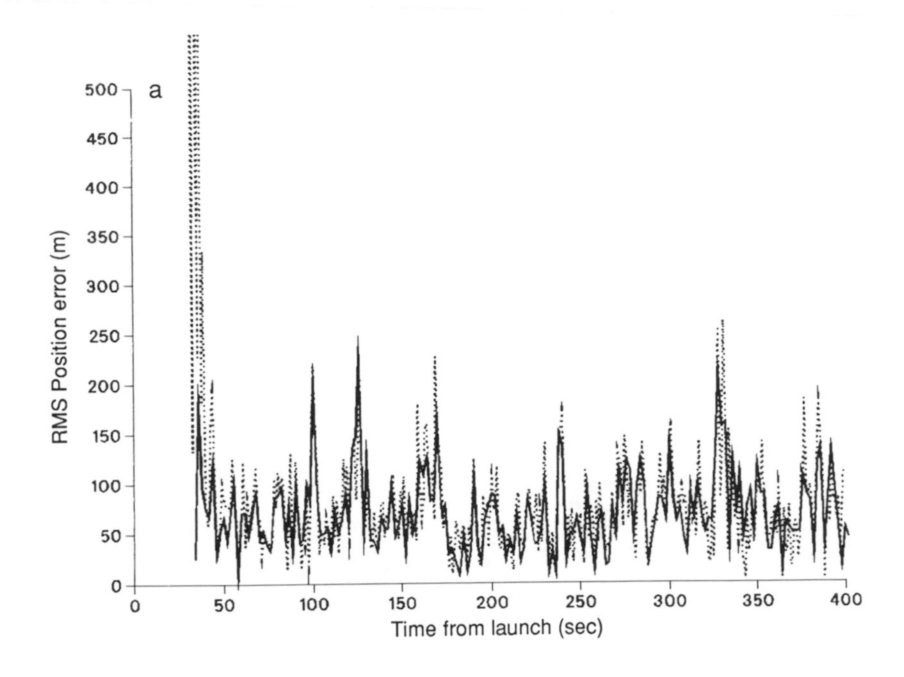

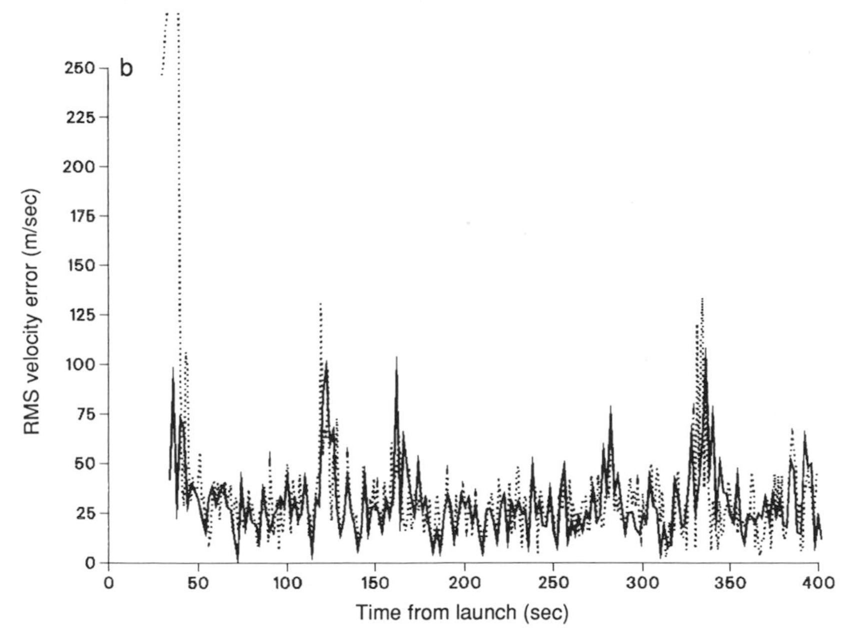

Fig. 20. TTT (solid line) versus DFKFT (dotted line) performance comparison—test 5 (2-sec frame time, 1-arcsec LOS): (a) position-tracking error and (b) velocity-tracking error.

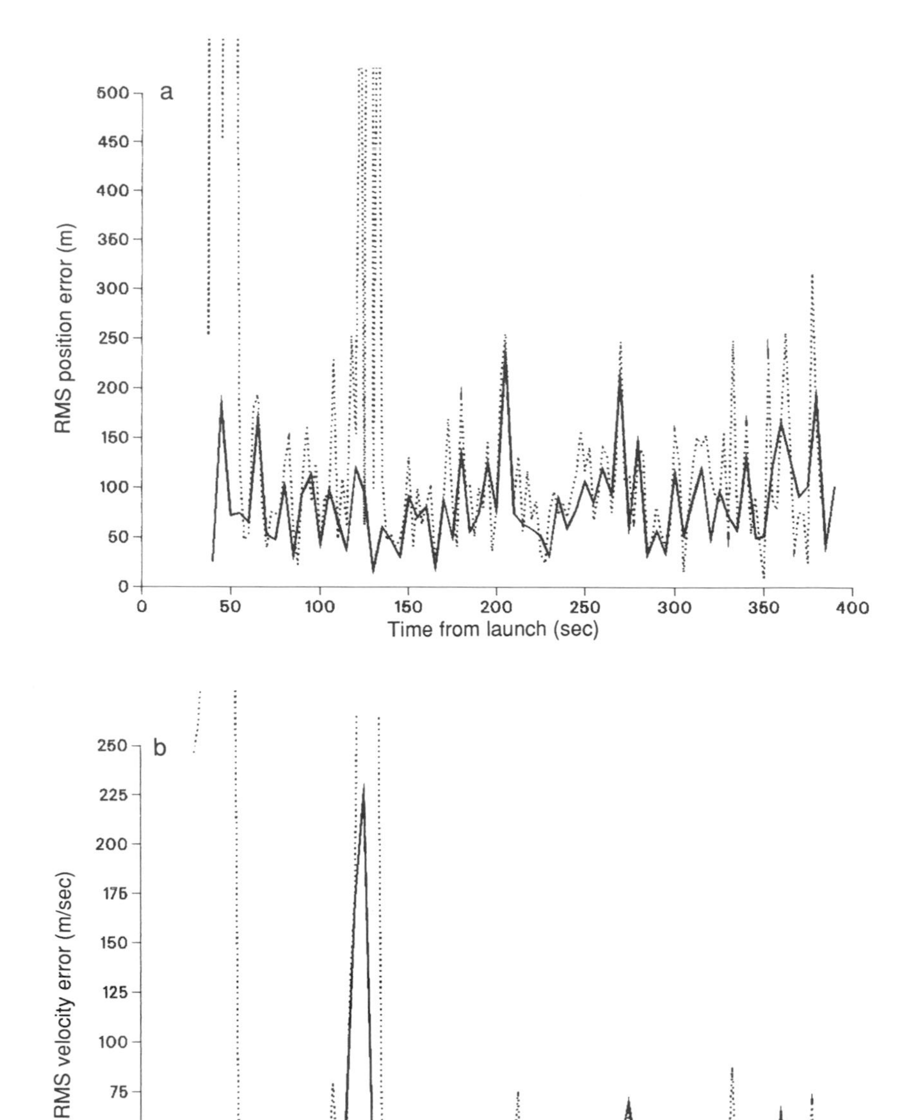

Fig. 21. TTT (solid line) versus DFKFT (dotted line) performance comparison—test 6 (5-sec frame time, 1-arcsec LOS): (a) position-tracking error and (b) velocity-tracking error.

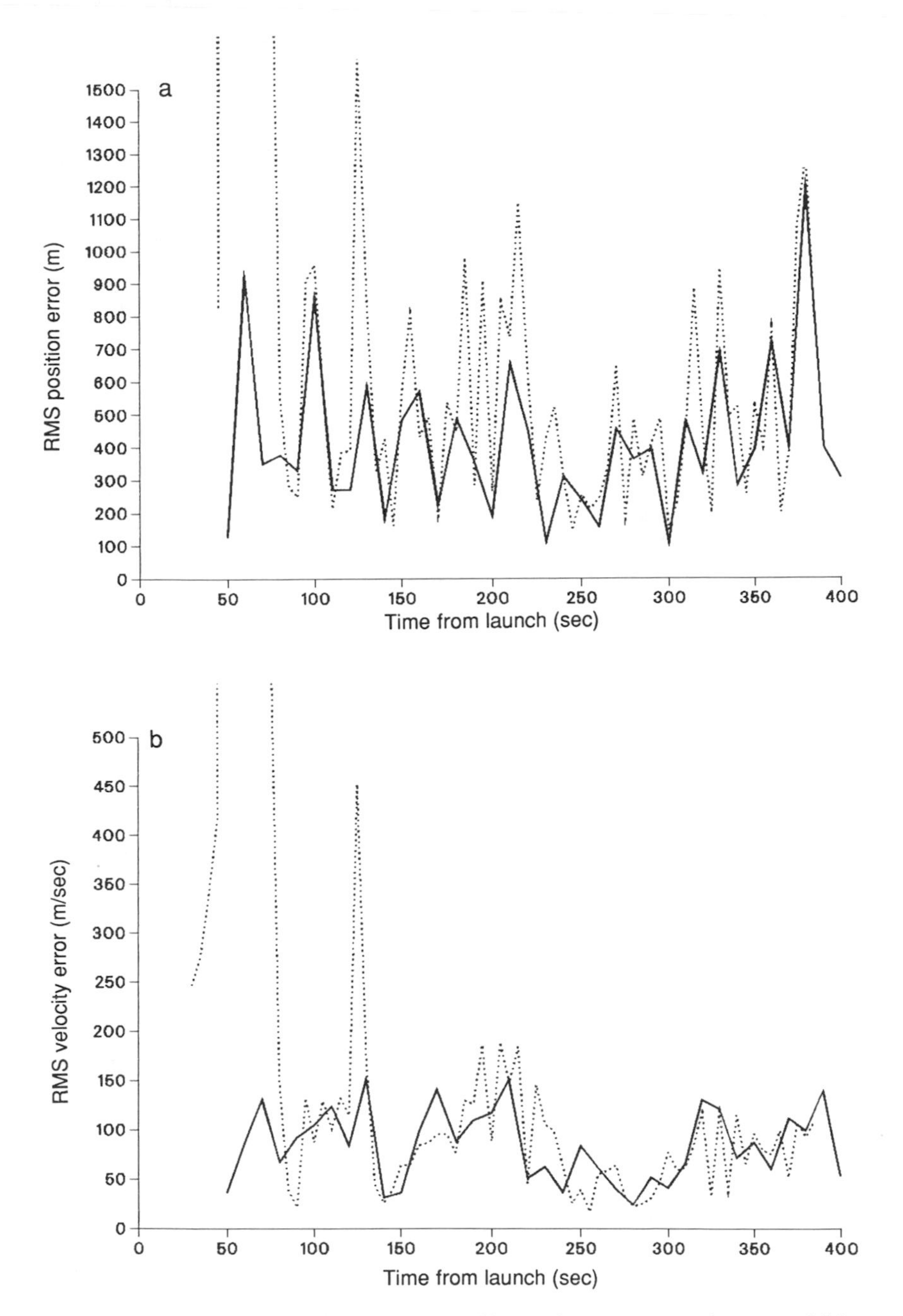

Fig. 22. TTT (solid line) versus DFKFT (dotted line) performance comparison—test 7 (10-sec frame time, 5-arcsec LOS): (a) position-tracking error and (b) velocity-tracking error.

tracking, no velocity information is directly measured. Instead, velocities (and accelerations) are obtained by position–time differencing, a process that is subject to round-off or truncation errors.

V. CONCLUSIONS

An algorithm for estimating ballistic missile trajectories with angles-only sensors has been developed. Certain assumptions made during the development of the algorithm were tested. Algorithm performance was investigated against two different missile types, for multiple variations of sensor parameters. The triangulation trajectory tracker algorithm was compared directly to the existing dynamics free Kalman filter tracker.

In terms of Kalman filter design, the most critical assumption made in the development of the TTT algorithm was to neglect the correlation between successive measurements. Recall that the correlation was introduced by the third-order interpolation scheme used to obtain sensor observations at a common time. Performance test results indicate (Figs 3a–4b) that the measurement correlation, although violating the definition of the Kalman filter, does not detract from the algorithm's performance. Again, this hypothesis is only valid for the test cases performed and needs further investigation. The allowable range of frame times should be evaluated for a given range of sensor LOS accuracies.

The decision to include an adaptive mechanism in the formulation of the TTT algorithm was clearly correct. Although the test results indicate that adaptive filtering is not required in all cases, they do indicate (Figs. 5a and 5b) that significant performance improvements can be obtained by including this type of adaptive compensation. Tests on other missile types and with other sensor parameters should be conducted to determine the range of performance of adaptive filtering.

The dynamics free state-transition matrix used in the TTT algorithm proved to perform well under most circumstances. It would be interesting to redevelop the TTT algorithm using the equations of motion instead of the kinematic relationships when defining the state-transition properties. In doing so, the performance of the dynamics-free approach could be evaluated against the true, nonlinear dynamics.

Most importantly, the test results prove that the concept of target triangulation can be successfully implemented in a trajectory-tracking algorithm. The performance of the algorithm was very good over the range of cases tested. The comparison between the TTT and DFKFT filters showed that the performance of the TTT algorithm is comparable to that required by many of today's ballistic missile-tracking applications.

REFERENCES

1. C. B. CHANG and K. P. DUNN, "Angle-Only Tracking Algorithms and Their Performance," Proj. Rep. No. RMP-183, Lincoln Laboratory, MIT, October 1979.
2. M. R. SALAZAR, "State Estimation of Ballistic Trajectories With Angle Only Measurements," *in* "Advances in the Techniques of the Application of Nonlinear Filters and Kalman Filters" (C. T. Leondes, ed.), Advisory Group For Aerospace Research and Development, AGARDogr. No. 256 (1982).
3. S. C. NARDON and V. J. AIDALA, "Observability Criteria for Bearings-Only Target Motion Analysis," *IEEE Trans. Aerosp. Electron. Syst.* **AES-17**, 162–166 (1981).
4. R. G. BROWN, "Not Just Observable, But How Observable," *Proc. Natl. Electron. Conf.* **22**, 709–714 (1966).
5. H. F. DODGE, R. DANCHICK and D. W. ESTES, *Space Defense Scenario Simulation*, Project Report No. 85353154, TRW Inc., Redondo Beach, Ca., 31 March 1986.
6. R. L. BURDEN and J. D. FAIRES, *Numerical Analysis 3rd Edition*, Prindle, Weber, and Schmidt Publishers, Boston, Massachusetts, 1985.
7. G. M. LERNER, "Covariance Analysis," *Spacecraft Attitude Determination and Control*, D. Reidel Publishing Company, Dordrecht, Holland, J. R. Wertz (ed.), 1985.
8. J. S. MEDITCH, *Stochastic Optimal Linear Estimation and Control*, McGraw-Hill Book Company, New York, New York, 1969.
9. P. S. MAYBECK, R. L. JENSEN and D. A. HARNLY, "An Adaptive Extended Kalman Filter for Target Image Tracking," *IEEE Trans. on Aerospace and Electronic Systems*, Vol. AES-17, No. 2, pp. 173–180, March 1981.
10. R. L. MOOSE, "An Adaptive State Estimation Solution to the Maneuvering Target Problem," *IEEE Trans. on Automatic Control*, Vol. AC-20, pp. 359–362, June 1975.
11. R. L. MOOSE, and N. H. GHOLSON, "Adaptive Tracking of Abruptly Maneuvering Targets," *Proc. 1976 IEEE Conf. Decision and Control*, pp. 804–808, Dec. 1976.
12. G. A. KORN and T. M. KORN, *Mathematical Handbook for Scientists and Engineers*, McGraw-Hill Book Company, New York, New York, 1968.

A PERTURBATION APPROACH TO THE MANEUVERING AND CONTROL OF SPACE STRUCTURES

L. MEIROVITCH

Department of Engineering Science and Mechanics
Virginia Polytechnic Institute and State University
Blacksburg, Virginia 24061

Y. SHARONY

Department of Electrical Engineering
Virginia Polytechnic Institute and State University
Blacksburg, Virginia 24061

I. INTRODUCTION

This study is concerned with the problem of simultaneous maneuvering and vibration control of a flexible space structure (Fig. 1). Fast maneuvering of highly flexible structures tends to excite elastic vibration, where the vibration is caused by the inertial forces resulting from the maneuver. The amplitude of the vibration increases as the ratio between the maneuver angular velocity and the natural frequencies of the structure increases. Quite often, the ratio is sufficiently high for the elastic motions to exceed the linear range, and even to cause damage to the structure. If the structure is to perform a maneuver as closely as possible to minimum time, with tight accuracy demands at the termination of the maneuver, the vibration control during the maneuver is essential.

The equations of motion describing the slewing of a flexible space structure constitute a hybrid set in the sense that the rigid-body motions of the space structure are described by nonlinear ordinary differential equations and the elastic motions, assumed to remain in the linear range, by linear partial

Copyright © 1990 by Academic Press, Inc.
All rights of reproduction in any form reserved.

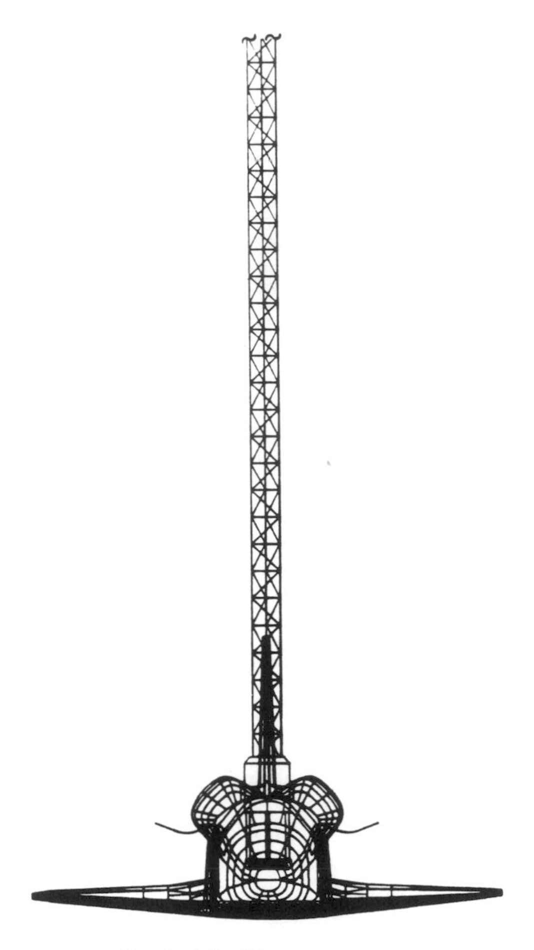

Fig. 1. A flexible space structure.

differential equations. The on-line control must be implemented by means of a relatively low-order compensator. Hence, the original model, which is nonlinear and infinite dimensional, must be controlled by a finite-order compensator. The design of an optimal compensator based on a nonlinear infinite-dimensional model is not feasible, so that the hybrid set of equations must be first discretized in space and then truncated. The discretization of the model is ordinarily carried out by representing the elastic motions as linear combinations of space-dependent admissible functions multiplied by time-dependent generalized coordinates [1]. By taking the linear combinations as finite, the system is truncated. It is assumed that the order of the truncated model is sufficiently high that it is unlikely that the ignored degrees of freedom, possessing higher inherent damping, will destabilize the closed-loop distributed parameter system. The outcome of the discretization process is a set of nonlinear ordinary differential equations that are still of relatively high order. Then, the problem consists of designing a reduced-order compensator to control the motion of the high-order, nonlinear model from one rest position to another in minimum time.

In recent years, there has been considerable interest in control of maneuvering space structures. Single-axis slewing maneuver about a principal axis of a space structure consisting of a rigid hub and a number of flexible appendages has been investigated in [2–4] by means of a reduced-order nonlinear model. The solution, satisfying prescribed final conditions, minimizes a performance measure in the form of a combination of control effort and energy, where the final time is fixed but otherwise arbitrary. The solution of the nonlinear two-point boundary-value problem (TPBVP) encountered in the minimum-time problem of the nonlinear reduced-order model is approximated by means of the continuation method and yields a closed-loop solution for the reduced-order model. In [5], the control is restricted in that only three switching times can be selected, where the final time is prescribed *a priori*. These three switching points are determined so as to minimize the post-maneuver elastic energy, whereas the minimization problem is constrained by prescribed angular impulse and its time integral. The solution results in on–off, opcn-loop control. The problem in which the cost is a combination of fuel and transition time was investigated in [6] by means of a nonmaneuvering reduced-order linear model, and expressions for the switching points were obtained by an approximate method. This also is on–off, open-loop control. Near time-optimal control minimizing the time and a certain measure of residual energy was investigated in [7]. The control is again open-loop. The slewing problem in which the cost is the transition time only was investigated in [8–10]. In these investigations, a time-optimal problem is solved for a nonmaneuvering reduced-order model, resulting in a linear model. In [10], this solution serves as an initial guess for the solution of the nonlinear TPBVP describing the maneuvering of the reduced-order model. The solution represents on–off, open-loop control in which the switching curve is determined by a TPBVP solver. The design of pure minimum-time control for a nonlinear model involves the solution of a nonlinear TPBVP, which in the case of structures has the added complexity of high dimensionality. The solution is very involved, even for off-line computers, and is not implementable on an on-line computer. Moreover, in general the solution represents open-loop control, which is highly sensitive to parameter inaccuracies ordinarily encountered in structures.

To circumvent the above difficulties, a perturbation approach was developed in [11, 12]. The motion of the structure consists of six rigid-body degrees of freedom of a set of body axes plus the elastic vibration relative to the body axes. Because the nonlinearity enters through the rigid-body motions, while the elastic motions tend to be small, the rigid-body maneuvering can be regarded as a zero-order problem, and the elastic motions and deviations from the rigid-body maneuvering can be regarded as a first-order problem. The zero-order problem is generally nonlinear, but of relatively low order, and the control is according to a minimum-time policy, which results in bang-bang control. The first-order problem represents a linear, time-varying model, albeit of high order, and is subjected to kinematical disturbances caused by inertial forces produced by the zero-order maneuver. The problem can be solved within the framework of linear time-varying feedback control. Hence, the method is less sensitive to parameter uncertainty and more able to cope with the effects of the residual model.

The first-order model consists of a rigid-body perturbed model describing the perturbations in the translations and rotations of a set of body axes and an elastic model describing the elastic displacements of the flexible appendages relative to the perturbed body axes. The problem of synthesizing feedback control for the first-order model during a minimum-time maneuver has been considered in [12, 13]. In both references a reduced-order model is considered. Reference [12] considers a general model of a maneuvering flexible spacecraft. The maneuver period is divided into several time intervals. A time-invariant reduced-order model corresponding to each time interval is then controlled according to a modal control law designed by ignoring kinematic disturbances caused by the zero-order maneuver. Two control techniques are considered: independent modal-space control (IMSC) and direct output feedback.

Single-axis slewing maneuver about a principal axis of a flexible space structure consisting of a rigid hub and a flexible appendage has been investigated in [13]. An optimal linear quadratic regulator (LQR) based on the time-varying reduced-order model, including integral feedback, is designed to accommodate disturbances resulting from the minimum-time maneuver of the rigid body and at the same time to shape the dynamic characteristics of the reduced-order model within the finite-time interval of the maneuver. The control minimizes a finite-time quadratic performance measure including the derivative of the control and an exponential term designed to force convergence within the finite-time interval.

A third approach to the control of the first-order model was developed in [14, 15] and is the main topic of this article. A single-axis slewing maneuver about a principal axis of a flexible spacecraft consisting of a rigid hub and a flexible appendage (Fig. 2) is considered. The approach is designed to stabilize a reduced-order model within a finite-time interval and at the same time to

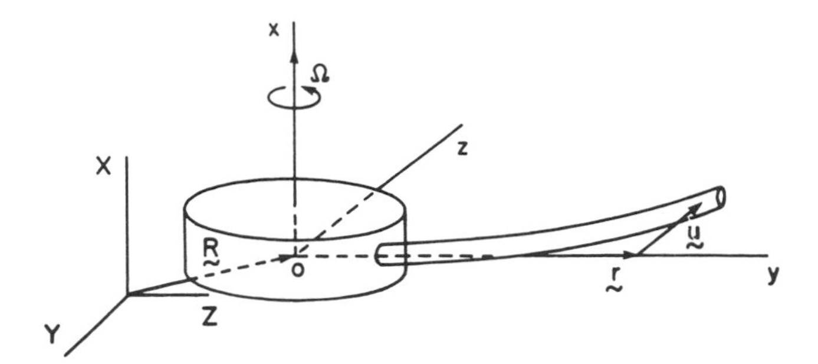

Fig. 2. A mathematical model of a flexible space structure.

optimize several characteristics of the entire first-order model. The control policy is based on a reduced-order compensator, which consists of a controller and a Luenberger observer. The observer estimates part of the modeled states, which include the controlled states and the observed but uncontrolled states, as well as part of the disturbance vector. The controller is divided into two parts, where the first part is designed to enhance the dynamic characteristics of the elastic model and the second part accommodates the persistent disturbances. The main features of this method are

1. Concepts of finite-time stability are defined and serve as design criteria for the control of both the reduced-order model and the first-order model.
2. The feedback control is designed not only to stabilize the reduced-order model but also to achieve quasi-constant conditions for the entire elastic model toward the end of the finite-time interval and to minimize the supremum time constant of the elastic model. This implies that the first-order model is finite-time stable.
3. A weighted norm of the disturbance response of the entire first-order model is minimized in quasi-constant conditions.

The approach is demonstrated by means of a numerical example.

II. PSEUDOMODAL EQUATIONS OF MOTION

We consider a space structure consisting of a rigid body and a flexible appendage (Fig. 3). The inertia of the rigid part is assumed to be appreciably larger than the inertia of the flexible part. It is convenient to describe the motion of the space structure in terms of a reference frame *xyz* embedded in

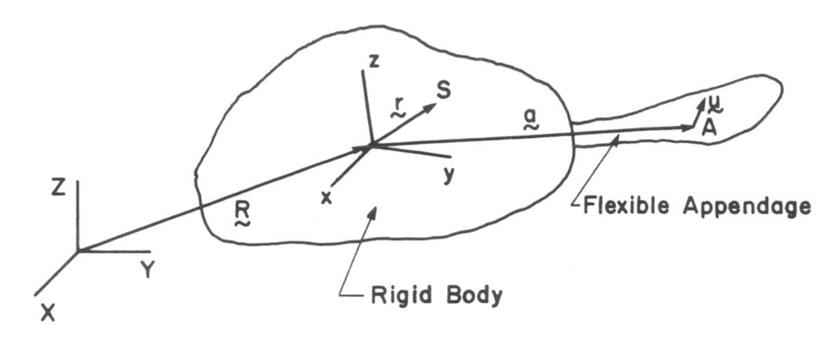

Fig. 3. The space structure with various reference frames.

the rigid body and referred to as the body axes. Then, the motion can be regarded as consisting of six rigid-body degrees of freedom, three translations and three rotations, associated with the motion of the body axes xyz and the elastic motion measured relative to the body axes. We assume that the origin 0 of the body axes is located at the center of mass of the undeformed space structure, which lies on the rigid body in view of the much larger inertia of the rigid part compared with that of the flexible part. We denote the position of 0 relative to the inertial axes XYZ by the radius vector $\mathbf{R}$ and the position of a point S on the rigid body relative to 0 by the radius vector $\mathbf{r}$. Moreover, we denote the radius vector from 0 to a nominal point A on the flexible appendage by $\mathbf{a}$ and the elastic displacement vector of A by $\mathbf{u}$. The elastic motion is governed by partial differential equations, which implies an infinite number of degrees of freedom for each displacement component. For practical reasons, the system must be discretized in space. To this end, we express the elastic displacement vector in the form

$$\mathbf{u} = \Phi\mathbf{q} \tag{1}$$

where Φ is a $3 \times 2N$ matrix of space-dependent admissible functions [1] and $\mathbf{q}$ is a $3N$ vector of time-dependent generalized coordinates.

From [11], the equations of motion for the space structure in the absence of gravitational effects are

$$m\ddot{\mathbf{R}} + C^T\bar{\Phi}\ddot{\mathbf{q}} + 2C^T\tilde{\omega}^T\bar{\Phi}\dot{\mathbf{q}} + C^T(\tilde{\omega}^2 + \dot{\tilde{\omega}}^T)\bar{\Phi}\mathbf{q} = C^T\mathbf{F} \tag{2a}$$

$$I_0\dot{\omega} + \tilde{\omega}^T I_0\omega + \psi^T\ddot{\mathbf{q}} + \{\widetilde{[C\ddot{\mathbf{R}}]}\bar{\Phi} + J(\dot{\omega}) + \tilde{\omega}^T J(\omega)\}\mathbf{q} + [\tilde{\omega}^T\psi^T + J(\omega)]\dot{\mathbf{q}} = \mathbf{M} \tag{2b}$$

$$M_A\ddot{\mathbf{q}} + \bar{\Phi}^T C\ddot{\mathbf{R}} + \psi\dot{\omega} + \int_{m_A} \Phi^T\tilde{\omega}^T\tilde{a}\omega\, dm_A + [D + \tilde{L}_A(\omega)]\dot{\mathbf{q}} + [\bar{L}_A(\omega) + \tilde{L}_A(\dot{\omega}) + K_A]\mathbf{q} = \mathbf{Q}, \tag{2c}$$

where m is the total mass of the space structure, m_A the mass of the flexible part, C a rotation matrix from axes XYZ to axes xyz, I_0 the total mass moment of inertia about 0 of the undeformed space structure, and $\boldsymbol{\omega}$ the angular velocity vector of the axes xyz. Moreover, we used the relation

$$\dot{C} = \tilde{\omega}C \tag{3}$$

as well as the notation

$$\bar{\Phi} = \int_{m_A} \Phi \, dm_A, \qquad \psi = \int_{m_A} \Phi^T \tilde{a} \, dm_A, \tag{4a}$$

$$J(\omega) = \int_{m_A} (\tilde{a}\tilde{\omega} + \widetilde{[\tilde{a}\omega]})\Phi \, dm_A, \tag{4b}$$

$$\tilde{L}_A(\omega) = \int_{m_A} \Phi^T \tilde{\omega}^T \Phi \, dm_A, \qquad \bar{L}_A(\omega) = \int_{m_A} \Phi^T \tilde{\omega}^2 \Phi \, dm_A, \tag{4c}$$

$$M_A = \int_{m_A} \Phi^T \Phi \, dm_A, \tag{4d}$$

$$D = \int_{D_A} \Phi^T \mathscr{D} \Phi \, dD_A, \tag{4e}$$

$$K_A = [\Phi, \Phi], \tag{4f}$$

in which $\mathscr{D}$ denotes a 3×3 distributed viscous damping matrix, [,] denotes an energy inner product [1], and a tilde over a symbol denotes a skew symmetric matrix of the form

$$\tilde{v} = \begin{bmatrix} 0 & v_z & -v_y \\ -v_z & 0 & v_x \\ v_y & -v_x & 0 \end{bmatrix}, \tag{5}$$

where v_x, v_y, v_z, are the Cartesian components of the vector $\mathbf{v}$. The terms on the right side of (2) are given by

$$\mathbf{F} = \int_{D_S} \mathbf{f}_S \, dD_S + \int_{D_A} \mathbf{f}_A \, dD_A, \tag{6a}$$

$$\mathbf{M} = \int_{D_S} (\tilde{r}^T \mathbf{f}_S + \mathbf{T}_S) \, dD_S + \int_{D_A} (\tilde{a}^T \mathbf{f}_A + \mathbf{T}_A) \, dD_A + \int_{D_A} \tilde{f}_A \Phi \, dD_A \mathbf{q}, \tag{6b}$$

$$\mathbf{Q} = \int_{D_A} (\Phi^T \mathbf{f}_A + \Phi'^T \mathbf{T}_A) \, dD_A, \tag{6c}$$

and they represent generalized force vectors in terms of components along axes xyz, in which D_S and D_A represent the rigid and flexible domains, respectively. Moreover, $\mathbf{f}_S$ and $\mathbf{T}_S$ denote distributed force and torque vectors acting on the rigid part and $\mathbf{f}_A$ and $\mathbf{T}_A$ are the counterparts for the flexible part. In addition, Φ' denotes the spatial derivative of Φ with the respect to the proper coordinate. It should be pointed out that the position vector $\mathbf{R}$, its time derivatives, and the Euler angles α_1, α_2, and α_3, defining the rotation of axes xyz, have been assumed to be of arbitrary magnitude, so that (2) is highly nonlinear. The angular velocity vector $\boldsymbol{\omega}$ is related to the Eulerian angular velocities by

$$\boldsymbol{\omega} = D_T(\boldsymbol{\alpha})\dot{\boldsymbol{\alpha}}, \tag{7}$$

where $\mathbf{D}_T(\boldsymbol{\alpha})$ is a transformation matrix depending nonlinearly on the angles α_1, α_2, and α_3, and we note that $\boldsymbol{\alpha}$ is not really a vector for finite angles.

As just pointed out, Eqs. (2) are highly nonlinear, where the nonlinearity enters through the rigid-body motions. Our interest is in perturbation equations, obtained by separating the rigid-body motions from perturbations caused by flexibility, where the latter include the elastic motions and perturbations in the rigid-body motions. Hence, we consider a first-order perturbation of $\mathbf{R}$ and $\boldsymbol{\alpha}$, or

$$\mathbf{R} = \mathbf{R}_0 + \mathbf{R}_1, \tag{8a}$$

$$\boldsymbol{\alpha} = \boldsymbol{\alpha}_0 + \boldsymbol{\alpha}_1, \tag{8b}$$

where the first-order terms $\mathbf{R}_1$ and $\boldsymbol{\alpha}_1$ are small compared to the zero-order terms $\mathbf{R}_0$ and $\boldsymbol{\alpha}_0$. Introducing (8) into the nonlinear equations of motion (2) and separating orders of magnitude, we obtain zero-order and first-order perturbation equations. The zero-order equations can be used for maneuvering the space structure, and the first-order equations for vibration suppression and rigid-body corrections. Before proceeding with this technique, we first develop some expressions relating the perturbation vector $\boldsymbol{\alpha}_1$ in the Euler angles to the small angular displacement vector $\boldsymbol{\beta}$ expressed in the body-fixed frame. This is done so as to permit expressing all the variables in the perturbation equations in terms of components along the body-fixed frame, the frame in which state measurements are taken and actuating forces are applied.

First, consider (7), which relates the velocity vector $\dot{\boldsymbol{\alpha}}$ in terms of the Euler angles to the body-fixed angular velocity vector $\boldsymbol{\omega}$. Introducing (8b) into Eq. (7) and neglecting higher-order terms, we obtain the perturbed angular velocity vector

$$\boldsymbol{\omega} = \boldsymbol{\omega}_0 + \boldsymbol{\omega}_1, \tag{9}$$

where

$$\boldsymbol{\omega}_0 = D(\boldsymbol{\alpha}_0)\dot{\boldsymbol{\alpha}}_0, \tag{10a}$$

$$\boldsymbol{\omega}_1 = D(\boldsymbol{\alpha}_1)\dot{\boldsymbol{\alpha}}_0 + D(\boldsymbol{\alpha}_0)\dot{\boldsymbol{\alpha}}_1. \tag{10b}$$

Using similar considerations, it can be shown that the body-fixed perturbation angles can be related to the perturbed Euler angles by the expression

$$\boldsymbol{\beta} = D(\boldsymbol{\alpha}_0)\boldsymbol{\alpha}_1. \tag{11}$$

Taking the time derivative of (11) and introducing the result into (10b), it can be shown, after much algebra, that the perturbed angular velocity vector $\boldsymbol{\omega}_1$ is related to the angular displacement vector $\boldsymbol{\beta}$ by the expression

$$\boldsymbol{\omega}_1 = \tilde{\omega}_0^T \boldsymbol{\beta} + \dot{\boldsymbol{\beta}}. \tag{12}$$

Taking the time derivative of (12), the perturbed angular acceleration becomes

$$\dot{\boldsymbol{\omega}}_1 = \dot{\tilde{\omega}}_0^T \boldsymbol{\beta} + \tilde{\omega}_0^T \dot{\boldsymbol{\beta}} + \ddot{\boldsymbol{\beta}}. \tag{13}$$

Next, we recall that the elements of the transformation matrix C from the body-fixed frame to the inertial frame consist of functions of the Euler angles, so that a perturbation of this matrix also involves the angles $\beta_1, \beta_2, \beta_3$. This relation can be derived using (12). Instead, we consider the frame $0'$, which differs from the 0 frame by the angles $\beta_1, \beta_2, \beta_3$. Then, recalling that β_1, β_2, and β_3 are small, the transformation from the $0'$ frame to the 0 frame is $[I + \tilde{\beta}]$, where I is the identity matrix. Letting C_0 be the transformation matrix from the 0 frame to the inertial frame, the total transformation matrix C from the $0'$ frame to the inertial frame can be expressed as

$$C \cong C_0 + C_1, \tag{14}$$

where

$$C_1 = \tilde{\beta} C_0. \tag{15}$$

In keeping with our objective of expressing the first-order perturbation equations in the body-fixed frame, (8a) is replaced by

$$\mathbf{R} = \mathbf{R}_0 + C_0^T \mathbf{R}_1, \tag{16}$$

where $\mathbf{R}_1$ is now a vector measured with respect to the 0 frame. The control forces and moments can also be expressed in first-order perturbed form as follows:

$$\mathbf{F} = \mathbf{F}_0 + \mathbf{F}_1, \qquad \mathbf{M} = \mathbf{M}_0 + \mathbf{M}_1. \tag{17}$$

Introducing (8)–(17) into (2) and ignoring second-order terms, we obtain the zero-order equations

$$m\ddot{\mathbf{R}}_0 = C_0^T \mathbf{F}_0, \tag{18a}$$

$$I_0 \dot{\boldsymbol{\omega}}_0 + \tilde{\omega}_0^T I_0 \boldsymbol{\omega}_0 = \mathbf{M}_0 \tag{18b}$$

and the first-order equations

$$M\ddot{\mathbf{x}} + (G + D_{\mathrm{v}})\dot{\mathbf{x}} + (K_{\mathrm{S}} + K_{\mathrm{NS}})\mathbf{x} = \mathbf{F}^*, \tag{19}$$

where

$$\mathbf{x}^T = [\mathbf{R}_1^T \mid \boldsymbol{\beta}^T \mid \mathbf{q}^T], \tag{20a}$$

$$F^{*T} = [\mathbf{F}_1^T \mid \mathbf{M}_1^T \mid \mathbf{Q}_0^T + \mathbf{Q}_1^T], \tag{20b}$$

$$M = \begin{bmatrix} M_0 & 0 & \bar{\Phi} \\ 0 & I_0 & \psi^T \\ \bar{\Phi}^T & \psi & M_{\mathrm{A}} \end{bmatrix}, \qquad M_0 = \begin{bmatrix} m & 0 & 0 \\ 0 & m & 0 \\ 0 & 0 & m \end{bmatrix}, \tag{20c}$$

$$G = \begin{bmatrix} 2M_0\tilde{\omega}_0^T & 0 & 2\tilde{\omega}_0^T\bar{\Phi} \\ 0 & I_0\tilde{\omega}_0^T + \tilde{\omega}_0^T I_0 + [\widetilde{I_0\tilde{\omega}_0}] & \tilde{\omega}_0^T\psi^T + J_0 \\ -2(\tilde{\omega}_0^T\bar{\Phi})^T & -[\tilde{\omega}_0^T\psi^T + J_0]^T & 2\tilde{L}_{\mathrm{A}} \end{bmatrix}, \tag{20d}$$

$$D_{\mathrm{v}} = \begin{bmatrix} 0 & 0 & 0 \\ 0 & 0 & 0 \\ 0 & 0 & D \end{bmatrix}, \tag{20e}$$

$$K_{\mathrm{S}} = \begin{bmatrix} M_0\tilde{\omega}_0^2 & \tilde{F}_0^T & [\tilde{\omega}_0^2 + \tilde{H}]\psi^T \\ \tilde{F}_0 & \tilde{\omega}_0 I_0 \tilde{\omega}_0 & \tilde{\omega}_0^T J_0 + \tilde{H}\bar{\Phi} \\ \psi[\tilde{\omega}_0^2 + \tilde{H}]^T & J_0^T\tilde{\omega}_0 + \bar{\Phi}^T\tilde{H}^T & \tilde{L}_{\mathrm{A}} + K_{\mathrm{A}} \end{bmatrix}, \tag{20f}$$

$$K_{\mathrm{NS}} = \begin{bmatrix} M_0\dot{\tilde{\omega}}_0^T & 0 & \dot{\tilde{\omega}}_0^T\bar{\Phi} \\ 0 & [I_0\dot{\tilde{\omega}}_0^T + [\widetilde{I_0\omega_0}]\tilde{\omega}_0^T & \dot{J}_0 \\ -(\dot{\tilde{\omega}}_0^T\bar{\Phi})^T & \psi\dot{\tilde{\omega}}_0^T & \dot{\tilde{L}}_{\mathrm{A}} \end{bmatrix}, \tag{20g}$$

in which

$$\tilde{H} = (\widetilde{C_0\ddot{\mathbf{R}}_0}) \tag{21a}$$

$$\mathbf{Q}_0 = -\left(\bar{\Phi}^T C_0 \ddot{\mathbf{R}}_0 + \psi\dot{\omega}_0 + \int_{m_{\mathrm{A}}} \Phi^T \tilde{\omega}_0^T \tilde{a} \omega_0 \, dm_{\mathrm{A}}\right). \tag{21b}$$

For any desired maneuver strategy $\mathbf{R}_0(t)$ and $\boldsymbol{\omega}_0(t)$, (18) yield the required controls $\mathbf{F}_0(t)$ and $\mathbf{M}_0(t)$. Because the rigid-body motions $\mathbf{R}_0(t)$ and $\boldsymbol{\omega}_0(t)$ are known functions of time, (19) represents a time-varying system. The stiffness matrix has been split into a symmetric matrix K_{S} and a nonsymmetric matrix K_{NS}. If desired, of course, K_{NS} can be separated into symmetric and skew symmetric matrices. As can be seen in (20g), the terms containing the angular acceleration $\dot{\boldsymbol{\omega}}_0$ of the body-fixed frame lead to nonsymmetric terms in the stiffness matrix. In fact, many of these terms are skew symmetric, and hence circulatory. Owing to the nonsymmetry of the stiffness matrices, the vibration problem is non-self-adjoint. The matrix G is due entirely to gyroscopic effects, and hence conservative in nature. On the other hand, the damping matrix is nonconservative and the stiffness matrix contains circulatory, nonconservative terms. During the maneuver, the centrifugal, tangential, and gravitational forces cause a change in the initial elastic deflection of the structure which, in turn, affects the orientation of the space structure.

The first-order model (19) represents a $(6 + 3N)$-degree-of-freedom time-varying system. Quite often, the number of degrees of freedom is so large that it is necessary to truncate the system. This can be effected by means of a linear transformation involving the premaneuver eigenvectors of the undamped structure. Equation (19) can be rewritten as

$$M\ddot{\mathbf{x}}(t) + G(t)\dot{\mathbf{x}}(t) + [K_0 + K_t(t)]\mathbf{x}(t) = \mathbf{F}^*(t), \tag{22}$$

where K_t consists of the time-varying terms of the stiffness matrix and K_0 contains the remaining constant terms. The premaneuver equation of motion of the undamped space structure can be expressed as

$$M\ddot{\mathbf{x}}(t) + K_0\mathbf{x}(t) = \mathbf{F}^*(t). \tag{23}$$

Hence, the corresponding eigenvalue problem can be written in the form

$$K_0\mathbf{x} = \lambda M\mathbf{x}, \tag{24}$$

where $\lambda = \omega^2$ is an eigenvalue, in which ω is a corresponding natural frequency and $\mathbf{x}$ is the associated eigenvector.

Owing to the nature of the discretization process, only the lower modes are accurate representations of the actual modes. Of course, by increasing the number of admissible functions in the original discretization process (1), we can increase the number of accurate modes. We retain $6 + 3n$ lower modes in the truncated model, where n is sufficiently large that it is unlikely that the ignored degrees of freedom will destabilize the actual closed-loop system. The modes can be arranged in a $(6 + 3N) \times (6 + 3n)$ matrix X of eigenvectors, which can be normalized so as to satisfy

$$X^T M X = I, \qquad X^T K_0 X = \Lambda, \tag{25}$$

where I is the identity matrix and Λ is the $(6 + 3n) \times (6 + 3n)$ diagonal matrix of eigenvalues. Using the expansion theorem [1], the solution of (22) can be approximated by a linear combination of the lower $(6 + 3n)$ modes as follows:

$$\mathbf{x}(t) = X\mathbf{v}(t), \tag{26}$$

where $\mathbf{v}(t)$ is a $(6 + 3n)$-vector of generalized coordinates. Inserting (26) into

(22), premultiplying by X^T, and considering (25), we obtain

$$\ddot{\mathbf{v}}(t) + [\bar{G}(t) + \bar{D}_v]\dot{\mathbf{v}}(t) + [\Lambda + \bar{K}(t)]\mathbf{v}(t) = \mathbf{f}(t), \tag{27}$$

where

$$\bar{G}(t) = X^T G(t) X, \qquad \bar{D}_v = X^T D_v X, \qquad \bar{K}(t) = X^T K_t(t) X \tag{28}$$

are the reduced-order gyroscopic, damping, and stiffness matrices, respectively. We assume that the off-diagonal terms in $\bar{D}_v$ are sufficiently small that they can be ignored, so that

$$\bar{D}_v \cong \text{diag}(0 \quad 0 \quad 0 \quad 0 \quad 0 \quad 0 \quad 2\zeta_i\omega_i), \qquad i = 7, 8, \ldots, 6 + 3n \tag{29}$$

in which ζ_i are viscous damping factors and ω_i natural frequencies of the premaneuvering undamped space structure. Moreover,

$$\mathbf{f}(t) = X^T \mathbf{F}^*(t) \tag{30}$$

is a generalized force vector. We refer to (27) as being in quasi-modal form.

III. CONTROL STRATEGY

The rigid-body equations of motion are described by (18). We assume that the position of point 0 is not altered during the angular maneuver, that is,

$$\ddot{\mathbf{R}}_0 \equiv \mathbf{0}, \qquad \mathbf{F}_0 = \mathbf{0}. \tag{31}$$

Moreover, we consider a single-axis maneuver about a principal axis. Assuming that x, y, z are principal axes, we pick the x axis as the rotation axis. Hence, (18b) reduces to the scalar equation

$$I_x \dot{\Omega}_0 = M_0 \tag{32}$$

where I_x, $\dot{\Omega}_0$, and M_0 are the moment of inertia, the slewing angular acceleration, and the control torque about the x axis, respectively.

The rigid-body slewing is controlled according to a minimum-time policy, which implies bang-bang control [16]. Assuming an ideal actuator, (32) represents a linear time-invariant system with two real eigenvalues. As a result, there is only one switching time, namely, halfway through the maneuver time interval. Hence, denoting by t_0 the initial time and by t_f the final time, the switching time is simply $t_1 = (t_f - t_0)/2$, so that the slewing angular acceleration and velocity are

$$\dot{\Omega}_0(t) = \begin{cases} c & \text{for} \quad t_0 \le t < t_1, \\ -c & \text{for} \quad t_1 \le t \le t_f, \end{cases} \tag{33a}$$

$$\Omega_0(t) = \begin{cases} c(t - t_0) & \text{for} \quad t_0 \le t < t_1, \\ -c(t - t_f) & \text{for} \quad t_1 \le t \le t_f. \end{cases} \tag{33b}$$

As a mathematical model, we consider a flexible space structure consisting of a rigid hub and a flexible appendage, where the appendage is in the form of a beam with one end attached to the hub and the other end free (Fig. 2). The perturbation equations (22) represent a set of high-order linear differential equations with time-varying coefficients and subjected to persistent disturbances caused by the rigid-body maneuver. Due to the configuration under consideration, and recalling (31), the inertial disturbances are proportional to the slewing angular acceleration, as can be concluded from (20b) and (21b). Moreover, we assume that there are two force actuators T_1 and T_2 and one torque actuator T_3 acting at the mass center of the space structure in undeformed state and p torque actuators T_{3+i} acting on the elastic appendage and mounted at the points $y = y_i (i = 1, 2, \ldots, p)$ of the appendage. The displacement and velocity measurements are denoted by the vectors $\mathbf{y}_d$ and $\mathbf{y}_v$. We assume that there are two pairs of translational displacement and velocity sensors and one pair of angular displacement and velocity sensors collocated with the actuators $T_i (i = 1, 2, 3)$, respectively. In addition, there are $m/2$ pairs of angular displacement and velocity sensors located throughout the flexible appendage at points $y = y_j (j = 1, 2, \ldots, m/2)$ of the appendage. The perturbed motion is the difference between the sensor data and the commanded motion expressed in body axes. Considering all this, we can rewrite (1), (4a), (4d), and (4f) as

$$\mathbf{u}(\mathbf{r}, t) = [0 \quad 0 \quad u_z(y, t)]^T, \qquad u_z(y, t) = \boldsymbol{\phi}^T(y)\mathbf{q}(t), \tag{34a}$$

$$\bar{\boldsymbol{\phi}} = \int_m \boldsymbol{\phi}\, dm_A, \qquad \boldsymbol{\psi} = \int_m y\boldsymbol{\phi}\, dm_A, \tag{34b}$$

$$M_A = \int_{m_A} \boldsymbol{\phi}(y)\boldsymbol{\phi}^T(y)\, dm_A, \qquad K_A = [\boldsymbol{\phi}, \boldsymbol{\phi}], \tag{34c}$$

where $\mathbf{q}(t)$ is an N-vector of generalized coordinates and $\boldsymbol{\phi}(y)$ is an N-vector of admissible functions. Then, defining $\mathbf{x} = [\mathbf{R}_1^T \mid \beta_x \mid \mathbf{q}^T]^T$, where $\mathbf{R}_1 = [R_y \quad R_z]^T$ is the vector of translational perturbations in the y- and z-directions, respectively, and β_x is the rotational perturbation about the x axis, we can rewrite (19), as well as the output equations, in the form

$$M\ddot{\mathbf{x}} + (G + D_v)\dot{\mathbf{x}} + K\mathbf{x} = E\mathbf{T} - \dot{\Omega}_0\boldsymbol{\psi}, \tag{35a}$$

$$\mathbf{y}_d = \tilde{C}\mathbf{x}, \qquad \mathbf{y}_v = \tilde{C}\dot{\mathbf{x}}, \tag{35b}$$

where

$$M = \left[\begin{array}{c|c|c} mI_{2\times 2} & \mathbf{0}_2 & \mathbf{e}_2\bar{\boldsymbol{\phi}}^T \\ \hline \mathbf{0}_2^T & I_x & \boldsymbol{\psi}^T \\ \hline \bar{\boldsymbol{\phi}}\mathbf{e}_2^T & \boldsymbol{\psi} & M_A \end{array}\right], \tag{36a}$$

$$D_v = \left[\begin{array}{c|c|c} 0_{2\times 2} & \mathbf{0}_2 & 0_{2\times N} \\ \hline \mathbf{0}_2^T & 0 & \mathbf{0}_N^T \\ \hline 0_{N\times 2} & \mathbf{0}_N & D \end{array}\right], \tag{36b}$$

in which $\mathbf{e}_1 = [1 \quad 0]^T$ and $\mathbf{e}_2 = [0 \quad 1]^T$ are standard unit vectors and $\mathbf{0}_2$ and $\mathbf{0}_N$ are null vectors of dimensions 2 and N, respectively. Moreover,

$$G = 2\Omega_0 G^*, \qquad K = K_c + \dot{\Omega}_0 G^* - \Omega_0^2 K_s, \tag{37}$$

where

$$G^* = \left[\begin{array}{c|c|c} mP & \mathbf{0}_2 & -\mathbf{e}_1\bar{\boldsymbol{\phi}}^T \\ \hline \mathbf{0}_2^T & 0 & \mathbf{0}_N^T \\ \hline \bar{\boldsymbol{\phi}}\mathbf{e}_1^T & \mathbf{0}_N & 0_{N\times N} \end{array}\right] = -(G^*)^T \tag{38a}$$

$$K_c = \left[\begin{array}{c|c|c} 0_{2\times 2} & \mathbf{0}_2 & 0_{2\times N} \\ \hline \mathbf{0}_2^T & 0 & \mathbf{0}_N^T \\ \hline 0_{N\times 2} & \mathbf{0}_N & K_A \end{array}\right] = K_c^T, \qquad K_s = \left[\begin{array}{c|c|c} mI_{2\times 2} & \mathbf{0}_2 & \mathbf{e}_2\bar{\boldsymbol{\phi}}^T \\ \hline \mathbf{0}_2^T & 0 & \mathbf{0}_N^T \\ \hline \bar{\boldsymbol{\phi}}\mathbf{e}_2^T & \mathbf{0}_N & M_A \end{array}\right] = K_s^T, \tag{38b}$$

$$\tilde{C} = \left[\begin{array}{ccc|c} & I_{3\times 3} & & 0_{3\times N} \\ \hline \mathbf{0}_{m/2} & \mathbf{0}_{m/2} & \mathbf{1}_{m/2} & C^* \end{array}\right], \qquad E = \left[\begin{array}{c|c} & \mathbf{0}_p^T \\ I_{3\times 3} & \mathbf{0}_p^T \\ & \mathbf{1}_p^T \\ \hline 0_{N\times 3} & E^* \end{array}\right], \qquad \boldsymbol{\Psi} = [\mathbf{0}_3^T \mid \boldsymbol{\psi}^T]^T, \tag{39}$$

where P is a 2×2 unit skew symmetric matrix, $\mathbf{1}_N$ is an N-vector of ones, $E^* = [\boldsymbol{\phi}'(y_1) \quad \boldsymbol{\phi}'(y_2) \quad \cdots \quad \boldsymbol{\phi}'(y_p)]$ is an $N \times p$ modal participation matrix, in which primes denote derivatives with respect to y, and $C^* = [\boldsymbol{\phi}'(y_1) \quad \boldsymbol{\phi}'(y_2) \quad \cdots \quad \boldsymbol{\phi}'(y_{m/2})]^T$ is an $m/2 \times N$ modal participation matrix. We notice that, in the case in which $m/2 = p$ and the sensors are collocated with the actuators $\tilde{C}^T = E$.

At this point, we wish to obtain the quasi-modal version of (35). Following the developments of Section II, we rewrite (25) in the form

$$U^T M U = I, \qquad U^T K_c U = \Lambda, \tag{40}$$

where U is a $(3 + N) \times (3 + n)$ matrix of eigenvectors and $\Lambda =$

$[0 \quad 0 \quad 0 \quad \omega_1^2 \quad \omega_2^2 \quad \cdots \quad \omega_n^2]$ is a diagonal matrix of eigenvalues, in which $\omega_i (i = 1, 2, \ldots, n)$ are distinct natural frequencies of the premaneuvering undamped space structure. Next, we introduce the linear transformation

$$\mathbf{x}(t) = U\mathbf{v}(t) \tag{41}$$

into (35), multiply on the left by U^T, consider (37), and obtain

$$\ddot{\mathbf{v}} + (2\Omega_0 \bar{G} + \bar{D}_v)\dot{\mathbf{v}} + (\Lambda + \dot{\Omega}_0 \bar{G} - \Omega_0^2 \bar{K})\mathbf{v} = \bar{E}\mathbf{T} - \dot{\Omega}_0 \bar{\mathbf{\Psi}} \tag{42a}$$

$$\mathbf{y}_d = \bar{C}\mathbf{v}, \qquad \mathbf{y}_v = \bar{C}\dot{\mathbf{v}}, \tag{42b}$$

where

$$\begin{aligned} &\bar{G} = U^T G^* U, \qquad \bar{D}_v = U^T D_v U, \qquad \bar{K} = U^T K_s U, \qquad \bar{E} = U^T E, \\ &\bar{C} = \tilde{C} U, \qquad \bar{\mathbf{\Psi}} = U^T \mathbf{\Psi}. \end{aligned} \tag{43}$$

Equations (42) represent a set of pseudomodal equations.

Due to the nature of the coefficient matrices and of the excitation vector, (42) possesses a relatively simple structure. To show this, we express the matrix M, as given by (36a), in the partitioned form

$$M = \begin{bmatrix} M_{11} & M_{12} \\ M_{12}^T & M_{22} \end{bmatrix}, \tag{44}$$

where

$$M_{11} = \begin{bmatrix} mI_{2\times 2} & \mathbf{0}_2 \\ \mathbf{0}_2^T & I_x \end{bmatrix}, \qquad M_{12} = \begin{bmatrix} \mathbf{e}_2 \bar{\boldsymbol{\phi}}^T \\ \boldsymbol{\psi}^T \end{bmatrix}, \qquad M_{22} = M_A \tag{45}$$

are 3×3, $3 \times N$, and $N \times N$ matrices, respectively. Accordingly, we partition the matrix of eigenvectors as follows:

$$U = \begin{bmatrix} U_{11} & U_{12} \\ U_{21} & U_{22} \end{bmatrix}. \tag{46}$$

Inserting (44)–(46) into (40), we conclude that

$$U_{11} = \begin{bmatrix} m^{-1/2} I_{2\times 2} & \mathbf{0}_2 \\ \mathbf{0}_2^T & I_x^{-1/2} \end{bmatrix}, \qquad U_{12} = -\begin{bmatrix} m^{-1}\mathbf{e}_2 \bar{\boldsymbol{\phi}}^T \\ I_x^{-1} \boldsymbol{\psi}^T \end{bmatrix} U_{22}, \qquad U_{21} = 0. \tag{47}$$

Moreover, the matrix U_{22} and the nonzero eigenvalues in Λ are the solution of the eigenvalue problem

$$K_A U_{22} = [M_A - (m^{-1}\bar{\boldsymbol{\phi}}\bar{\boldsymbol{\phi}}^T + 1_x^{-1}\boldsymbol{\psi}\boldsymbol{\psi}^T)] U_{22} \Lambda_e, \tag{48}$$

where $\Lambda_e = \text{diag}\ (\omega_1^2 \quad \omega_2^2 \quad \cdots \quad \omega_n^2)$. Then, introducing (46) in conjunction with (47) into the first five equations of (43) and considering (36) and

(39), we obtain

$$\bar{G} = \left[\begin{array}{ccc|c} 0 & -1 & 0 & \\ 1 & 0 & 0 & 0_{3\times n} \\ 0 & 0 & 0 & \\ \hline & 0_{n\times 3} & & 0_{n\times n} \end{array}\right], \qquad \bar{D}_v = \left[\begin{array}{c|c} 0_{3\times 3} & 0_{3\times n} \\ \hline 0_{n\times 3} & \bar{D}_e \end{array}\right] \tag{49a}$$

$$\bar{K} = \left[\begin{array}{ccc|c} 1 & 0 & 0 & \\ 0 & 1 & 0 & 0_{3\times n} \\ 0 & 0 & 0 & \\ \hline & 0_{n\times 3} & & \bar{K}_{22} \end{array}\right], \qquad \bar{E} = \left[\begin{array}{ccc|c} e_1 & 0 & 0 & \\ 0 & e_1 & 0 & 0_{2\times p} \\ 0 & 0 & e_2 & e_2\mathbf{1}_p^T \\ \hline \mathbf{0}_n & \bar{\mathbf{E}}_2 & \bar{\mathbf{E}}_3 & \bar{E}_e \end{array}\right] \tag{49b}$$

$$\bar{C} = \left[\begin{array}{ccc|c} e_1 & 0 & 0 & \mathbf{0}_{1\times n} \\ 0 & e_1 & 0 & \mathbf{E}_2^T \\ 0 & 0 & e_2 & \mathbf{E}_3^T \\ \hline \mathbf{0}_{m/2} & \mathbf{0}_{m/2} & e_2\mathbf{1}_{m/2} & \bar{C}_e \end{array}\right], \tag{49c}$$

where $\mathbf{1}_p^T = [1 \quad 1 \quad \cdots \quad 1]$. We assume that the off-diagonal terms of the damping matrix are sufficiently small that they can be ignored, so that

$$\bar{D}_e = \operatorname{diag}(2\zeta_1\omega_1 \quad 2\zeta_2\omega_2 \quad \cdots \quad 2\zeta_n\omega_n) \tag{50}$$

in which ζ_i $(i = 1, 2, \ldots, n)$ are viscous damping factors. Other terms entering into (49) are as follows:

$$\bar{K}_{22} = I_{n\times n} + I_x^{-1}U_{22}^T\boldsymbol{\psi}\boldsymbol{\psi}^T U_{22}, \qquad e_1 = m^{-1/2}, \qquad e_2 = I_x^{-1/2}, \tag{51a}$$

$$\bar{\mathbf{E}}_2 = -m^{-1}U_{22}^T\bar{\boldsymbol{\phi}}, \qquad \bar{\mathbf{E}}_3 = -I_x^{-1}U_{22}^T\boldsymbol{\psi}, \tag{51b}$$

$$\bar{E}_e = U_{22}^T(E^* - I_x^{-1}\boldsymbol{\psi}\mathbf{1}_p^T), \qquad \bar{C}_e = (C^* - I_x^{-1}\mathbf{1}_{m/2}\boldsymbol{\psi}^T)U_{22}. \tag{51c}$$

The bottom components of $\boldsymbol{\psi}$ tend to decrease rapidly relative to the top components. It follows that the bottom components of $\mathbf{v}$ are less affected by the maneuver velocity $\Omega_0(t)$ and acceleration $\dot{\Omega}_0(t)$, so that these components tend to behave as if the structure were not slewing. Furthermore, the inertia of the hub is relatively large, so that, considering (51b), it can be concluded that the components of $\bar{\mathbf{E}}_2$ and $\bar{\mathbf{E}}_3$ are relatively small. Equations (42) describe a linear time-varying system subjected to persistent disturbances. The design of a reduced-order compensator, as described later, can be based on this set of equations. It turns out, however, that the equations describing the translational perturbations preclude the achievement of quasi-constant gain matrices. Hence, to achieve suboptimal steady-state disturbance accommodation, it is necessary to control the translational motion independently. Furthermore, to decrease the control gains, so as to reduce the spillover effect, it is desirable to control the rotational motion independently as well. To this end, we let

$$T_3 = T_r - \sum_{j=4}^{p+3} T_j, \qquad y^*_{di} = y_{di} - y_{d3}, \qquad y^*_{vi} = y_{vi} - y_{v3},$$
$$i = 4, 5, \ldots, 3 + m/2, \tag{52}$$

and define

$$\tilde{P} = \begin{bmatrix} 0 & -1 & 0 \\ 1 & 0 & 0 \\ 0 & 0 & 0 \end{bmatrix}, \qquad \tilde{I} = \begin{bmatrix} 1 & 0 & 0 \\ 0 & 1 & 0 \\ 0 & 0 & 0 \end{bmatrix}, \qquad \bar{E}^*_e = \bar{E}_e - \bar{\mathbf{E}}_3 \mathbf{1}_p^T = U_{22}^T E^*, \tag{53}$$

so that (42) can be rewritten as

$$\ddot{\mathbf{v}}_R + 2\Omega_0(t)\tilde{P}\dot{\mathbf{v}}_R + [\dot{\Omega}_0(t)\tilde{P} - \Omega_0^2(t)\tilde{I}]\mathbf{v}_R = \bar{E}_R \mathbf{T}_R \tag{54a}$$

$$\ddot{\mathbf{v}}_e + \bar{D}_e\dot{\mathbf{v}}_e + [\Lambda_e - \Omega_0^2(t)\bar{K}_{22}]\mathbf{v}_e = \bar{E}^*_e\mathbf{T}_e + T_2\bar{\mathbf{E}}_2 + T_r\bar{\mathbf{E}}_3 - \dot{\Omega}_0(t)\bar{\psi} \tag{54b}$$

$$\mathbf{y}^*_d = \bar{C}^*\mathbf{v}, \qquad \mathbf{y}^*_v = \bar{C}^*\dot{\mathbf{v}} \tag{54c}$$

where $\mathbf{v}_R = [v_1 \quad v_2 \quad v_3]^T$ is a vector of rigid-body perturbations, in which v_1 is a translational perturbation in the y-direction, v_2 is a translational perturbation in the z-direction, and v_3 is the rotational perturbation, $\bar{E}_R = \mathrm{diag}(e_1 \quad e_1 \quad e_2)$ and $\mathbf{T}_R = [T_1 \quad T_2 \quad T_r]^T$. Moreover, $\mathbf{v}_e = [v_4 \quad v_5 \quad \cdots \quad v_{n+3}]^T$ is an n-vector of pseudomodal coordinates corresponding to the elastic motion, and $\mathbf{T}_e = [T_4 \quad T_5 \quad \cdots \quad T_{3+p}]^T$ is a vector of torques on the elastic appendage. The structure of the output matrix $\bar{C}^*$ is

$$\bar{C}^* = [C_1 \quad C_2] = \left[\begin{array}{ccc|c} e_1 & 0 & 0 & \mathbf{0}_{1\times n} \\ 0 & e_1 & 0 & \bar{\mathbf{E}}_2^T \\ 0 & 0 & e_2 & \bar{\mathbf{E}}_3^T \\ \hline & \mathbf{0}_{m/2\times 3} & & \bar{C}^*_e \end{array}\right] \tag{55}$$

where C_1 is an $(3 + m/2) \times (3 + m/2)$ matrix. We recognize that the rigid-body perturbations $\mathbf{v}_R$, which are not subjected to input disturbances caused by the maneuver, can be controlled independently of the elastic motions if $\mathbf{v}_R$ and $\dot{\mathbf{v}}_R$ are extracted from the solution of

$$\mathbf{y}^*_d = C_1\mathbf{v}_s, \qquad \mathbf{y}^*_v = C_1\dot{\mathbf{v}}_s, \tag{56}$$

respectively, where $\mathbf{v}_s^T = [\mathbf{v}_R^T \mid v_{e,1} \quad \cdots \quad v_{e,m/2}]$ provided that the sensors are placed such that C_1^{-1} exists. In this process, the coupling terms consisting of the higher elastic states v_{ei} and $\dot{v}_{ei}$ $(i = 1 + m/2, 2 + m/2, \ldots, n)$ and contaminating the rigid-body measurements are neglected. This can be justified by the fact that higher states are difficult to excite and their contribution to $\mathbf{E}_2$ and $\mathbf{E}_3$ is relatively small. A linear quadratic regulator is designed so as to produce fast decay in $\mathbf{v}_R$ in response to impulse disturbances, which implies that the controls T_1, T_2, and T_r are of relatively short duration. Then, because $\bar{\mathbf{E}}_2$ and $\bar{\mathbf{E}}_3$ are small, we can regard the terms $T_2\bar{\mathbf{E}}_2$ and $T_r\bar{\mathbf{E}}_3$ as small transient disturbances, so that the elastic motions can be controlled independently of the rigid-body perturbations. In view of this, we focus our attention on the control of the elastic motions under persistent disturbances caused by the maneuver.

IV. CONTROL OF THE ELASTIC MODEL

Introducing the state vector $\mathbf{z}_e(t) = [\mathbf{v}_e^T(t) \quad \dot{\mathbf{v}}_e^T(t)]^T$, as well as the coefficient matrices

$$A(t) = \left[\begin{array}{c|c} 0_{n\times n} & I_{n\times n} \\ \hline -\Lambda_e + \Omega_0^2(t)\bar{K}_{22} & -\bar{D}_e \end{array}\right], \qquad B = \left[\begin{array}{c} 0_{n\times p} \\ \hline \bar{E}_e^* \end{array}\right],$$
$$C = \left[\begin{array}{c|c} \bar{C}_e^* & 0_{m/2\times n} \\ \hline 0_{m/2\times n} & \bar{C}_e^* \end{array}\right] \tag{57}$$

and the vectors

$$\mathbf{R} = [\mathbf{0}_n^T \mid \bar{\psi}^T]^T, \tag{58a}$$

$$\mathbf{d}(t) = [\mathbf{0}_n^T \mid (T_2\bar{E}_2 + T_r\bar{E}_3)^T]^T, \tag{58b}$$

we can rewrite Eq. (54b,c) in the state form

$$\dot{\mathbf{z}}_e = A(t)\mathbf{z}_e + B\mathbf{T}_e - \dot{\Omega}_0\mathbf{R} + \mathbf{d}(t), \qquad \mathbf{y}_m = C\mathbf{z}_e. \tag{59}$$

We refer to (59) as the elastic model. Considering (33), we define $\tau = [t_i, t_h)$, where $t_i = t_0$ and $t_h = t_1$ for the first half of the maneuver and $t_i = t_1$ and $t_h = t_f$ for the second half, and recognize that $A(t)$ is analytic and bounded for all $t \in \tau$. Moreover, we assume that the homogeneous part of (59) represents a quasi-contractively stable system (see Appendix) during the maneuver.

The order $n_M = 2n$ of the elastic model is generally high, so that control implementation considerations dictate further truncation. It is typical of structures that inherent damping improves the robustness of higher modes and causes them to decay faster than lower modes. In addition, higher modes are more difficult to excite and the components of the disturbance vector $-\dot{\Omega}_0(t)\bar{\psi}$ corresponding to the higher states tend to have lower values. Hence, it is reasonable to retain the first $n_C/2$ components of $\mathbf{v}_e$ for control, and denote the corresponding state vector by $\mathbf{z}_C$. If we were to use an observer to estimate the controlled state only, we could experience observation spillover which in conjunction with control spillover can cause instability [17]. This effect can be eliminated from the significant part of the elastic model by increasing the order of the estimated state [18], thus shifting the observation spillover to more robust parts of the elastic model. Hence, we propose to estimate $n_0/2$ additional components without controlling them, and denote the corresponding part of the state vector by $\mathbf{z}_0$. Finally, we refer to the remaining $n_R/2$ components as residual, and denote the corresponding part of the state vector by $\mathbf{z}_R$. Hence, $\mathbf{z}_e = [\mathbf{z}_C^T \mid \mathbf{z}_0^T \mid \mathbf{z}_R^T]^T$ and $n_C + n_0 + n_R = n_M$. Considering the above definitions, (59) can be partitioned as follows:

$$\begin{bmatrix} \dot{\mathbf{z}}_C \\ \dot{\mathbf{z}}_0 \\ \dot{\mathbf{z}}_R \end{bmatrix} = \begin{bmatrix} A_{CC} & A_{C0} & A_{CR} \\ A_{0C} & A_{00} & A_{0R} \\ A_{RC} & A_{R0} & A_{RR} \end{bmatrix} \begin{bmatrix} \mathbf{z}_C \\ \mathbf{z}_0 \\ \mathbf{z}_R \end{bmatrix} + \begin{bmatrix} B_C \\ B_0 \\ B_R \end{bmatrix} \mathbf{T}_e - \dot{\Omega}_0 \begin{bmatrix} \mathbf{R}_C \\ \mathbf{R}_0 \\ \mathbf{R}_R \end{bmatrix} + \begin{bmatrix} \mathbf{d}_C(t) \\ \mathbf{d}_0(t) \\ \mathbf{d}_R(t) \end{bmatrix}, \qquad t \in \tau, \tag{60a}$$

$$\mathbf{y}_m = [C_C \mid C_0 \mid C_R] \begin{bmatrix} \mathbf{z}_C \\ \mathbf{z}_0 \\ \mathbf{z}_R \end{bmatrix}, \tag{60b}$$

where

$$A_{ii} = \begin{bmatrix} 0_{i \times i} & I_{i \times i} \\ -\Lambda_{ii} + \Omega_0^2(t)\bar{K}_{ii} & -\bar{D}_{ii} \end{bmatrix}, \qquad A_{ij} = \Omega_0^2(t) \begin{bmatrix} 0_{i \times j} & 0_{i \times j} \\ \bar{K}_{ij} & 0_{i \times j} \end{bmatrix} \tag{61}$$

in which $i,j = $ C, 0, R.

Our objective is to design a compensator exhibiting the following characteristics:

1. The controlled model must be stabilized according to a quadratic performance measure such that the closed-loop controlled model represents a strictly finite-time stabilizable system (see Definition 3 and Conclusion from Definition 3 in the Appendix).
2. A weighted norm of the constant part in the disturbance response of the elastic model must be minimized in quasi-constant conditions imposed on the elastic model toward the end of the finite-time interval. We consider a constant minimizing matrix because of computer limitations.
3. The supremum time constant (see Appendix) of the elastic model must be minimized, which implies that the elastic model is exponentially contractively stable (see Definition 1 in the Appendix), without compromising the quasi-constant conditions.

Equations (60) represent a linear, time-varying system subjected to a piecewise-constant disturbance and a transient disturbance. The commanded angular velocity $\Omega_0(t)$ and angular acceleration $\dot{\Omega}_0(t)$ are known *a priori*. On the other hand, we regard $\mathbf{R}$ as insufficiently accurate for direct compensation. Hence, we propose to treat the persistent disturbance as unknown, except that it is piecewise constant. Moreover, we assume that the number of sensors is not sufficient to measure all the states, so that we must use an observer to estimate the controlled state $\mathbf{z}_C$ and part of the disturbance, where that part is denoted by $\mathbf{f}_E$.

Due to the nature of the disturbance vector $\dot{\Omega}_0\mathbf{R}$, as described by (33a), we propose to carry out the control and estimation over one half of the maneuver period τ at a time. We divide the control into two parts as follows [19, 20]:

$$\mathbf{T}_e(t) = \mathbf{u}_C(t) + \mathbf{u}_D(t), \tag{62}$$

where $\mathbf{u}_C$ represents feedback control in the absence of persistent disturbances and is designed so as to exhibit characteristics (1) and (3) and to impose quasi-constant conditions on the elastic model within the finite-time interval. On the other hand, $\mathbf{u}_D$ represents the disturbance accommodation part of the control and is designed so as to possess characteristic (2).

V. DESIGN OF A REDUCED-ORDER COMPENSATOR

To design the feedback control for the stabilization of the reduced-order model, we consider the controlled model in absence of disturbances. From (60) and (62), we write

$$\dot{\mathbf{z}}_C(t) = A_{CC}(t)\mathbf{z}_C(t) + B_C\mathbf{u}_C(t), \ \mathbf{z}_C(t_i) = \mathbf{z}_{Ci}, \qquad t \in \tau, \tag{63}$$

where the dimension of $\mathbf{z}_C$ is n_C and that of $\mathbf{u}_C$ is p, where $p \le n_C/2$. It is shown in [21] that the pair $[A_{CC}(t), B_C]$ is completely controllable in every arbitrary subinterval of τ, i.e., totally controllable [22] for every $t \in t$. To achieve the control objectives described above, we consider the performance measure

$$J = \int_{t_i}^{t_h} e^{2\alpha t}(\mathbf{z}_C^T Q \mathbf{z}_C + \mathbf{u}_C^T R \mathbf{u}_C)\,dt + e^{2\alpha t_h}\mathbf{z}_C^T(t_h)S_1\mathbf{z}_C(t_h), \tag{64}$$

where $\alpha \ge 0$ is a convergence factor and

$$Q = Q^T > 0, \qquad R = R^T > 0, \qquad S_1 = S_1^T \ge 0. \tag{65}$$

The problem defined by (63) and (64) can be reduced to a standard LQR problem by introducing the transformations [23]

$$\mathbf{z}_C^*(t) = e^{\alpha t}\mathbf{z}_C(t), \qquad \mathbf{u}_C^*(t) = e^{\alpha t}\mathbf{u}_C(t), \qquad \alpha \ge 0. \tag{66}$$

Inserting (66) into (63) and rearranging, we obtain

$$\dot{\mathbf{z}}_C^*(t) = A_{CC}^*(t)\mathbf{z}_C^*(t) + B_C\mathbf{u}_C^*(t), \tag{67}$$

where

$$A^*_{CC}(t) = A_{CC}(t) + \alpha I. \tag{68}$$

It is shown in [21] that the total controllability of the original system (63) also implies the total controllability of the modified system defined by (67). The control law minimizing J is

$$\mathbf{u}_C(t) = K_C(t)\mathbf{z}_C(t), \qquad K_C(t) = -R^{-1}B_C^T S(t), \tag{69}$$

where $S(t)$ is the solution of the matrix Riccati equation

$$\dot{S}(t) = -Q - S(t)A^*_{CC}(t) - A^{*T}_{CC}(t)S(t) + S(t)B_C R^{-1}B_C^T S(t), \qquad S(t_h) = S_1. \tag{70}$$

Next, we wish to develop an explicit sufficiency condition guaranteeing the strictly finite-time stabilizability of the closed-loop system. To this end, we consider the sufficiency condition for finite-time stabilizability (see Appendix) and define the Lyapunov function

$$V(\mathbf{z}^*_C, t) = \mathbf{z}^{*T}_C S(t)\mathbf{z}^*_C. \tag{71}$$

Differentiating (71) with respect to time, evaluating the result along solutions of (67), and considering (70), we obtain

$$\dot{V}(\mathbf{z}^*_C, t) = -\mathbf{z}^{*T}_C C(t)\mathbf{z}^*_C, \tag{72}$$

where

$$C(t) = Q + S(t)B_C^T R^{-1}B_C S(t) \tag{73}$$

is a positive definite matrix for all $t \in \tau$, because $Q > 0$ and $S(t)B_C^T R^{-1}B_C S(t) \geq 0$ for all $t \in \tau$. This proves the first part of the sufficiency condition. The second part of the sufficiency condition (173) is satisfied if [21]

$$\lambda_M[S(t_i, \alpha + \Delta\alpha)]/\lambda_M[S(t_i, \alpha)] < \exp[2\,\Delta\alpha(t - t_i)] \qquad \text{for all} \quad t \in \tau_1$$
$$\text{and} \qquad \alpha \geq \bar{\alpha}(\varepsilon) \geq 0, \qquad \Delta\alpha > 0, \tag{74}$$

where $\tau_1 = (T_1, t_h)$, $t_i < T_1 < t_h$. We recognize that condition (74) is expressed in terms of the Riccati solution $S(t)$ evaluated at the initial time t_i, where the solution is likely to reach its steady-state value if α is sufficiently large. Hence, we can state that, if inequality (74) is satisfied, then the system represented by (63), (69), and (70) is strictly finite-time stabilizable. Hence, to every ε and $\mathbf{z}_{Ci}$ corresponds an $\tilde{\alpha}(\varepsilon, \mathbf{z}_{Ci})$ such that for every $\alpha > \tilde{\alpha}(\varepsilon, \mathbf{z}_{Ci})$ the response to $\mathbf{z}_{Ci}$ decays to a value below ε within the finite-time interval τ.

The observer is expected to estimate not only part of the state vector but also part of the persistent disturbance vector. Hence, let us introduce the notation

$$-\dot{\Omega}_0\mathbf{R} = \mathbf{W} = [\mathbf{W}_C^T \mid \mathbf{W}_0^T \mid \mathbf{W}_R^T]^T, \tag{75}$$

where $\mathbf{W}_C$, $\mathbf{W}_0$, and $\mathbf{W}_R$ are the parts of $-\dot{\Omega}_0\mathbf{R}$ corresponding to $\mathbf{z}_C$, $\mathbf{z}_0$, and $\mathbf{z}_R$, respectively. Considering the disturbance dynamics (33a) and the structure of $\mathbf{R}$, (58a), we denote the part of $-\dot{\Omega}_0\bar{\boldsymbol{\psi}}$ to be estimated by $\mathbf{f}_E$ and introduce the relations

$$\dot{\mathbf{f}}_E = \boldsymbol{\delta}_E(t), \quad \mathbf{W}_E = H_E\mathbf{f}_E \tag{76}$$

where $\mathbf{f}_E$ is an $n_W/2$ vector, $n_W \leq n_C + n_0$, $\boldsymbol{\delta}_E(t)$ is a vector of Dirac delta functions of the same dimension, and H_E is an $(n_C + n_0) \times n_W/2$ matrix of zeros and ones. For convenience, we rewrite (75) as

$$-\dot{\Omega}_0\mathbf{R} = [(\mathbf{W}_E^{**})^T \quad \mathbf{W}_R^T]^T + \begin{bmatrix} H_E \\ \hline 0 \end{bmatrix}\mathbf{f}_E, \tag{77}$$

where 0 denotes a null matrix. Defining the vector to be estimated as $\boldsymbol{\eta}_E = [\mathbf{z}_C^T \mid \mathbf{z}_0^T \mid \mathbf{f}_E^T]^T$, (60a), (76), and (77) can be combined into

$$\begin{bmatrix} \dot{\boldsymbol{\eta}}_E \\ \dot{\mathbf{z}}_R \end{bmatrix} = \begin{bmatrix} A_{EE} & A_{ER} \\ A_{RE} & A_{RR} \end{bmatrix}\begin{bmatrix} \boldsymbol{\eta}_E \\ \mathbf{z}_R \end{bmatrix} + \begin{bmatrix} B_E \\ B_R \end{bmatrix}(\mathbf{u}_C + \mathbf{u}_D) + \begin{bmatrix} I_E^* \\ 0 \end{bmatrix}\boldsymbol{\delta}_E(t) + \begin{bmatrix} \mathbf{W}_E^* \\ \mathbf{W}_R \end{bmatrix} + \begin{bmatrix} \mathbf{d}_E^*(t) \\ \mathbf{d}_R(t) \end{bmatrix}, \tag{78}$$

where

$$A_{EE} = \left[\begin{array}{c|c|c} A_{CC} & A_{C0} & \multirow{2}{*}{H_E} \\ \hline A_{0C} & A_{00} & \\ \hline 0 & 0 & 0 \end{array}\right], \qquad A_{ER} = \begin{bmatrix} A_{CR} \\ \hline A_{0R} \\ \hline 0 \end{bmatrix}, \qquad A_{RE} = [A_{RC} \quad A_{R0} \quad 0] \tag{79}$$

$$B_E = \begin{bmatrix} B_C \\ \hline B_0 \\ \hline 0 \end{bmatrix}, \qquad I_E^* = \begin{bmatrix} 0 \\ \hline 0 \\ \hline I_E \end{bmatrix}, \qquad \mathbf{W}_E^* = \begin{bmatrix} \mathbf{W}_E^{**} \\ \hline \mathbf{0} \end{bmatrix}, \qquad \mathbf{d}_E^* = \begin{bmatrix} \mathbf{d}_C \\ \hline \mathbf{d}_0 \\ \hline \mathbf{0} \end{bmatrix},$$

in which I_E is an $n_W/2 \times n_W/2$ identity matrix; the various submatrices in (79) can be deduced from (57) and (58). Consistent with the above, the output vector can be written in the form

$$\mathbf{y}_m(t) = C_E\boldsymbol{\eta}_E(t) + C_R\mathbf{z}_R(t), \tag{80}$$

where, in accordance with the last of Eqs. (57),

$$C_E = \begin{bmatrix} \bar{C}_C^* & 0 & \bar{C}_0^* & 0 & 0 \\ 0 & \bar{C}_C^* & 0 & \bar{C}_0^* & 0 \end{bmatrix}, \qquad C_R = \begin{bmatrix} \bar{C}_R^* & 0 \\ 0 & \bar{C}_R^* \end{bmatrix}. \tag{81}$$

Introducing the estimated vector $\hat{\boldsymbol{\eta}}_E = [\hat{\mathbf{z}}_C^T \mid \hat{\mathbf{z}}_0^T \mid \hat{\mathbf{f}}_E^T]^T$, we consider an observer described by

$$\dot{\hat{\boldsymbol{\eta}}}_E(t) = A_{EE}(t)\hat{\boldsymbol{\eta}}_E(t) + B_E[\mathbf{u}_C(t) + \mathbf{u}_D(t)] + K_E(t)[\hat{\mathbf{y}}_m(t) - \mathbf{y}_m(t)], \tag{82}$$

where $K_E(t)$ is the observer gain matrix and

$$\hat{\mathbf{y}}_m(t) = C_E\hat{\boldsymbol{\eta}}_E(t). \tag{83}$$

Defining the observer error vector as

$$\mathbf{e}(t) = \boldsymbol{\eta}_E(t) - \hat{\boldsymbol{\eta}}_E(t), \tag{84}$$

we obtain the homogeneous part of the error equation

$$\dot{\mathbf{e}}(t) = [A_{EE}(t) + K_E(t)C_E]\mathbf{e}(t). \tag{85}$$

It is shown in [21] that the pair $[A_{EE}(t), C_E]$ is completely observable in every subinterval of τ, i.e., totally observable [22] for every $t \in \tau$. Introducing the transformation

$$\mathbf{e}^*(t) = e^{\beta t}\mathbf{e}(t), \qquad \beta \geq 0 \tag{86}$$

and inserting (86) into (85), we obtain

$$\dot{\mathbf{e}}^*(t) = [A_{EE}^*(t) + K_E(t)C_E]\mathbf{e}^*(t) \tag{87}$$

where

$$A_{EE}^*(t) = A_{EE}(t) + \beta I. \tag{88}$$

Using the approach suggested in [24], we consider the gain matrix

$$K_E(t) = -\tfrac{1}{2}P(t)C_E^T \tag{89}$$

in which $P(t)$ is the solution of the Riccati equation

$$\begin{aligned} \dot{P}(t) &= A_{EE}^*(t)P(t) + P(t)A_{EE}^{*T}(t) - P(t)C_E^T C_E P(t) + Q_E, \\ P(t_i) &= P_0, \end{aligned} \tag{90a}$$

where

$$Q_E = Q_E^T > 0, \qquad P_0 = P_0^T > 0. \tag{90b}$$

Next, we wish to develop a sufficiency condition guaranteeing the strictly finite-time stabilizability of the system defined by (87) and (89). To this end, we define the Lyapunov function

$$V(\mathbf{e}^*, t) = \mathbf{e}^{*T}P^{-1}(t)\mathbf{e}^*. \tag{91}$$

Then, by a completely dual process to the one described earlier in this section, it is shown in [21] that properties (90b) and

$$\lambda_M[P(t, \beta + \Delta\beta)]/\lambda_M[P(t,\beta)] < \exp[2\,\Delta\beta(t - t_i)] \qquad \text{for all} \qquad t \in \tau_1 \\ \text{and} \qquad \beta \geq \bar{\beta}(\varepsilon) \geq 0, \qquad \Delta\beta > 0, \tag{92}$$

where $\tau_1 = (T_1, t_h)$, $t_i < T_1 < t_h$, are sufficient to guarantee a strictly finite-time stabilizable system. Hence, to every ε and $\mathbf{e}(t_i)$ corresponds a $\tilde{\beta}[\varepsilon, \mathbf{e}(t_i)]$ such that for every $\beta > \tilde{\beta}[\varepsilon, \mathbf{e}(t_i)]$ the response of the system to $\mathbf{e}(t_i)$ decays to a value below ε within the finite-time interval τ.

Finally, we propose to derive the closed-loop equations for the modeled system incorporating the observer described previously. To this end, we write the feedback control and the persistent disturbance control in the form

$$\mathbf{u}_C(t) = K_C(t)\hat{\mathbf{z}}_C(t), \qquad \mathbf{u}_D(t) = K_D\hat{\mathbf{f}}_E(t). \tag{93}$$

Then, denoting the uncontrolled state vector by $\mathbf{z}_U = [\mathbf{z}_0^T \mid \mathbf{z}_R^T]^T$ and considering (78) and (82), we obtain

$$\begin{bmatrix} \dot{\mathbf{z}}_C \\ \dot{\mathbf{e}} \\ \dot{\mathbf{z}}_U \end{bmatrix} = \begin{bmatrix} \tilde{A}_{CC} & \tilde{A}_{CE} & \tilde{\mathbf{A}}_{CU} \\ 0 & \tilde{\mathbf{A}}_{EE} & \tilde{\mathbf{A}}_{EU} \\ \tilde{A}_{UC} & \tilde{A}_{UE} & \tilde{A}_{UU} \end{bmatrix} \begin{bmatrix} \mathbf{z}_C \\ \mathbf{e} \\ \mathbf{z}_U \end{bmatrix} + \begin{bmatrix} \mathbf{W}_C + B_C K_D \mathbf{f}_E \\ \mathbf{W}_E^* \\ \mathbf{W}_U + B_U K_D \mathbf{f}_E \end{bmatrix} + \begin{bmatrix} 0 \\ I_E \\ 0 \end{bmatrix} \boldsymbol{\delta}_E(t) + \begin{bmatrix} \mathbf{d}_C(t) \\ \mathbf{d}_E^*(t) \\ \mathbf{d}_U(t) \end{bmatrix}, \tag{94}$$

where

$$\begin{aligned} &\tilde{A}_{CC} = A_{CC} + B_C K_C, \qquad \tilde{A}_{CE} = -[B_C K_C \mid 0 \mid B_C K_D], \\ &\tilde{A}_{CU} = [A_{C0} \mid A_{CR}], \qquad \tilde{A}_{EE} = A_{EE} + K_E C_E, \\ &\tilde{A}_{EU} = \begin{bmatrix} 0 & A_{CR} + K_{EC} C_R \\ 0 & A_{0R} + K_{E0} C_R \\ 0 & K_{Ef} C_R \end{bmatrix}, \qquad \tilde{A}_{UC} = \begin{bmatrix} A_{0C} + B_0 K_C \\ A_{RC} + B_R K_C \end{bmatrix}, \\ &\tilde{A}_{UE} = -\begin{bmatrix} B_0 K_C & 0 & B_0 K_D \\ B_R K_C & 0 & B_R K_D \end{bmatrix}, \qquad \tilde{A}_{UU} = \begin{bmatrix} A_{00} & A_{0R} \\ A_{R0} & A_{RR} \end{bmatrix}, \end{aligned} \tag{95}$$

in which we introduced the notation

$$K_E = [K_{EC}^T \mid K_{E0}^T \mid K_{Ef}^T]^T, \qquad B_U = \begin{bmatrix} B_0 \\ B_R \end{bmatrix}, \qquad \mathbf{W}_U = \begin{bmatrix} \mathbf{W}_0 \\ \mathbf{W}_R \end{bmatrix}, \\ \mathbf{d}_U = \begin{bmatrix} \mathbf{d}_0 \\ \mathbf{d}_R \end{bmatrix}. \tag{96}$$

In the future, we will refer to the coefficient matrix in (94) as $\tilde{A}(t)$.

VI. DISTURBANCE ACCOMMODATION

The object of disturbance accommodation is to minimize the response of an output vector associated with the disturbance. To this end, we define an n_M-dimensional output vector

$$\mathbf{y}_1(t) = L_y \mathbf{z}_e(t), \tag{97}$$

where we recall that $\mathbf{z}_e = [\mathbf{z}_C^T \mid \mathbf{z}_0^T \mid \mathbf{z}_R^T]^T$. For the minimization process contemplated, we rearrange the state vector in (94) and retain only the terms related to the observed disturbance vector $\hat{\mathbf{f}}_E(t)$. Hence, we rewrite (94) in the form

$$\begin{bmatrix} \dot{\mathbf{z}}_e \\ \hline \dot{\mathbf{e}} \end{bmatrix} = \left[\begin{array}{c|c} F_{11} & F_{12} \\ \hline F_{21} & F_{22} \end{array}\right] \begin{bmatrix} \mathbf{z}_e \\ \hline \mathbf{e} \end{bmatrix} + \begin{bmatrix} H_F + BK_D \\ \hline 0 \end{bmatrix} \hat{\mathbf{f}}_E, \tag{98}$$

where

$$F_{11} = \left[\begin{array}{c|c} \tilde{A}_{CC} & \tilde{A}_{CU} \\ \hline \tilde{A}_{UC} & \tilde{A}_{UU} \end{array}\right], \qquad F_{12} = [-BK_C \mid 0 \mid H_F], \qquad F_{21} = [0 \mid \tilde{A}_{EU}],$$
$$F_{22} = \tilde{A}_{EE}, \qquad H_F = \begin{bmatrix} H_E \\ \hline 0 \end{bmatrix}, \qquad B = \begin{bmatrix} B_C \\ \hline B_U \end{bmatrix}. \tag{99}$$

Then, denoting the response to $\hat{\mathbf{f}}_E(t)$ by $\mathbf{y}_f(t)$, we obtain

$$\mathbf{y}_f(t) = \int_{t_i}^{t} [L_y \mid 0] \Phi_F(t, \sigma) \begin{bmatrix} H_F + BK_D \\ \hline 0 \end{bmatrix} \hat{\mathbf{f}}_E(\sigma)\, d\sigma, \qquad t \in \tau, \tag{100}$$

where $\Phi_F(t, \sigma)$ is the transition matrix corresponding to the coefficient matrix $F(t)$ in (98). With due regard to on-line computer limitations, we seek a constant gain matrix K_D minimizing the performance measure

$$J_D = \lim_{t \to t_h} \|\mathbf{y}_f(t)\|^2. \tag{101}$$

The solution of (101) has meaning only if quasi-constant conditions exist for $t \in \tau$. Hence, we assume now, and justify later, that the coefficient matrix can be written in the form

$$F(t) = F_0 + \varepsilon F_1(t), \qquad \varepsilon \ll 1, \qquad t \in \tau, \tag{102}$$

where F_0 is a constant matrix. Then, the transition matrix can be shown to have the expression [25]

$$\Phi_F(t, t_i) = e^{F_0(t - t_i)} + \varepsilon e^{F_0 t} \int_{t_i}^{t} e^{-F_0 \tau} F_1(\tau) e^{F_0 \tau}\, d\tau + 0(\varepsilon^2), \tag{103}$$

so that, using (100) and (103), retaining the zero-order term only, and denoting by $\bar{\hat{\mathbf{f}}}_E$ the constant part of $\hat{\mathbf{f}}_E(t)$, we obtain

$$\mathbf{y}_{f0}(t) = -[L_y \quad 0]F_0^{-1}[I - e^{F_0(t-t_i)}]\begin{bmatrix} H_F + BK_D \\ \hline 0 \end{bmatrix}\bar{\hat{\mathbf{f}}}_E, \qquad t \in \tau. \quad (104)$$

At this point, we assume that F_0 is a stable matrix, that is,

$$\operatorname{Re}\lambda_i(F_0) < 0, \qquad i = 1, 2, \ldots, n_A; \qquad n_A = n_M + n_C + n_0 + n_W/2, \quad (105)$$

which guarantees the existence of F_0^{-1}. Next, we denote the constant part of $\mathbf{y}_{f0}(t)$ by $\bar{\mathbf{y}}_{f0}$ and determine the constant gain matrix K_D minimizing the performance measure

$$\bar{J}_D = \|\bar{\mathbf{y}}_{f0}\|^2. \quad (106)$$

The transient part of $\mathbf{y}_{f0}(t)$ and $\hat{\mathbf{f}}_E(t)$, which are not included in the minimization process, decay below small threshold values as $t \to t_h$, provided the homogeneous part of (94) is quasi-contractively stable (see Definition 2 in the Appendix) with $\rho \ll 1$ and

$$\lim_{t \to t_h} \|e^{F_0 t}\| \ll 1. \quad (107)$$

To produce the gain matrix K_D, we partition F_0 in the same way as in (98) and introduce the notation

$$F_0 = \begin{bmatrix} \bar{F}_C & \bar{F}_{12} \\ \bar{F}_{21} & \bar{F}_E \end{bmatrix}. \quad (108)$$

Inserting (108) into (104), we obtain

$$\bar{\mathbf{y}}_{f0} = -L_y\Gamma^{-1}(H_F + BK_D)\bar{\hat{\mathbf{f}}}_E = -L_y\Gamma^{-1}(H_F\bar{\hat{\mathbf{f}}}_E + B\mathbf{u}_D), \quad (109)$$

where

$$\Gamma = \bar{F}_C - \bar{F}_{12}\bar{F}_E^{-1}\bar{F}_{21}. \quad (110)$$

Equation (110) requires that $\bar{F}_E$ be nonsingular, which implies the assumption

$$\operatorname{Re}\lambda_i(\bar{F}_E) < 0, \qquad i = 1, 2, \ldots, n_A. \quad (111)$$

In addition, Γ itself must be nonsingular. Using (108), (110), and (111), it can be shown that $\Gamma^{-1} = (F_0^{-1})_{11}$, where F_0^{-1} is partitioned in the same way as F_0 in (108). At this point, we seek the value of $\mathbf{u}_D$ minimizing $\bar{J}_D$. This value must satisfy

$$\partial\bar{J}_D/\partial\mathbf{u}_D = B^T\Gamma^{-T}W\Gamma^{-1}(H_F\bar{\hat{\mathbf{f}}}_E + B\mathbf{u}_D) = \mathbf{0}, \quad (112)$$

where $\Gamma^{-T} = (\Gamma^{-1})^T$ and $W = L_y^T L_y$, so that recalling the second of

Eqs. (93), the minimizing gain matrix has the expression

$$K_D = -(B^T\Gamma^{-T}W\Gamma^{-1}B)^{-1}B^T\Gamma^{-T}W\Gamma^{-1}H_F. \tag{113}$$

A sufficient condition for the existence of (113) minimum is

$$B^T\Gamma^{-T}W\Gamma^{-1}B > 0. \tag{114}$$

Considering inequalities (105), this condition is guaranteed by choosing $W > 0$, and by placing the actuators so as to ensure a full-rank matrix B.

In the following, we propose to validate two of the necessary conditions implied by (113), namely assumptions (102) and (111). The other necessary condition is assumption (105), which represents a design requirement and is analyzed in the next section.

Let us consider the system described by (66)–(70). The discussion can be generalized to some extent by replacing the matrix $A^*_{CC}(t)$ by

$$A^*(t) = \left[\begin{array}{c|c} \alpha I & I \\ \hline \varphi(t)\bar{K}_{22} - \Lambda_e & \alpha I - \bar{D}_e \end{array}\right], \tag{115}$$

where $0 \le \alpha \le \alpha_M$ and $\varphi(t) = \varphi_0 t^k$, $k \ge 2$. We assume that φ_0 is such that $\varphi(t)\|\bar{K}_{22}\| \gg \|\Lambda_e\|$ for $t \in \tau$, where t is sufficiently large. As in the case of the original matrix $A^*_{CC}(t)$, the pair $[A^*(t), B_C]$ is totally controllable for every $t \in \tau$. Then, it is verified numerically that the Riccati matrix $S(t)$ possesses the following characteristics:

1. By increasing the parameter α, the steady-state solution of the Riccati equation, (70), can be reached within τ.
2. The above steady-state solution can be expressed as

$$\bar{S}(t) = \bar{S}_0 + \varepsilon_1(\alpha)S_1(t), \tag{116}$$

where $\bar{S}_0$ is the solution of the algebraic Riccati equation using the matrix A^*_0, obtained from (115) by setting $\varphi(t) \equiv 0$, and $\varepsilon_1(\alpha)$ is a positive scalar decreasing monotonically with increasing α.

3. Choosing the end condition $S_1 = \bar{S}_0$ for (70), we obtain

$$S(t) = \bar{S}(t) + \varepsilon_2(\alpha)S_2(t) \qquad \text{for all} \quad t \in \tau, \tag{117}$$

where $\varepsilon_2(\alpha)$ is a positive scalar. Then, in view of (116),

$$S(t) = \bar{S}_0 + \varepsilon_3(\alpha)S_3(t) \quad \text{for all} \quad t \in \tau, \tag{118}$$

where $\varepsilon_3(\alpha)$ decreases monotonically with increasing α.

These three characteristics hold for the solution of the observer Riccati equation, (90a). Indeed, choosing the initial condition $P_0 = \bar{P}_0$, we obtain

$$P(t) = \bar{P}_0 + \varepsilon_4(\beta)P_1(t) \qquad \text{for all} \quad t \in \tau, \tag{119}$$

where $\varepsilon_4(\beta)$ is a positive scalar decreasing monotonically with increasing β and $\bar{P}_0$ is the solution of the algebraic Riccati equation with $A^*_{\mathrm{EE}}(\Omega_0 \equiv 0)$ replacing $A^*_{\mathrm{EE}}(t)$ in (90a). We refer to solutions (118) and (119) as quasi-constant.

Next, we introduce (118) into the second equation of (69) and obtain the quasi-constant optimal control gain matrix

$$K_{\mathrm{C}}(t) = \bar{K}_{\mathrm{C}} + \varepsilon K_{\mathrm{C}1}(t), \qquad \varepsilon \ll 1, \tag{120}$$

where

$$\bar{K}_{\mathrm{C}} = -R^{-1}B_{\mathrm{C}}^T\bar{S}_0, \tag{121a}$$

$$K_{\mathrm{C}1}(t) = -R^{-1}B_{\mathrm{C}}^T S_1(t). \tag{121b}$$

Similarly, inserting (119) into Eq. (89), we obtain the quasi-constant observer gain matrix

$$K_{\mathrm{E}}(t) = \bar{K}_{\mathrm{E}} + \varepsilon K_{\mathrm{E}1}(t), \qquad \varepsilon \ll 1, \tag{122}$$

where

$$\bar{K}_{\mathrm{E}} = -\tfrac{1}{2}\bar{P}_0 C_{\mathrm{E}}^T, \tag{123a}$$

$$K_{\mathrm{E}1}(t) = -\tfrac{1}{2}P_1(t)C_{\mathrm{E}}^T. \tag{123b}$$

Recalling (95) and (99) and introducing (121) into (98), $F(t)$ can be split into a constant and a time-varying part, as indicated by (102). Then, in light of the relatively high gains needed to achieve characteristics 1 and 2 (see Section IV), if we consider the first of Eqs. (57) and assume that the order of the controlled model is such that $\|\Lambda_{ei}\| > \Omega_0^2(t)\|\bar{\mathbf{K}}_i\|$ for all $t \in \tau$ and for all $i > n_{\mathrm{C}}/2$, where i denotes the row number, it is clear that the terms $B_{\mathrm{C}}\bar{K}_{\mathrm{C}}$, $\bar{K}_{\mathrm{E}}C_{\mathrm{E}}$, and Λ_{ei} $(i = 1 + n_{\mathrm{C}}/2, 2 + n_{\mathrm{C}}/2, \ldots, N)$ dominate the matrix $F(t)$. Hence, ε in (102) must be regarded as small.

Next, we verify the assumption implied by inequalities (111). Introducing (123) into the fourth of Eqs. (99) and recalling the fourth of Eqs. (95), we obtain

$$\bar{F}_{\mathrm{E}} = \bar{A}_{\mathrm{EE}} - \tfrac{1}{2}\bar{P}_0 C_{\mathrm{E}}^T C_{\mathrm{E}}, \tag{124}$$

where $\bar{A}_{\mathrm{EE}} = \tilde{A}_{\mathrm{EE}}(\Omega_0 \equiv 0)$ and $\bar{P}_0$ is the solution of the algebraic Riccati equation

$$(\bar{A}_{\mathrm{EE}} + \beta I)\bar{P}_0 + \bar{P}_0(\bar{A}_{\mathrm{EE}}^T + \beta I) - \bar{P}_0 C_{\mathrm{E}}^T C_{\mathrm{E}}\bar{P}_0 + Q_{\mathrm{E}} = 0. \tag{125}$$

Inserting $\bar{A}_{\mathrm{EE}}$ from (124) into (125), we obtain the Lyapunov equation

$$(\bar{F}_{\mathrm{E}} + \beta I)\bar{P}_0 + \bar{P}_0(\bar{F}_{\mathrm{E}}^T + \beta I) = -Q_{\mathrm{E}}. \tag{126}$$

In view of the fact that $\bar{P}_0$ and Q_E are symmetric and positive definite, we have [26]

$$\operatorname{Re} \lambda_i(\bar{F}_E + \beta I) < 0, \qquad i = 1, 2, \ldots, n_C + n_0 + n_W/2. \tag{127}$$

VII. FINITE-TIME STABILITY OF THE ELASTIC MODEL

The third stage of the design process is to determine the optimal values for the convergence factors α and β minimizing the supremum time constant of the elastic model, which implies that the elastic model is exponentially contractively stable (see Definition 1 in the Appendix), and at the same time to preserve the exponential stability of the matrix F_0 so as to agree with assumption (105). This optimization process is constrained by the minimal values of α and β, say α_0 and β_0, that provide quasi-constant Riccati solutions according to (118) and (119), and by a proper threshold ε, (159), on the response of the controlled model.

The closed-loop elastic model is defined by (94). Introducing the n_A-dimensional state vector $\boldsymbol{\eta}_A = [\mathbf{z}_C^T \mid \mathbf{e}^T \mid \mathbf{z}_U^T]^T$, we can rewrite the homogeneous part of (94) as

$$\dot{\boldsymbol{\eta}}_A(t) = \tilde{A}(t)\boldsymbol{\eta}_A(t), \qquad \boldsymbol{\eta}_A(t_0) = \boldsymbol{\eta}_{A0} \qquad \text{for all} \quad t \in \tau. \tag{128}$$

The design process is based on the fact that system (128), in which $\tilde{A}(t, \alpha, \beta) = \tilde{A}(t, \alpha_0, \beta_0)$, is exponentially contractively stable. Indeed, the convergence characteristics of system (128), as well as those of F_0, are dominated by the main diagonal submatrices, where the dominance of the main diagonal submatrices can be enhanced by adding critical uncontrolled states to the observed model. The diagonal submatrices $\tilde{A}_{CC}(t)$ and $\tilde{A}_{EE}(t)$ of $\tilde{A}(t)$, (94), are finite-time stable by design and the open-loop submatrix $\tilde{A}_{UU}(t)$ is finite-time stable by assumption (Section IV). Similarly, the various submatrices in F_0 can be shown to have the structure

$$\begin{aligned} \bar{F}_C &= \bar{A} + B[\bar{K}_C \mid 0 \mid 0], \qquad & \bar{F}_{12} &= -[B K_C \mid 0 \mid H_F], \\ \bar{F}_{21} &= \bar{K}_E[0 \mid 0 \mid C_R], \qquad & \bar{F}_E &= \bar{A}_{EE} + \bar{K}_E C_E, \end{aligned} \tag{129}$$

in which

$$\bar{A} = \begin{bmatrix} \bar{A}_{CC} & 0 & 0 \\ 0 & \bar{A}_{00} & 0 \\ 0 & 0 & \bar{A}_{RR} \end{bmatrix}, \qquad \bar{A}_{EE} = \begin{bmatrix} \bar{A}_{CC} & 0 & \multirow{2}{*}{H_E} \\ 0 & \bar{A}_{00} & \\ 0 & 0 & 0 \end{bmatrix},$$

$$\bar{A}_{jj} = \begin{bmatrix} 0 & I \\ -\Lambda_{ej} & -\bar{D}_{ej} \end{bmatrix}, \qquad j = C, 0, R. \tag{130}$$

Hence, the diagonal submatrices $\bar{F}_C$ and $\bar{F}_E$ of F_0 are exponentially stable.

At this point, we wish to determine a single Lyapunov function, (163), capable of achieving both goals specified above. Rearranging the state vector in (94) to correspond to the state vector in (98), the constant part of $\tilde{A}(t)$, denoted by $\tilde{A}_0$, can be rewritten as

$$\tilde{A}_0^f = \left[\begin{array}{c|c} \bar{F}_C & \bar{A}_{12} \\ \hline \bar{F}_{12} & \bar{F}_E \end{array}\right], \tag{131}$$

where

$$\bar{A}_{12} = -[BK_C \mid 0 \mid BK_D]. \tag{132}$$

Then, comparing the matrix F_0, (108) and (129), to the matrix $\tilde{A}_0^f$, (131), we conclude that the difference between the eigenvalues of these two matrices is due to the spillover matrix $\bar{A}_{12}$, (132). This difference can be reduced by adding more states to the observed model. Furthermore, usually $\|\bar{K}_C\| > \|K_D\|$, so that comparing $\bar{A}_{12}$, (132), to $\bar{F}_{12}$, obtained from the second of Eqs. (99) by replacing K_C with $\bar{K}_C$, we can expect the difference between the eigenvalues of these two matrices to be small. Hence, it is essential to define a Lyapunov function guaranteeing both the exponential contractive stability of the system defined by (128) and that

$$\operatorname{Re} \lambda_i(\tilde{A}_0) < 0, \quad i = 1, 2, \ldots, n_A. \tag{133}$$

To this end, we introduce the following Lyapunov function [27]

$$V(\boldsymbol{\eta}_A, t) = \boldsymbol{\eta}_A^T(t) B_a(t) \boldsymbol{\eta}_A(t), \tag{134}$$

where $B_a(t)$ is a Hermitian matrix given by

$$B_a(t) = [U_a(t) U_a^H(t)]^{-1}, \tag{135}$$

in which the superscript H denotes the complex conjugate transpose of a matrix and $U_a(t)$ is the matrix of instantaneous eigenvectors of $\tilde{A}(t)$. It is assumed that the eigenvalues of $\tilde{A}(t)$ are distinct for all $t \in \tau$, so that the matrix $U_a(t)$ is always nonsingular. It can be shown [27] that

$$\dot{V}(\boldsymbol{\eta}_A, t) = \boldsymbol{\eta}_A^T(t) U_a^{-H}(t) C_a(t) U_a^{-1}(t) \boldsymbol{\eta}_A(t), \tag{136}$$

where $[\quad]^{-H}$ is the inverse of $[\quad]^H$ and

$$C_a(t) = 2 \operatorname{Re}[\Lambda_a(t)] + T(t), \tag{137}$$

in which

$$T(t) = U_a^H(t) \dot{B}_a(t) U_a(t), \tag{138}$$

where $\Lambda_a(t)$ is the diagonal matrix of the eigenvalues of $\tilde{A}(t)$. Following the developments in the Appendix, we can write

$$\dot{\boldsymbol{\eta}}_{\mathrm{A}}^{*}(t) = \tilde{A}^{*}(t)\boldsymbol{\eta}_{\mathrm{A}}^{*}(t), \tag{139}$$

where

$$\boldsymbol{\eta}_{\mathrm{A}}^{*}(t) = e^{\gamma t}\boldsymbol{\eta}_{\mathrm{A}}(t), \qquad \tilde{A}^{*}(t) = \tilde{A}(t) + \gamma I, \qquad \gamma \geq 0, \tag{140}$$

and γ is yet to be defined. Then, by analogy with (137), we obtain

$$C_{\mathrm{a}}^{*}(t) = 2\,\mathrm{Re}[\Lambda_{\mathrm{a}}^{*}(t)] + T^{*}(t), \tag{141}$$

where the terms denoted with an asterisk correspond to the eigensolution of $\tilde{A}^{*}(t)$. Considering the first of Eqs. (140) and the definitions of $\Lambda_{\mathrm{a}}(t)$ and $T(t)$, we have

$$\Lambda_{\mathrm{a}}^{*}(t) = \gamma I + \Lambda_{\mathrm{a}}(t), \qquad T^{*}(t) = T(t), \qquad C_{\mathrm{a}}^{*}(t) = 2\gamma I + C_{\mathrm{a}}(t). \tag{142}$$

Finally, using (177) with $C(t)$ replaced by $C_{\mathrm{a}}^{*}(t)$, we obtain the supremum time constant of the elastic model

$$\pi = 2\left\{\min_{t\in\tau}|\lambda_{\mathrm{m}}[C_{\mathrm{a}}(t)]|\right\}^{-1} \tag{143}$$

provided that

$$\lambda_{\mathrm{m}}[C_{\mathrm{a}}(t)] < 0 \qquad \text{for all } t \in \tau, \tag{144}$$

where $\lambda_m[\ \ ]$ denotes the minimum eigenvalue of $[\ \ ]$. Considering (137), we recognize that

$$\lambda_i[C_{\mathrm{a}}(t_{\mathrm{A}})] = 2\,\mathrm{Re}[\lambda_i(\tilde{A}_0)], \qquad i = 1, 2, \ldots, n_{\mathrm{A}}, \tag{145}$$

where $t_{\mathrm{A}} = t_0$ for the first half of the maneuver and $t_{\mathrm{A}} = t_{\mathrm{f}}$ for the second half of the maneuver. Hence, inequality (144) also implies conditions (133). Moreover, the minimum supremum time constant is given by

$$\pi_{\mathrm{m}} = \min_{\alpha,\beta}[\pi(\alpha, \beta)]. \tag{146}$$

The minimum supremum time constant can be approximated by an expression involving the eigensolutions of a constant matrix by keeping the zero-order terms only in the expressions of the optimal gains. Hence, inserting (121a) and (123a) into (94) and considering the first of Eqs. (57), we can rewrite the matrix $\tilde{A}(t)$ as

$$\tilde{A}(t) = \tilde{A}_0 + \Omega_0^2(t)A_1, \tag{147}$$

where A_1 is also a constant matrix. Then, recalling the relatively high norm of $\bar{K}_{\mathrm{C}}$ and $\bar{K}_{\mathrm{E}}$, we can assume that

$$\|A_0\| \gg \Omega_0^2(t)\|A_1\| \qquad \text{for all } t \in \tau. \tag{148}$$

At this point, we can obtain the eigensolutions of $\tilde{A}(t)$ by means of a perturbation technique [1], where $\tilde{A}(t)$ is defined by (147). Retaining first-order terms only, we obtain

$$\Lambda_a(t) = \Lambda_0 + \Omega_0^2(t)\Lambda_1, \tag{149a}$$

$$U_a(t) = U_0 + \Omega_0^2(t)U_1, \tag{149b}$$

where Λ_0 is the diagonal matrix of eigenvalues of $\tilde{A}_0$, and U_0 is the matrix of right eigenvectors of $\tilde{A}_0$. Then, defining

$$E_P = V_0^T A_1 U_0, \tag{150}$$

where V_0 is the matrix of left eigenvectors of $\tilde{A}_0$, the perturbations in the eigensolutions can be expressed as

$$\Lambda_1 = \text{diag}(e_{Pii}), \qquad U_1 = U_0 \bar{E}_P, \tag{151}$$

where e_{Pii} are the diagonal elements of E_P and $\bar{E}_P$ has the entries

$$\bar{e}_{Pij} = \frac{e_{Pij}}{e_{Pjj} - e_{Pii}}, \qquad i = j; \qquad \bar{e}_{Pii} = 0; \qquad i, j = 1, 2, .., n_A. \tag{152}$$

Inserting (149b) into (138) and considering (33b), (135), and the first of Eqs. (151), we obtain

$$T(t) = -2(t - t_A)\dot{\Omega}_0^2[M_a + M_a^H] \tag{153}$$

in which

$$M_a = [I_{n_A \times n_A} + (t - t_A)^2 \dot{\Omega}_0^2 \bar{E}_P]^{-1} \bar{E}_P. \tag{154}$$

The existence of the inverse is guaranteed by the assumption implied by the perturbation technique [28], as reflected in inequality (148). Using the inverse formula for a perturbed matrix [28], inserting (153) and (154) into (137) and considering (33b), we finally obtain

$$\bar{C}_a(t) = 2\,\text{Re}[\Lambda_0] + 2\dot{\Omega}_0^2\{(t - t_A)^2\,\text{Re}[\Lambda_1] - (t - t_A)[\bar{E}_P + \bar{E}_P^H]\}. \tag{155}$$

It follows that the approximate supremum time constant is given by (143) with $C_a(t)$ replaced by $\bar{C}_a(t)$.

VIII. NUMERICAL EXAMPLE

The above developments were applied to the flexible space structure shown in Fig. 2. The mathematical model consists of a rigid hub and a flexible appendage in the form of a uniform beam 24-ft long. The mass of the hub is $m_h = 27$ slugs, the density of the beam is $\rho_b = 0.01$ slug ft^{-1}, and the mass moment of inertia of the hub about its own mass center is 264 slug ft^2. The perturbed model, described by (54), is assumed to possess two rigid-body translational, one rigid-body rotational, and seven elastic degrees of freedom, for a total of ten degrees of freedom. The first ten natural frequencies of the structure are 0, 0, 0, 0.32, 1.8, 5.0, 9.8, 16.2, 24.2, and 33.8 (Hz). All damping factors are assumed to have the same value, $\zeta_i = 0.01$ ($i = 4, 5, \ldots, 10$).

The "rigid-body" slewing of the space structure is carried out according to a minimum-time control policy, which implies bang-bang control. A 180° maneuver is shown in Fig. 4. The maneuver is relatively fast, so that in the absence of vibration control the elastic deformations tend to be large, as implied by Fig. 5, which compares the uncontrolled response and the controlled response. There are up to seven controlled degrees of freedom; they consist of three rigid-body motions and up to four elastic motions. As indicated in Section III, the rigid-body perturbations are controlled independently of the elastic motions by three sets of collocated actuators and sensors, one for each degree of freedom. All actuators and sensors for control of the rigid-body perturbations are located at the mass center of the space structure, assumed to lie on the hub. The elastic motions are controlled by up to four sets of collocated torque actuators and angular displacement and angular velocity sensors, where the collocation is advised for spillover reduction. These actuators and sensors are placed on the flexible beam at $y_i = iL/4$ ($i = 1, 2, 3, 4$). The feedback control $\mathbf{u}_C$ is applied during and after the maneuver, while the disturbance accommodation $\mathbf{u}_D$ operates only during the maneuver.

The observer is designed to estimate the controlled state and a two-dimensional disturbance vector; additional observed but uncontrolled states are not needed in this example. During the maneuver the full observer model, (82), is operational. After the maneuver is completed, $\hat{\mathbf{f}}_E$ must be removed from the observer state, as the persistent disturbance no longer exists. Otherwise, significant transient estimation errors arising from the disturbance estimation errors can plague the controlled state, causing an increase in the settling time after the termination of the maneuver. Hence, the observer model used after the completion of the maneuver is the one defined by (82), in which $\hat{\mathbf{f}}_E$ and H_E are omitted.

From now on, we focus our attention on the 14-order elastic model consisting of the elastic degrees of freedom, although the simulation results are based on the entire perturbed model (54), where the control of the rigid-body perturbed model is based on (54a) and (56). In every case, the response represents the deflection at the tip of the beam, at which point the deflection tends to be the largest.

The response of the controlled model is bounded according to inequality (171). As argued in Section V, by increasing the convergence factor α, the

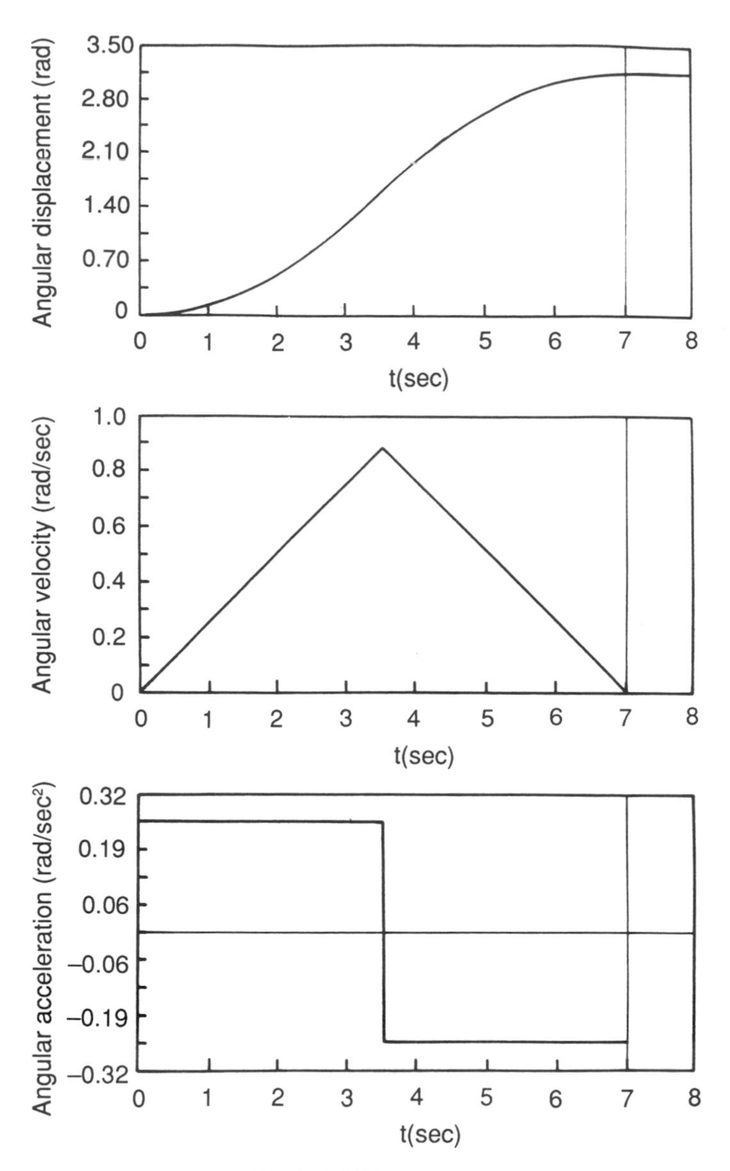

Fig. 4. A 180° maneuver.

response to initial conditions can be forced to drop below any desired threshold value as $t \to t_h$. This process is demonstrated in Fig. 6, where it is shown that the term $r_{\mathrm{B}}^{1/2}(t)\, e^{-\alpha t}$ is dominated by the exponential part for $t \in (T_1, t_h)$, $t_i < T_i < t_h$, for sufficiently large α.

To obtain a satisfactory steady-state disturbance accommodation, it is necessary that the Riccati equation for both the controller and observer

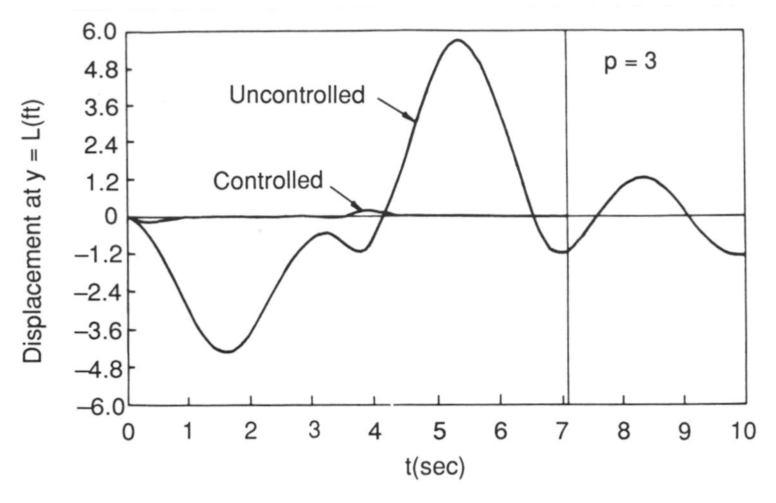

Fig. 5. Displacement at the tip; maneuver angle = 180°.

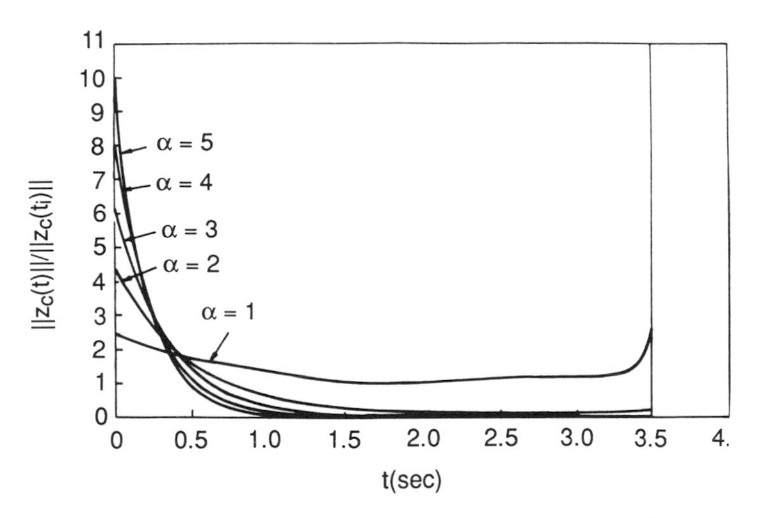

Fig. 6. Convergence of the controlled response upper bound.

possess quasi-constant solution according to (118) and (119), respectively. To demonstrate this, we consider an extreme case. With reference to (115), we use a second-order model, where $\varphi(t) = 0.25t^2$ $(\text{rad/sec})^2$, $t \in [0, 2.5]$, $\omega_1 = 0.2$ rad/sec, $\|\bar{K}_{22}\| \cong 1$, so that $\varphi(t)\|\bar{K}_{22}\| > \omega_1^2$, for $t > 0.4$ sec. Figures 7a and 7b show plots of $s_{11}(t)/\bar{s}_{011}$ versus t for $\alpha = 4$ and $\alpha = 8$, respectively, where $s_{11}(t)$ is the entry of $S(t)$, (70), exhibiting the largest deviation relative to the corresponding entry $\bar{s}_{011}$ of the algebraic Riccati matrix $\bar{S}_0$.

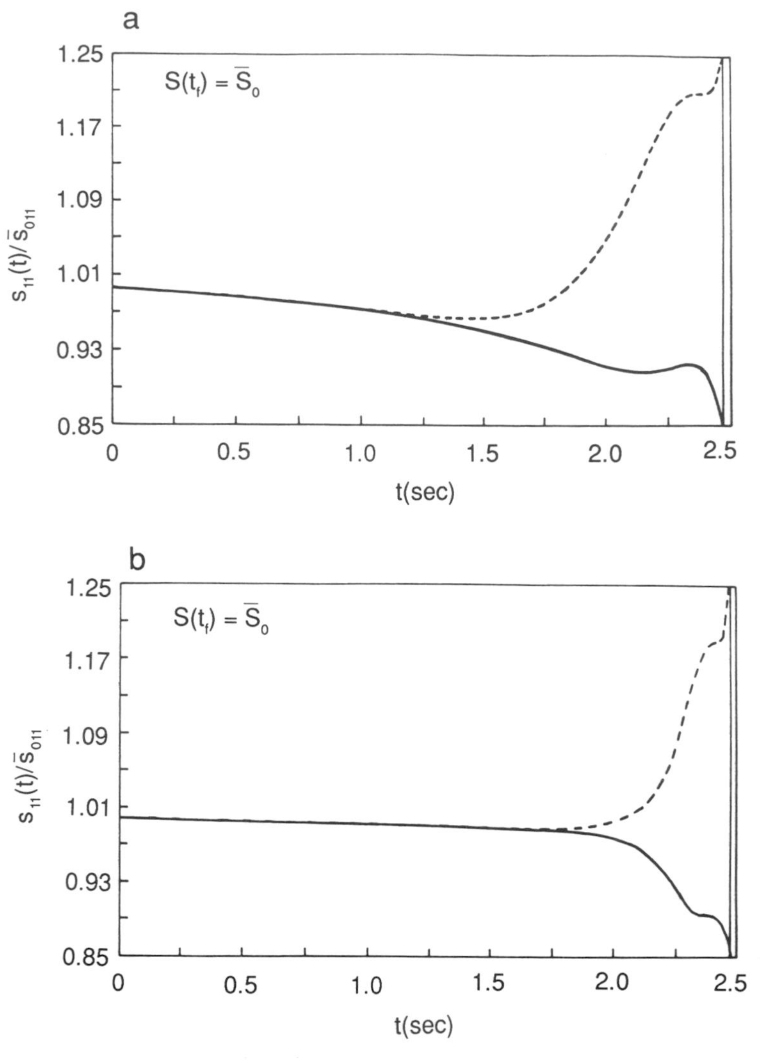

Fig. 7. Convergence of the controller steady-state Riccati solution for (a) $\alpha = 4$ and (b) $\alpha = 8$.

Of course, $\bar{S}_0$ satisfies the algebraic version of (70). It is obvious that convergence to the steady-state solution is faster for increasing α. Figure 8a demonstrates the decrease in $\varepsilon_1(\alpha)$, as defined by (116), for increasing α. Finally, Fig. 8b shows that, by choosing $S(t_f) = \bar{S}_0$, because $\varepsilon_3(\alpha)$ approaches zero as α increases, the transient Riccati solution approaches the algebraic solution for $t \in \tau$, as can be concluded from (118). From Fig. 8b, we see that the maximum deviation from the algebraic solution is smaller than 5% for $\alpha = 4$ and smaller than 0.5% for $\alpha = 12$.

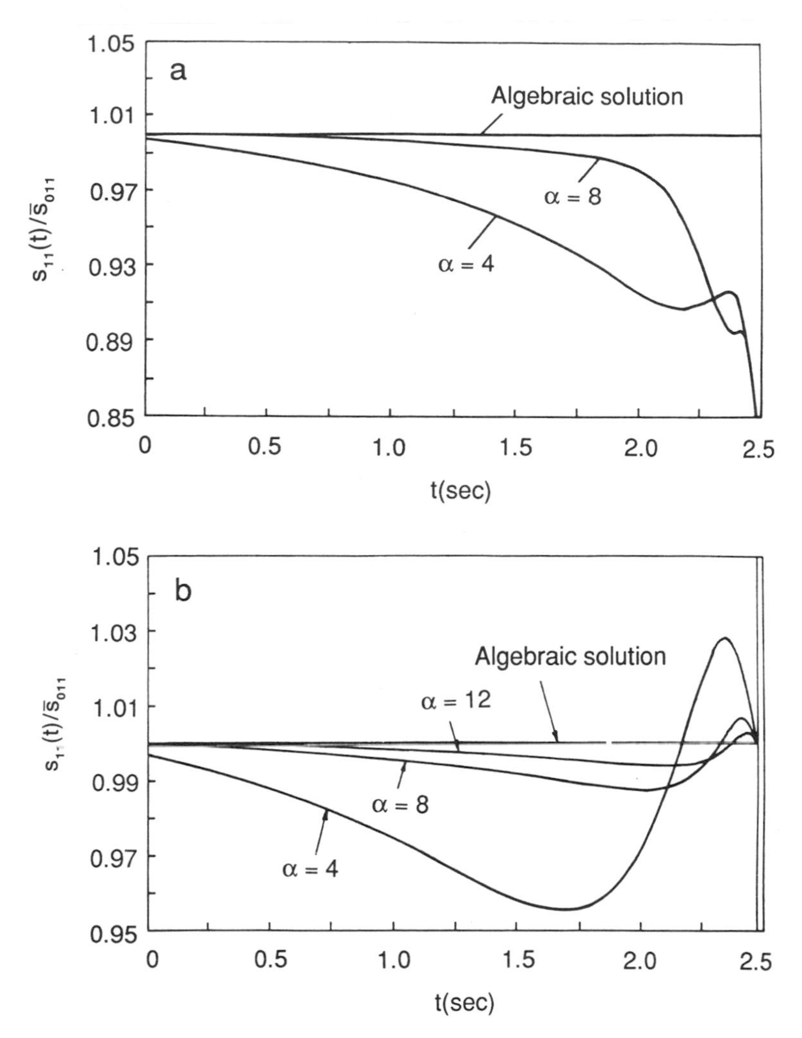

Fig. 8. Deviation of the time-varying Riccati solution from the algebraic solution for (a) $S(t_f) = 0.85\bar{S}_0$ and (b) $S(t_f) = \bar{S}_0$.

The design policy is demonstrated next. A sixth-order controlled model is examined, so that $n_C = 6$. The gain matrices $K_C(t)$, $K_E(t)$, and K_D are calculated according to the second of Eqs. (69), (89), and (113), respectively, the convergence factor β is taken as $\beta = 3\alpha$, and the weighting matrices Q and R in (64) and Q_E in (90a) are taken as unit matrices. Moreover, the boundary condition in the Riccati equations, (70) and (90a), are taken as $S_1 = \bar{S}_0$ and $P_0 = \bar{P}_0$, respectively, where $\bar{S}_0$ and $\bar{P}_0$ are the solutions of the corresponding algebraic equations. There are $p = m/2 = 3$ collocated torque actuators and pairs of displacement and velocity sensors located along the beam at $y_j = jL/4\ (j = 1, 2, 3)$. The process begins with the determination of the minimum convergence factor guaranteeing a satisfactory controlled model dynamics and quasi-constant optimal gains. This value is found to be $\alpha = 1.5$. According to Fig. 9, the displacement error drops to a value below 0.38×10^{-4} ft as $t \to t_h$. Then, the convergence factor is raised until the minimum supremum time constant is found. Figure 10 shows the plot π^{-1} versus α.

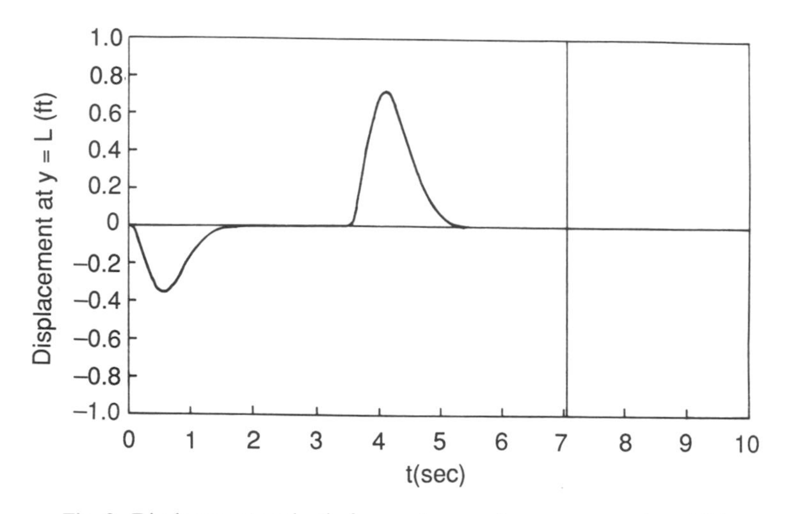

Fig. 9. Displacement at the tip for $p = 3$, $\alpha = 1.5$; maneuver angle = 180°.

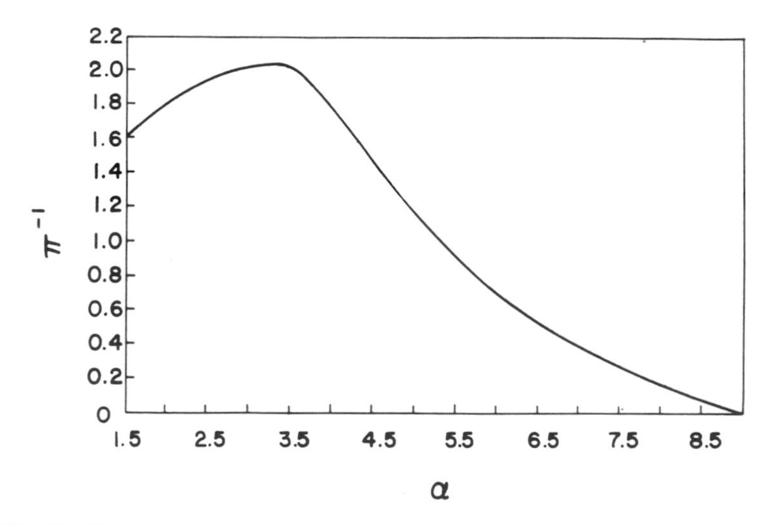

Fig. 10. The reciprocal of the supremum time constant versus the convergence factor.

The minimum supremum time constant is calculated according to (143) and (146), where $C_a(t)$ is defined by (137) exactly and by (155) approximately. The exact value is $\pi_m = 0.49$ sec and is obtained for $\alpha = 3.5$, whereas the approximate value is $\pi_m = 0.506$ sec and is obtained for $\alpha = 3.0$. It is obvious that the difference between the two values is small, but the reduction in computation effort using the approximate value is significant. We choose $\alpha = 3.0$ as the optimal value. The satisfaction of inequalities (105) by using a Lyapunov function guaranteeing the satisfaction of inequalities (133) is based on the small difference between the eigenvalues of the matrices F_0 and $\tilde{A}_0$. Table I shows the two sets of eigenvalues. The disturbance accommodation policy is meaningful as long as the optimal gains are quasi-constant. Figure 11 shows the deviation of four controller optimal gains from the algebraic values representing the most extreme cases out of the 15 existing distinct gains. We conclude that the maximum deviation is less than 5% and it drops below 1% for k_{Ci} ($i = 4, 5, \ldots, 15$). Finally, Fig. 12a shows the displacement error for the controlled model and for the perturbed model. As $t \to t_h$, the error drops below 0.4×10 and 0.6×10^{-2} ft, respectively.

The displacement error of the perturbed model in the case of a fourth-order controlled model and $p = 2$ is shown in Fig. 12b and in the case of an eighth-order controlled model and $p = 1$ is shown in Fig. 12c. The optimal values of α are 2.2 and 2.0, respectively. As $t \to t_h$, the displacement error drops below 0.8×10^{-2} and 1.7×10^{-2}, respectively. As expected, we conclude that an increased controlled model enhances the quasi-constant conditions imposed on the elastic model. Hence, the minimization process associated with the disturbance is closer to the ideal one, whose performance measure is defined

TABLE I. EIGENVALUES OF $\tilde{A}_C^f$ AND F_0 COMPARED[a]

$\tilde{A}_C^f$	F_0
$-6.491 \pm j261.917$	$-5.502 \pm j242.167$
$-3.691 \pm j172.799$	$-2.893 \pm j161.150$
$-2.376 \pm j117.189$	$-2.528 \pm j114.038$
$-1.272 \pm j64.597$	$-1.124 \pm j63.116$
$-5.879 \pm j33.457$	$-5.790 \pm j33.441$
$-11.892 \pm j30.647$	$-11.856 + j30.595$
$-5.796 \pm j15.333$	$-6.488 \pm j15.497$
$-4.533 \pm j9.666$	$-5.286 \pm j10.594$
$-6.059 \pm j9.619$	$-6.796 \pm j9.687$
$-5.226 \pm j1.930$	$-5.255 \pm j1.885$
$-9.550 \pm j0.0$	$-9.597 \pm j0.187$
$-10.190 \pm j0.0$	

[a] Parameter values: $n_M = 14$, $n_C = 6$, $p = 3$, $\alpha = 3$.

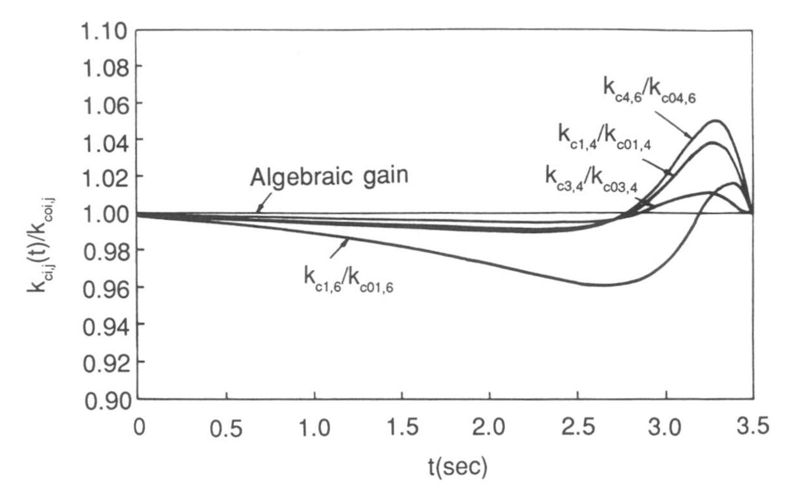

Fig. 11. Deviation of controller optimal gains from the algebraic values for $S(t_f) = \bar{S}_0$, $\alpha = 3$.

by (101). Finally, Fig. 13 represents the angular displacement of the space structure at the tip of the beam for an 180° maneuver. Deviations from the ideal bang-bang maneuver are well below the resolution of the graph.

IX. SUMMARY AND CONCLUSIONS

The equations of motion describing the slewing of a flexible space structure, following discretization in space and truncation, consist of a set of nonlinear ordinary differential equation that can still be of relatively high order. To circumvent the difficulties involved in solving the nonlinear two-point boundary-value problem of high order encountered in a minimum-time problem, a perturbation technique is adopted. This yields a low-order nonlinear zero-order model describing the rigid-body motions and a linear time-dependent first-order model representing the elastic vibrations. The zero-order model is controlled according to a minimum-time control policy, whereas the first-order model is stabilized simultaneously by a linear feedback control law.

A design approach is developed for the case of a single-axis slewing maneuver about a principal axis of a space structure consisting of a rigid-hub and a flexible appendage. The first-order model consists of a rigid-body perturbed model and an elastic model. Using decoupling control inputs and measurements and neglecting small coupling terms due to relatively high elastic modes contaminating the measurements of the rigid-body perturbations, it is possible to design independent control for the rigid-body

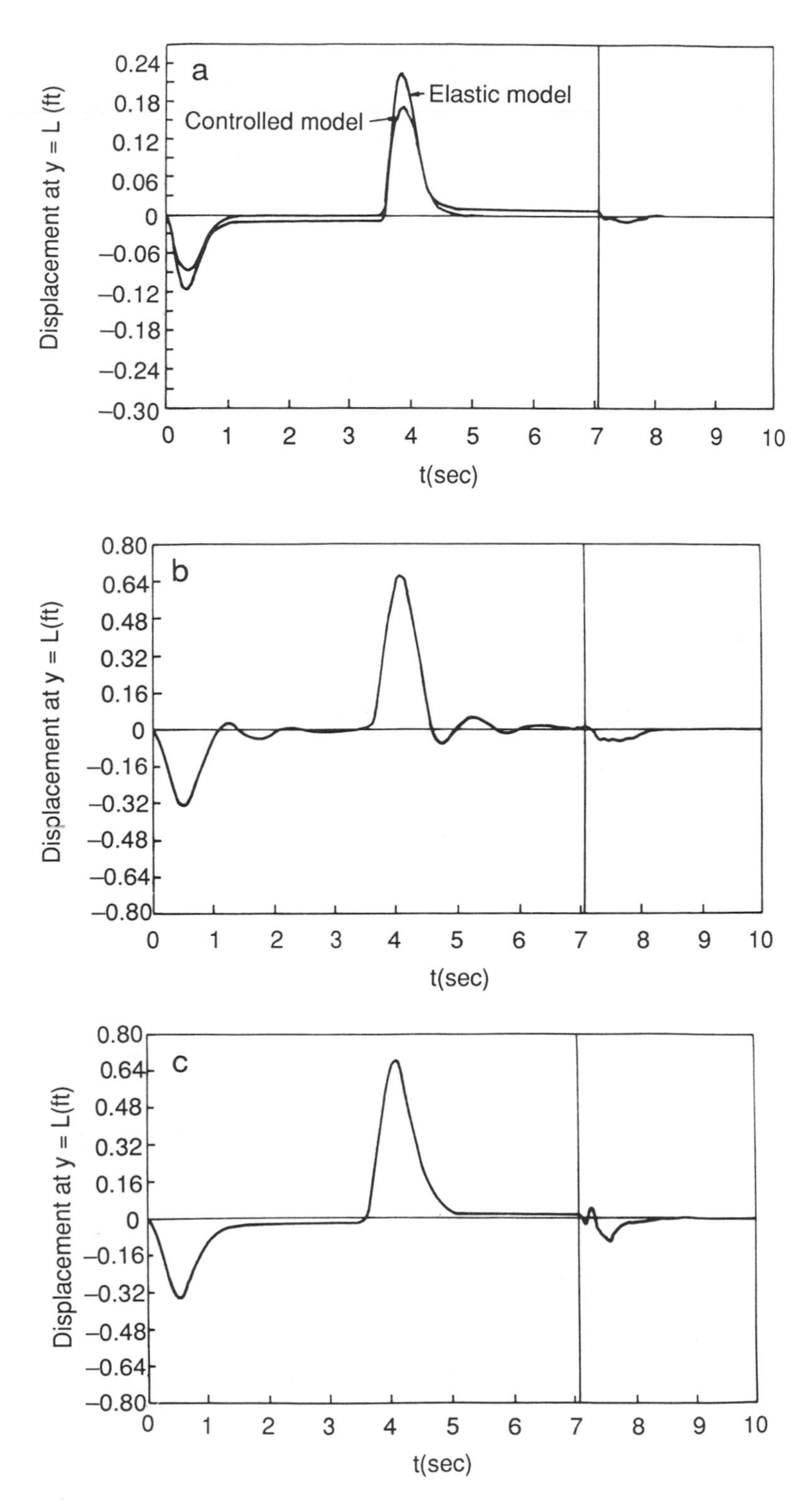

Fig. 12. Displacement at the tip for (a) $p = 3$, $\alpha = 3$, maneuver angle = 180°; (b), $p = 2$, $\alpha = 2.2$, maneuver angle = 180°; (c) $p = 1$, $\alpha = 2$, maneuver angle = 180°.

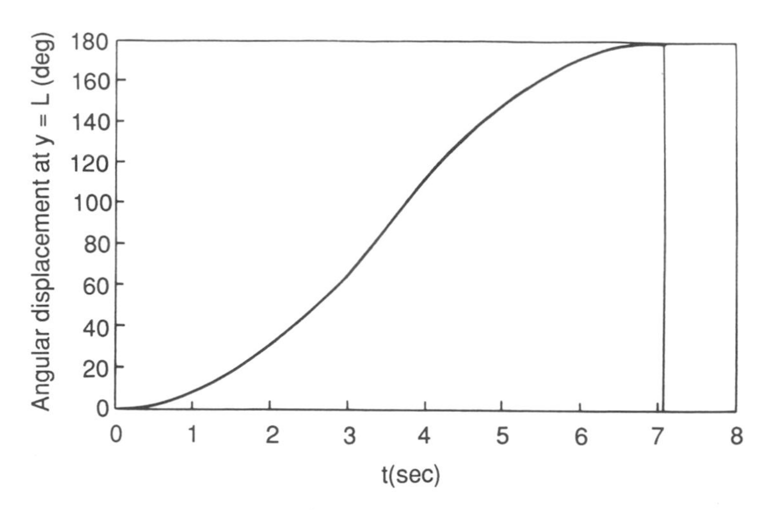

Fig. 13. Angular displacement at the tip for $p = 3$.

model and for the elastic model. The rigid-body perturbations, which are not subjected to persistent disturbances caused by the maneuver, are controlled by a standard linear quadratic regulator, whereas the control of the elastic perturbations consists of a reduced-order compensator including feedback control and disturbance accommodation control.

The main features of the control are:

1. A reduced-order model is stabilized during a finite-time interval according to a quadratic performance measure. A convergence factor is included in the controller performance measure and in the observer Riccati equation. Then, by increasing these convergence factors, the response of the reduced-order model to initial conditions can be forced to decay below any desired threshold value.

2. The disturbance accommodation control minimizes a weighted norm spanning the constant part of the perturbed state, provided that a time-invariant coefficient matrix is exponentially stable. The minimization is meaningful only if quasi-constant conditions exist. These quasi-constant conditions can be achieved toward the end of the finite-time interval if the optimal solutions of both the controller and the observer Riccati equation achieve quasi-constant values toward the end of the time interval. The elastic model is characterized by the fact that, as the convergence factors increase, the optimal solutions reach steady-state values. These steady-state solutions converge monotonically to the corresponding algebraic solutions as the convergence factors increase. Hence, choosing the boundary conditions in the Riccati equations equal to the corresponding algebraic solutions, it is possible to obtain quasi-constant values throughout the entire time interval.

3. A Lyapunov function based on the eigensolution of the elastic model and a perturbation technique are combined to yield a sufficiency condition guaranteeing the exponential contractive stability of the elastic model, as well as the exponential stability of the coefficient matrix associated with the disturbance accommodation process. Moreover, an expression for the supremum time constant of the elastic model is developed and the convergence factors are determined so as to minimize this supremum.

The above developments were demonstrated by means of a numerical example.

APPENDIX. FINITE-TIME STABILITY

The ordinary stability definitions relate to systems operating over an infinite-time interval. Hence, for finite-time stability, we need different definitions. Under consideration is the linear system

$$\dot{\mathbf{z}} = A(t)\mathbf{z}, \qquad \mathbf{z}(t_0) = \mathbf{z}_0 \tag{156}$$

defined over the time interval $v = [t_0, t_f)$, where $A(t)$ is an $n \times n$ matrix continuous on $R^n \times v$ and bounded for all $t \in v$. Then, we denote by $\bar{\mathbf{z}}(t^*, t, \mathbf{z})$ a trajectory of (156) evaluated at time t^*, which takes on the value $\mathbf{z}$ at time t, and by $\| \quad \|$ a norm on $R^n \times v$.

Definition 1. The system (156) is exponentially contractively stable (ECS) with respect to $[\delta, \psi, \alpha, \mu(t), v, \| \quad \|]$, where α is a positive constant, $\mu(t)$ is a continuous positive function, and $\psi \geq \delta$, if every trajectory $\bar{\mathbf{z}}(t^*, t_0, z_0)$ for which $\|\mathbf{z}_0\| < \delta$ is such that

$$\|\bar{\mathbf{z}}(t, t_0, \mathbf{z}_0)\| < \psi \exp\left[-\int_{t_0}^{t} \mu(s)\, ds\right] \exp[-\alpha(t - t_0)], \qquad t \in \tau. \tag{157}$$

Definition 2 [29]. The system (156) is quasi-contractively stable (QCS) with respect to $[\delta, \rho, v, \| \quad \|]$ if for every trajectory $\bar{\mathbf{z}}(t^*, t_0, \mathbf{z}_0)$, where $\|\mathbf{z}_0\| < \delta$, there exists a $T_1 \in (t_0, t_f)$ such that

$$\|\bar{\mathbf{z}}(t^*, t_0, \mathbf{z}_0)\| < \rho, \qquad \rho < \delta, \qquad t^* \in (T_1, t_f). \tag{158}$$

Definition 3. The system (156) is strictly finite-time stabilizable (SFTS) with respect to $[\delta, \varepsilon, v, \| \quad \|]$ if for any positive ε there exists a $T_1 \in (t_0, t_f)$ and every trajectory $\bar{\mathbf{z}}(t^*, t_0, \mathbf{z}_0)$ for which $\|\mathbf{z}_0\| < \delta$ has the property

$$\|\bar{\mathbf{z}}(t^*, t_0, \mathbf{z}_0)\| < \varepsilon \qquad \text{for all } t^* \in (T_1, t_f). \tag{159}$$

At this point, we wish to formalize sufficiency conditions for ECS and SFTS systems and to introduce a definition of the supremum time constant of the system in terms of a convergence factor. With reference to (156), we define the new variable

$$\mathbf{z}^* = e^{\gamma t}\mathbf{z}, \qquad \gamma \geq 0 \tag{160}$$

so that (156) yields

$$\dot{\mathbf{z}}^* = A^*(t)\mathbf{z}^*, \qquad \mathbf{z}^*(t_0) = \mathbf{z}_0^*, \tag{161}$$

where

$$A^*(t) = A(t) + \gamma I. \tag{162}$$

Next, we introduce the Lyapunov function

$$V(\mathbf{z}^*, t) = \mathbf{z}^{*T}B(t)\mathbf{z}^*, \tag{163}$$

where $B(t)$ is a symmetric and differentiable matrix chosen such that

$$0 \leq c_1 \leq \|B(t)\| \leq c_2 < \infty, \qquad t \in v. \tag{164}$$

The implication of the above is that $V(\mathbf{z}^*, t)$ is a positive definite function for all $t \in v$. Differentiating (163) with respect to time and evaluating along a trajectory of (161), we obtain

$$\dot{V}(\mathbf{z}^*, t) = -\mathbf{z}^{*T}C(t)\mathbf{z}^*, \tag{165}$$

where $C(t)$ is a symmetric differentiable matrix such that

$$\dot{B}(t) = -B(t)A^*(t) - A^{*T}(t)B(t) - C(t). \tag{166}$$

Next, we introduce the definition

$$\mu(t) = \min_i \lambda_i[C(t)]/\max_i \lambda_i[B(t)], \tag{167}$$

so that, considering (163) and (165), we obtain

$$\dot{V}(\mathbf{z}^*, t) \leq -\mu(t)V(\mathbf{z}^*, t) \qquad \text{for all} \quad t \in v. \tag{168}$$

Then, integrating both sides of inequality (168), we obtain

$$V(\mathbf{z}^*, t) \leq V(\mathbf{z}_0^*, t_0)\exp\left[-\frac{1}{2}\int_{t_0}^{t} \mu(s)\, ds\right]. \tag{169}$$

We consider now the following property of positive definite quadratic forms

$$\min_i \lambda_i(B(t))\|\mathbf{z}^*\|^2 \leq \|\mathbf{z}^{*T}B(t)\mathbf{z}^*\| \leq \max_i \lambda_i(B(t))\|\mathbf{z}^*\|^2 \tag{170}$$

so that, in view of (160), (163), (169), and (170), we obtain

$$\|\mathbf{z}(t)\| < \|\mathbf{z}_0\| r_{\mathrm{B}}^{1/2}(t)\exp\left[-\frac{1}{2}\int_{t_0}^{t} \mu(s)\, ds\right]\exp[-\gamma(t - t_0)], \tag{171}$$

where

$$r_B(t) = \max_i \lambda_i[B(t_0)]/\min_i \lambda_i[B(t)]. \tag{172}$$

Hence, recalling inequality (164) and (167), we conclude that $\mu(t) > 0$ for all $t \in v$ if and only if $C(t) > 0$ for $t \in v$. The above developments, in conjunction with Definitions 1 and 3, permit us to state the following sufficiency conditions:

1. If $C(t) > 0$ for $t \in v$, Eq. (156) represents an ECS system.
2. If $C(t) > 0$ for $t \in v$ and in addition

$$\delta r_B^{1/2}(t)\exp[-\gamma(t - t_0)] < \varepsilon \qquad \text{for all } t \in (T_i, t_f), \qquad t_0 < T_1 < t_f, \tag{173}$$

then (156) represents an SFTS system.

Conclusion from Definition 3. In a finite-time stabilizable system, to every ε and $\mathbf{z}_0$ corresponds an $\tilde{\gamma}(\varepsilon, \mathbf{z}_0)$ such that, for every $\gamma > \tilde{\gamma}(\varepsilon, \mathbf{z}_{C0})$, the response to $\mathbf{z}_0$ decays to a value below ε within the finite-time interval v.

Finally, defining

$$\bar{r}_B = \max_{t \in \tau}[r_B(t)], \tag{174}$$

we can rewrite inequality (171) in the form

$$\|\mathbf{z}(t)\| < \|\mathbf{z}_0\| {r_B}^{1/2} \exp\left[-\frac{1}{2}\int_{t_0}^{t} \mu(s)\,ds\right]\exp[-\gamma(t - t_0)]. \tag{175}$$

It is clear that inequality (175) gives an estimate of how fast equilibrium is approached. In fact, if $\mu(t) > 0$ for all $t \in \tau$, then $\|\mathbf{z}(t)\|$ converges to zero not slower than $e^{-\gamma t}$. Therefore, $1/\gamma$ can be interpreted as an upper bound on the time constant of system (156). We denote the least upper bound of $1/\gamma$ by π as the *supremum time constant* (STC). It is clear that the STC corresponds to the value of γ for which $\mu(\gamma, t) = 0$ for some $t \in \tau$. Hence, we can define π as

$$\pi = 1/\bar{\gamma}, \tag{176}$$

where $\bar{\gamma}$ is the maximum value of γ for which

$$\min_{t \in v} |\lambda_m[C(\bar{\gamma}, t)]| = 0. \tag{177}$$

REFERENCES

1. L. MEIROVITCH, "Computational Methods in Structural Dynamics." Sijthoff & Noordhoff, Alphen aan den Rijn, Netherlands, 1980.
2. J. D. TURNER and J. L. JUNKINS, "Optimal Large-Angle Single Axis Rotational Maneuvers of Flexible Spacecraft," *AIAA J. Guidance Control* **3**, 578–585 (1980).

3. J. D. TURNER and H. M. CHUN, "Optimal Distributed Control of a Flexible Spacecraft During a Large Angle Maneuver," *AIAA J. Guidance, Control Dyn.* **7**, 257–264 (1984).
4. J. N. JUANG, J. D. TURNER, and H. M. CHUN, "Closed Form Solution for Feedback Control with Terminal Constraints," *J. Guidance, Control Dyn.* **8**, 39–43 (1985).
5. S. B. SKAAR, L. TANG, and Y. YALDA-MOOSHABAD, "On–Off Attitude Control of Flexible Satellites," *AIAA J. Guidance, Control Dyn.* **9**, 507–510 (1986).
6. W. E. VANDER VELDE and J. HE, "Design of Space Structures Control Systems Using On–Off Thruster," *AIAA J. Guidance, Control Dyn.* **6**, 53–60 (1983).
7. R. C. THOMPSON, J. L. JUNKINS, and S. R. VADALI, "Near-Minimum Time Open-Loop Slewing of Flexible Vehicles," *J. Guidance, Control Dyn.* **12**, 82–89 (1989).
8. N. RAJAN, "Minimum-Time Slewing of the SIRTF Spacecraft," *Proc. AIAA Guidance, Navig. Control Conf., Monterey*, Calif., pp. 1222–1227 (1987).
9. T. E. BAKER and E. POLAK, "Computational Experiment in the Optimal Slewing of Flexible Structure," Electronics, Research Laboratories, Memo. No. UCB/ERL M87/72, University of California, Berkeley, September 1981.
10. J. Z. BEN-ASHER, J. A. BURNS, and E. M. CLIFF, "Time Optimal Slewing of Flexible Spacecraft," *Proc. 28th IEEE Conf. Decision Control, Los Angeles, Calif.*, pp. 524–528 (1987).
11. L. MEIROVITCH and R. D. QUINN, "Equation of Motion for Maneuvering Flexible Spacecraft," *AIAA J. Guidance, Control Dyn.* **10**, 453–465 (1987).
12. L. MEIROVITCH and R. D. QUINN, "Maneuvering and Vibration Control of Flexible Spacecraft," *J. Astronaut. Sci,* **35**, 301–328 (1987).
13. L. MEIROVITCH and Y. SHARONY, "Optimal Maneuvering of Flexible Spacecraft," *Proc. 6th VPI&SU/AIAA Symp. Dyn. Control Large Struct., Blacksburg, Va.*, pp. 579–602 (1987).
14. Y. SHARONY and L. MEIROVITCH, "Accommodation of Kinematic Disturbances During a Minimum-Time Maneuver of a Flexible Spacecraft," *AIAA/AAS Astrodyn. Conf. Minneapolis, Minn.*, Paper No. AIAA-88-4253-CP (1988).
15. Y. SHARONY and L. MEIROVITCH, "A Perturbation Approach to the Minimum-Time Slewing of a Flexible Spacecraft." Presented at the *1989 IEEE Int. Conf. Control Appl., Jerusalem* (1989).
16. D. E. KIRK, "Optimal Control Theory." Prentice-Hall, Englewood Cliffs, New Jersey, 1970.
17. M. J. BALAS, "Active Control of Flexible Systems," *J. Optim. Theory Appl.* **2**, 415–436 (1978).
18. L. MEIROVITCH and H. BARUH, "On the Problem of Observation Spillover in Self-Adjoint Distributed-Parameter Systems," *J. Optim. Theory Appl.* **39**, 269–291 (1983).
19. C. D. JOHNSON, "Accommodation of External Disturbances in Linear Regulator and Servomechanism Problems," *IEEE Trans. Autom. Control* **AC–16**, 335–343 (1971).
20. C. D. JOHNSON, "Accommodation of Disturbances in Optimal Control Problems," *Int. J. Control* **15**, 209–231 (1972).
21. Y. SHARONY, "Control of Flexible Spacecraft During a Minimum-Time Maneuver," Ph.D. Dissertation, Electrical Engineering Department, Virginia Polytechnic Institute and State University, Blacksburg (1988).
22. L. M. SILVERMAN and H. E. MEADOWS, "Controllability and Observability in Time-Variable Linear Systems," *SIAM J. Control* **5**, 64–73 (1967).
23. B. D. O. Anderson and J. B. Moore, "Linear Optimal Control." Prentice-Hall, Englewood Cliffs, New Jersey, 1971.
24. G. W. JOHNSON, "A Deterministic Theory of Estimation and Control," *IEEE Trans. Autom. Control* Aug., 380–384 (1969).
25. R. BELLMAN, "Perturbation Techniques in Mathematics, Physics and Engineering." Holt, Rinehart & Winston, New York, 1964.

26. R. E. KALMAN and J. E. BERTRAM, "Control System Analysis and Design via the Second Method of Lyapunov," *J. Basic Eng.* June, 371–393 (1960).
27. E. J. DAVISON, "Some Sufficient Conditions for the Stability of a Linear Time-Varying Systems," *Int. J. Control* **7**, 377–380 (1968).
28. G. W. STEWART, "Introduction to Matrix Computations." Academic Press, New York, 1973.
29. L. WEISS and E. F. INFANTE, "On the Stability of Systems Defined Over a Finite Time Interval," *Proc. Natl. Acad. Sci. USA* **54**, 44–48 (1965).

INDEX

N

O

R

S

T

U

V